HOW TO FIND OUT IN
MATHEMATICS

HOW TO FIND OUT IN

MATHEMATICS

A Guide to Sources of Information

SECOND REVISED EDITION

BY

JOHN E. PEMBERTON, BA, FLA

PERGAMON PRESS

Oxford · London · Edinburgh · New York
Toronto · Sydney · Paris · Braunschweig

Pergamon Press Ltd., Headington Hill Hall, Oxford
4 & 5 Fitzroy Square, London W.1

Pergamon Press (Scotland) Ltd., 2 & 3 Teviot Place, Edinburgh 1

Pergamon Press Inc., Maxwell House, Fairview Park, Elmsford,
New York 10523

Pergamon of Canada Ltd., 207 Queen's Quay West, Toronto 1

Pergamon Press (Aust.) Pty. Ltd., 19a Boundary Street, Rushcutters Bay,
N.S.W. 2011, Australia

Pergamon Press S.A.R.L., 24 rue des Écoles, Paris 5ᵉ

Vieweg & Sohn GmbH, Burgplatz 1, Braunschweig

First edition 1964
Second revised edition 1969

Library of Congress Catalog Card No. 74-94058

Printed in Great Britain by A. Wheaton & Co., Exeter

191946

08 006823 5 (flexicover)
08 006824 3 (hard cover)

TO MY WIFE

Contents

Chapter 1

CAREERS FOR MATHEMATICIANS   1

General survey. Guides to careers. Employment registers. Teaching and women. Research. Statistics. Operational research. Actuarial science.

Chapter 2

THE ORGANIZATION OF MATHEMATICAL INFORMATION   15
Dewey classes: 510, 510.1, 510.2

Guides to using libraries. Guides to libraries and special collections. Inter-library co-operation. Photocopying and microcopying. Library catalogues. Classification of mathematics. Philosophy and foundations of mathematics.

Chapter 3

MATHEMATICAL DICTIONARIES, ENCYCLOPEDIAS AND THESES   28
Dewey classes: 510.3, 510.4

General dictionaries. Mathematical dictionaries. Signs and symbols. Translations. Interlingual mathematical dictionaries. Encyclopedias: general, mathematical. Theses.

Chapter 4

MATHEMATICAL PERIODICALS AND ABSTRACTS   47
Dewey class: 510.5

Importance of periodicals. Guides to periodicals. Different kinds of periodicals: universities, societies, commercial houses. Foreign-language periodicals. Locating and using periodicals: searching periodical literature. Abstracting and indexing publications. Cover-to-cover translations.

List of Illustrations

Preface to the Second Edition

FOR this second edition every section of the book has been revised and updated. New titles have been added, in particular to the chapters on probability and statistics, dictionaries and encyclopedias, computing and mathematical education—in all, over 200 new works and new editions have been incorporated throughout the text. Account has been taken of developments of interest to students, teachers, and practitioners, for example the modifications to the content of professional actuarial examinations, the assimilation of modern mathematics into the school curriculum, and the establishment of government departments to administer financial support for mathematical research. Also recorded are the latest efforts to improve communication between mathematicians, for instance the inception of the Mathematical Offprint Service and the publication of *Contents of Contemporary Mathematical Journals* by the American Mathematical Society. And perhaps special mention should be made of the specimen pages from important reference works—three times as many as in the first edition—which, it is hoped, will encourage the reader to acquire a greater familiarity with his subject's bibliographical apparatus.

The purpose of this book is to give those studying, teaching, or applying mathematics an insight into the vast amount of information which is available and to act as a guide to its exploitation. To this end a balance has been sought between descriptions of actual sources and the principles of biblio-science—the effective use of books and libraries. The text has been designed to facilitate continuous reading. Its chapters are arranged according to the Mathematics class of the Dewey Decimal Classification, which provides a convenient and progressive grouping of material, and a number of exercises are included under each heading. At the same time, a detailed contents list and a comprehensive index enable the book to be used for reference purposes.

Sources of Russian mathematical information are treated separately in an appendix because of their unique interest, whilst references to the works of other countries are incorporated in the body of the book. Both actuarial science and the role of government departments are similarly dealt with as appendixes.

The individuals, societies, and publishers who have willingly given both advice and factual information are now too numerous to mention by name, and I am obliged to make my acknowledgement to them collectively. I owe a special debt, however, to Dr. Chandler, the series editor, for his unfailing advice and encouragement.

University of Warwick   JOHN E. PEMBERTON

CHAPTER 1

Careers for Mathematicians

"With the post-war eruption in mathematics the problem of supply [of mathematicians] has become a serious one, and in some countries has taken on the character of a political as well as a social crisis."

E. R. Duncan, in the *Year Book of Education*, Evans, London, 1961.

General Survey

The past 5 years have witnessed a tremendous upsurge in the number of qualified mathematicians. In the United States alone the number of earned bachelor's and first professional degrees in mathematics and statistics has doubled, and the annual production of PhD's has more than trebled. Demand, however, continues to increase, and the purpose of this chapter is to draw attention to the main areas of employment, giving at the same time a brief guide to sources of relevant information.

Before the question of careers can be discussed it is necessary to define the word *mathematician*. Mathematics is an essential training for many scientific and technical as well as commercial and business occupations, but to attempt to cover all the fields in which some mathematical knowledge is required is beyond the scope of the present work. The mathematician is here considered as a person who has specialized in mathematics at degree level.

Approximately one-half of each year's product of mathematicians find employment in teaching of one form or another. The other half are spread through a wide range of situations. Qualified mathematicians are indeed to be found in almost every

1

sector of industry, though, as might be expected, many are grouped in electronics, aeronautics, and space exploration.

Two developments which have in recent years opened up bright new horizons for mathematicians, and which will continue to offer good prospects for many years to come, are high-speed computers and operational research.

The now established concept of automation has affected many industries and brought in its wake new openings for those with the necessary skills. Industries which are particularly concerned in this development include chemical processing, nuclear engineering, iron and steel and, of course, those which produce the necessary control systems.

The specialized field of actuarial science is facing growing competition for new recruits, but continues to offer good prospects for suitably qualified men and women.

Patterns of mathematical employment are constantly changing, but they are accurately reflected in the reports of the Mathematical Sciences Section of the National Register of Scientific and Technical Personnel. This is maintained by the American Mathematical Society in co-operation with the National Science Foundation. Ten other societies, representing such fields as statistics, electronic computation, actuarial science, and operational research, assist in the work. Data covering the year 1960 are analysed in a report entitled: *Professional Characteristics of Mathematical Scientists*, American Mathematical Society, Providence, RI, 1962. Among the questions with which the report deals are employment trends among younger mathematicians, patterns of degree attainment, the proportion of qualified persons absorbed by the individual subfields of mathematics, and salaries. Two especially interesting inferences are drawn which are equally valid for many countries outside the United States. The first is the "enormous growth of opportunities for non-academic employment of mathematicians"; and the second is that "mathematicians in industry, business and non-profit organizations are getting higher salaries at earlier ages than those in other types of employment". The discernible shift away from teaching is largely due to a strong attraction from new specialities which do not demand as high a level of academic attainment.

It will be stated on several occasions in this book that the work of the mathematician is often geared to the task of helping to solve the problems of workers in apparently unrelated fields. The contribution which the application of mathematical techniques can make in terms of efficiency and hard cash has created an unprecedented demand for the services of competent mathematicians. They are now being taken on in industry for their knowledge of mathematics itself rather than of the fields to which it is applied. It is significant that in the United States the number of mathematicians employed in private industry increased by nearly 100 per cent between 1954 and 1960. Organizations both large and small have found that their volume of experimentation can be considerably reduced by the elaboration of mathematical formulations. The aircraft industry is one which has particularly benefited in this respect. However, many less obvious areas of research are also directly concerned in this movement.

Guides to Careers

Guides to careers in mathematics appear under three main headings: general guides to occupations for scientists, particularly graduates; those dealing with the entire range of mathematical opportunities; and very specific guides which restrict themselves to individual mathematical professions. The following examples represent the first and second categories. More detailed information on particular professions is given in succeeding sections.

Careers Encyclopedia, Cleaver-Hume, 5th edn., 1967.

Directory of Opportunities for Graduates, Cornmarket Press, London, Annual.

FORRESTER, GERTRUDE, *Occupational Literature: an Annotated Bibliography*, Wilson, New York, rev. edn., 1964.

GEHMAN, HARRY MERILL, *Opportunities in Mathematics Careers*, Vocational Guidance Manuals, New York, 1964.

Jobs in Mathematics, Science Research Associates, Chicago, 1959.

The Mathematician, HMSO, London, 2nd edn., 1965 (Choice of Careers, 109).

NATIONAL COUNCIL OF TEACHERS OF MATHEMATICS AND NATIONAL ACADEMY OF SCIENCES—NATIONAL RESEARCH COUNCIL, *Careers in Mathematics*, Washington, 1961.

Professional Opportunities in Mathematics, Mathematical Association of America, Buffalo. (Prepared by the Association's Committee on Advisement and Personnel. The booklet is an outgrowth of an article which first appeared in the *American Mathematical Monthly*, and has gone through several editions.)

SHEDD, A. NEAL *et al.*, *Careers in Engineering, Mathematics, Science and Related Fields: a Selected Bibliography*. US Office of Education, 1961.

TURNER, NURA D., A bibliograghy for careers in mathematics, in *Science Education*, Oct. 1961, and *Mathematics Teacher*, Nov. 1961.

US BUREAU OF LABOR STATISTICS, *Occupational Outlook Handbook*. Published every 2 years.

A pamphlet on careers in mathematics has been prepared by and is available from the Conference Board of the Mathematical Sciences, 1515 Massachusetts Avenue, NW, Washington 5, DC, USA. The first section deals with teaching careers and the second with industrial and government careers in mathematics. Also, an eight-page leaflet entitled *How About a Career with Mathematics* is published by the Committee on High School Contests and is avalaible from the Mathematical Association of America.

In 1962 the US Bureau of Labor Statistics published *Employment in Professional Mathematical Work in Industry and Government; Report on a 1960 Survey Prepared for the National Science Foundation in Co-operation with the Mathematical Association of America*. The Bureau has also published *Math. and Your Career*, a *Monthly Labor Review* reprint, 1968.

There are a number of periodicals on careers which may be consulted as they frequently contain articles on mathematical openings. They include the *Journal of College Placement*, *Graduate Careers* and *Occupational Outlook Quarterly*.

Employment Registers

An employment register is jointly maintained by the American Mathematical Society, the Mathematical Association of America and the Society for Industrial and Applied Mathematics. The Mathematical Sciences Employment Register (PO Box 6248, Providence, Rhode Island 02904, USA), whose services are not limited to members of these bodies, contains a list of positions available to mathematicians in industry, government, and academic institutions, and a list of applicants. Its booklet on *Finding Employment in the Mathematical Sciences* gives the young entrant to the profession some general guidance on the role of the mathematician in teaching, research and computing as well as sources of practical information on employment. The *National Directory of Employment Services* (Gale, Detroit) gives details of over 5000 agencies, including university placement bureaux.

In Britain, the national advisory body is the Department of Employment and Productivity which maintains a Professional and Executive Register (Atlantic House, Farringdon St., London EC 4) and regionally based Occupational Guidance Units.

Most, if not all, universities and colleges offer positive assistance in finding suitable employment to those completing their studies. Often, the placement officer is well known to local employers who notify him of vacancies. Those institutions which have a high reputation for producing good mathematicians have valuable employment contacts on a national, even an international basis. It is a good plan to peruse the advertisements in mathematical and other journals for some months before graduation in order to get the feel of the market. Such reading can give a helpful stimulus to a student preparing for his final examinations.

Teaching and Women

Teaching mathematics will be dealt with in Chapter 6, but a word may be added here on the employment of women. Married women who hold academic qualifications are being increasingly sought to help meet the present-day shortage of teachers. Several countries have launched campaigns to attract married women back into the profession.

Among newly qualified women teaching is a favoured occupation. In the United States in 1957, 1166 women were graduated with an undergraduate major in mathematics. Of these, 86 per cent took jobs, and within this group 42 per cent were engaged as teachers. By way of comparison, another 42 per cent were engaged as mathematicians or statisticians, and 16 per cent in a variety of other positions. The National Register of Scientific and Technical Personnel for 1956–8 showed 1277 women in mathematical occupations. This was over 10 per cent of the total mathematical recordings. About 50 per cent were employed by colleges and universities, 30 per cent by industry, and 20 per cent by government agencies and other bodies. The figure for 1960 showed a leap to 1633 women mathematicians, most of whom did not have higher degrees. It is reasonable to assume that an increasingly high proportion are finding employment as computer programmers.

Part-time abstracting work can often be an attractive proposition. It is quite well paid, particularly when the abstractor is capable of handling material in one or more foreign languages in her field of specialization. The Management Services Group of the National Federation of Science Abstracting and Indexing Services (301 East Capitol Street, Washington 3, DC) and Aslib (3 Belgrave Square, London, SW 1) can give advice on opportunities.

Research

In the field of research, mathematics is often referred to as *the language of science*. This is because it is through mathematical formulations used in the different branches of science that workers in one can understand the problems and progress of others, and can in turn communicate their own. Mathematics is the common medium through which research can progress. The mathematician performs the necessary function of formulating problems, of rationalizing them in a manner which will most readily render them capable of solution. In this he frequently uses the technique of the mathematical model.

Recently, a leading company in the propulsion field advertised for a mathematician "to develop programmes that transform design intent into the definitions required for numerically controlled tools. Will mathematically define rocket engine compounds. A degree in mathematics is required." This is typical of the current industrial demand.

As industrial developments shift their emphasis and change their course, the one constant need will be for men soundly versed in the theoretical disciplines—especially mathematics. In 1961, UNESCO published a volume entitled *Current Trends in Scientific Research* by the consultant Pierre Auger. The first section of this authoritative work gives a concise account of the development of mathematics in the first half of the twentieth century. Auger surveys the links between mathematics and other sciences and technologies. He draws attention to a "trend towards duality in the training of research workers and technicians" which he defines as "a tendency to separate, in some degree, the theoretical and abstract from the practical and concrete". He then confirms that "the relative importance of theory, and of its most extreme form— mathematics—is undoubtedly increasing as time goes on".

The pattern of research has been considerably affected by the use of electronic computers, and mathematicians are employed at all levels of computer work. Highly trained mathematicians contribute to the actual design and development of the machines. Computers may, in fact, be said to be the result of a happy marriage between mathematics and electronics. Whilst programming work does not necessarily demand advanced mathematical training, the basic formulation of problems to be fed to the machines certainly does.

The US Civil Service Commission publishes a guide entitled

New Directions: career opportunities for computer specialists in Federal Government (Pamphlet No. 71, 1968). If computers provide a service to mathematicians by performing calculations which might otherwise be impracticable, they have also created a demand for suitably qualified personnel. At the present time they absorb mathematicians at a rate second only to that of teaching.

Employers in the field of research are listed in such publications as the *Research Centers Directory* (Gale, Detroit) which is updated by a periodic supplement, *New Research Centers*. As far as government work is concerned, the US Department of Defense is easily the largest employer, though most agencies employ some mathematicians. High figures are, for example, reported by the National Aeronautics and Space Agency, and the Department of Commerce of which the National Bureau of Standards forms part. For those primarily interested in basic research, NASA offers the most promising prospects.

Information of this kind is available for other countries. A comprehensive British work is *Industrial Research in Britain*, published by Harrap Research Publications, London. The sixth edition appeared in 1968.

Statistics

Mathematical statisticians are required in all branches of science and technology, in government departments, and in business concerns. The demand for them grows as new applications for their work are explored. Business management, quality control, biological research and agriculture are examples of areas in which the use of statistical techniques is currently increasing. It is obviously necessary for the statistician to acquire a working knowledge of each particular subject to which he applies his skills. He must work in close collaboration with subject experts in other fields and must therefore be able to understand their language and their problems. This demands the quality of adaptability and, equally important, a high degree of competence in using bibliographical tools.

Training in statistics requires a good level of mathematical knowledge as a prerequisite. In Britain, only the University of London offers a degree course in statistics. The subject is, however, included in the degree courses for mathematics and economics in other universities. Postgraduate diploma courses are available in such universities as Aberdeen, Birmingham, Cambridge, Manchester, Oxford, and Wales (Swansea). Details are given in the calendars of the respective institutions. The Association of Incorporated Statisticians (55 Park Lane, London, W 1) holds examinations at two levels: intermediate and final.

The Royal Statistical Society issues a booklet entitled *The Career of Statistician.*

United States universities which offer a master's degree in statistics are listed in the American Mathematical Society's annual survey of *Assistantships and Fellowships in Mathematics.* The American Statistical Association and the Institute of Mathematical Statistics' joint booklet, *Careers in Statistics,* describes what statisticians do by means of individual case histories, and includes useful sections on education and salaries. The American Statistical Association itself "facilitates the placement of its members through special arrangements with the United States Employment Service and by other means, including the numerous professional contacts of its members".

Operational Research

The role of the mathematician in OR is as a member of a team. The team comprises members having a variety of basic skills. This is reflected in the pre-entry requirements to diploma and degree courses in the subject. The Imperial College of Science and Technology (University of London) stipulates, for example, that "Prospective entrants should possess an engineering or science degree or equivalent, together with some industrial experience". The diploma course offered by the London School of Economics and Political Science "is open to men and women who hold a university degree in *any* subject, but a knowledge of Mathematics, Statistics and Economics to the level of the Part I examination of the BSc(Econ) degree will be assumed".

The methods used in OR are largely based on advanced mathematical techniques. The mathematical background to the theory of games, linear programming, and the theory of queues are important facets of the work.

More and more large companies are establishing OR departments. These departments vary greatly in size, but it is significant that new departments which have proved their effectiveness tend to grow quite rapidly. Not every industrial company or commercial undertaking is large enough to warrant the creation of its own OR department. Operational research services are, however, offered by research organizations and consultants. Attractive openings exist here for mathematicians, and those who decide to specialize in this work still have an opportunity of "getting in on the ground floor".

Salaries are relatively high both for new entrants and those with experience. In the Mathematical Sciences Section of the National Register of Scientific and Technical Personnel, practitioners of OR are grouped in the subfield "Mathematics of resource use" which has a high percentage of non-academic employment and the highest median salary of all. Whilst this group also includes the highly paid actuarial profession, the fact that its salary rating is equally representative of the OR field is borne out by advertisements in the professional press.

To meet the demand for organized instruction in OR, new courses are being offered by colleges and universities, and there is in consequence a growing need for lecturers in the field. Operational research societies such as the Operational Research Society (London) undertake to arrange for suitably qualified lecturers to provide on-the-job instruction within industry, and can give advice on opportunities for work of this kind.

Formal training in OR at recognized educational institutions is normally given at post-graduate level. The validity of the subject as an academic discipline has now been recognized. The number of courses offered on a regular basis is growing rapidly. In Britain, the courses are usually of one year's duration, except in the case of the London School of Economics and Political Science, where normally a full-time course of instruction extending over one academic year is followed by a further year spent in practical work. The first British course was that introduced by the University of Birmingham in 1958, leading to the Degree of Master of Science in Operational Research. Other full-time courses are offered by the College of Aeronautics (Cranfield), Hull University, the Imperial College of Science and Technology (University of London), the University of Wales Institute of Science and Technology, and the University of Warwick.

The *Bulletin of the Operations Research Society of America* publishes information on education for OR. An extensive listing of courses was given in vol. 15, supplement 2, Summer 1967, and this has since been augmented. The Society has also produced a brochure entitled *Careers in Operations Research.*

The breadth of subject coverage in courses of instruction in OR and cognate techniques may be exemplified by listing the main areas of the syllabus of a typical course. They are statistics, scientific method, industrial psychology, industrial economics, management accounting, data processing, stochastic processes, mathematical programming, planning, and control of production and inventories.

Apart from full-time courses, many short courses are organized by various bodies. They are usually concerned with specific OR techniques, and the best way of keeping informed about them is to consult the OR periodical literature listed in Chapter 12. Furthermore, the national societies are always able to provide such information upon request.

Practical experience is an essential part of OR training. This is recognized by those organizations which utilize its techniques, and many of them engage young graduates in appropriate subjects and offer them training within working groups before financing their attendance at academic courses.

Actuarial Science

Actuaries are among the highest paid mathematicians. They belong to a relatively small profession. In Great Britain at any time there are only about a thousand Fellows of the Institute of Actuaries. The training is exacting, but the rewards are stimulating. The primary work of an actuary is concerned with long-term financial contracts, particularly in the field of life insurance. It involves, among other things, the balancing of assets and liabilities, the rating of premiums, and the investment of accrued funds. It is responsible work demanding integrity of a high order, and from a mathematical point of view, a thorough grounding in probability theory and statistics, and their practical application. Actuaries are extending their sphere of operation. They have, for some time, found employment in government service. For example, in 1917 the post of Government Actuary was created in Britain. This was followed by the foundation of the Government Actuary's Department, which was intended to deal with national social security activities. Local government and private pension schemes similarly require the participation of actuaries. Statistical activities in which actuaries find work include vital statistics, the construction of life tables, and demography. Stock exchanges are another source of employment. But it is in commerce that opportunities are most rapidly increasing. Actuarial skills are required in the manipulation of investments in large organizations.

An analysis of occupations of Fellows and Associates of the Institute of Actuaries (London) indicates that most are employed in assurance offices. The next highest figure is for whole-time consulting practice. Others are in central and local government service, stock exchange, industry and commerce. It is of interest to note that in 1925 there were no entries under the last heading.

In Britain, the Commonwealth, and South Africa the necessary qualifications are those awarded by the Institute of Actuaries and the Faculty of Actuaries in Scotland. Tuition is organized by the Actuarial Tuition Service which is administered by a Joint Committee of the two bodies. A list of textbooks and other useful study literature is given in the *Year Book* of the Institute. This list has the advantage of being graded according to the qualification being studied for. In 1969 the examinations were rearranged to form two stages—Intermediate and Final. Changes have been made in the syllabus intended to keep the examinations up to date and in line with current university practice; and it is of interest to students to note that these have the effect of:

(a) reducing the emphasis on finite differences and increasing that on numerical analysis;

(b) including the basic principles of computer programming at an early stage of the examinations so that the candidate becomes aware of the power, and also the limitations of computers as a means of performing actuarial computations;

(c) revising the approach to probability and statistics so as to give a more unified treatment of probability and small sample theory, leaving large sample theory as a limiting case of the latter;

(d) introducing the subject of operational research into the examination in more advanced statistics, with a reduction in the treatment of mathematical statistics.

Those wishing to qualify should obtain a copy of *Institute of Actuaries: the examinations, syllabus and course of reading*. More general information is given in its 20-page booklet, *The Actuarial Profession*. The Institute does everything possible to ensure a healthy profession. It has set up an Appointments Board with three stated aims: "To provide authoritative guidance to prospective employers of actuaries who seek such guidance; to provide recognized means of bringing suitable vacancies to the notice of members of the Institute; and to provide an opportunity for members who wish to do so, to discuss matters affecting their careers with a senior member of the profession."

In America, the examining body is the Society of Actuaries, 208 South La Salle Street, Chicago, Illinois 60604. There are two parts to the Preliminary Actuarial Examinations. The first is the General

Mathematics Examination which is based on the first 2 years of college mathematics; and the second is the Probability and Statistics Examination. The actual examinations are held at various centres in the United States and Canada. Another American insight into the profession is given in George E. Immerwahr, A career as an actuary, *J. College Placement*, **21,** 43 *et seq.* (1961).

Sources of information in actuarial science are treated separately in Appendix III as they involve subjects beyond the purely mathematical field.

CHAPTER 2

The Organization
of Mathematical Information

510: Mathematics, general
510.1: Philosophy of mathematics
510.2: Handbooks and outlines

THE science of mathematics is a product of human intelligence. Its development has been recorded through history in many forms, the most important of which has undoubtedly been the printed word. The invention of printing from movable type in the fifteenth century gave man a ready means of recording and communicating his thoughts and ideas, and has contributed significantly to the attainment of the highly sophisticated mathematical techniques of today.

In this process of communication by means of printed words and other symbols librarians play an active role. Not only do they collect and preserve the vital records, but they analyse and organize them for future use. Libraries are an essential complement to formalized courses of study and an important part of every mathematician's training is that devoted to the efficient use of libraries and the exploitation of all forms of recorded knowledge and sources of information.

Sources of mathematical information may be divided into two main categories: bibliographical and non-bibliographical. Books, periodicals, and other printed documents are bibliographical sources. The second category includes professional societies, research organizations, and subject specialists. It should, however, be noted that the *guides* to these do themselves take a bibliographical form. Non-bibliographical sources generally receive too

little attention, particularly in relation to their growing importance. The relationship between them and the more conventional
bibliographical sources has therefore been made a feature of
succeeding chapters.

The need for adequate training in exploiting information
sources becomes more strongly felt when formal education ceases.
It increases with the degree of specialization one achieves. This is
particularly true in today's pattern of scientific research where
specialization often takes the form of interdisciplinary synthesis,
producing such subjects as mathematical chemistry, mathematical
economics, and mathematical linguistics. Whilst, therefore, it is
important to know in detail the sources in one's own particular
subject field, it is equally important to acquire an understanding
of the basic principles of information techniques.

The first essential is a good working knowledge of libraries:
what they contain, and the services they provide.

Guides to Using Libraries

A useful introductory work is Jean Kay Gates's *Guide to the Use
of Books and Libraries* (McGraw-Hill, New York, 1962). Written
by an American librarian, it deals especially with college libraries
and their use. It is a textbook for students covering the different
kinds of library materials and how they are organized. Ella V.
Aldrich's *Using Books and Libraries* (Prentice-Hall), has gone
through several editions since 1940. It was originally written for
the freshman course in library use at Louisiana State University.

The series of which the present work forms part guides inquirers
in all the major subject fields through the relevant literature and
other sources of information. The first volume is the compact but
comprehensive *How to Find Out* by George Chandler (Pergamon
Press, Oxford), now in its third edition.

Guides to Libraries and Special Collections

The mathematician needs to use the resources of many different
kinds of libraries. From student days onwards the university or

college library and the public library will be his standby. On joining the appropriate professional society he will become aware of the resources of its library and information services. Should he find employment in an industrial firm he will in all probability discover that it has an efficient specialized library, for industrial libraries form one of the most rapidly developing segments of the library profession. Particularly if he has recourse to official statistical publications, he will come into contact, either directly or indirectly, with the libraries of givernment departments. The more specialized his work becomes the greater will be his need to know which libraries further afield will be likely to supply the information he requires.

There are national directories in existence for most countries. The *American Library Directory* (Bowker, New York) gives information on libraries in the United States and Canada. *Specialized Science Information Services in the United States* (National Science Foundation, 1961) is a directory of information centres which indicates subject specializations, services provided, user qualifications, and publications. Another Bowker compilation, now in its third edition, 1967, is *Subject Collections: a guide to special book collections and subject emphases as reported by university, college, public, and special libraries in the United States and Canada.* Prepared by Lee Ash and Denis Lorenz, it contains about 20,000 entries arranged in alphabetical order of subjects.

In Britain, the 2-volume *Aslib Directory* provides particulars of several hundred libraries arranged geographically, with a complete index to their subject specialities. The second edition of vol. 1, *Information Sources in Science, Technology and Commerce* (1968), has over 200 index references under mathematics and statistics. A more recent British guide is *Special Library and Information Services in the United Kingdom,* edited by J. Burkett and published by the Library Association (2nd edn., 1965). Separate chapters deal with the different types of libraries.

The libraries of professional societies are mentioned in a later chapter, but it is appropriate to quote here a publication of a professional society for documentation, the International Federation for Documentation. Its *Bibliography of Directories of Sources of Information* (The Hague, 1960) lists national and international guides to sources of scientific and technical information in no less than thirty-five countries. In 1965 UNESCO published a *World Guide to Science Information and Documentation Centres*; and, more specifically, mathematics collections in Europe are listed in Richard C. Lewanski's *Subject Collections in European Libraries: a directory and bibliographical guide* (Bowker, New York, 1965), in which the entries are arranged according to the Dewey Decimal Classification—there are fifty-five items under the mathematics heading.

Inter-library Co-operation

No library is ever complete. Even the great libraries of the world such as the British Museum, the Library of Congress, and the Lenin Library have their limitations. But librarians have for many years sought ways to increase the accessibility of recorded information, largely through formal and informal schemes of inter-lending. These are organized at local, national, and international levels. Every public library in Great Britain, for example, is linked through Regional Bureaux to the National Central Library which maintains vast union catalogues of library stocks. A request for the loan of a particular book not available in a local library will if necessary be circulated throughout the country until a library is found which has a copy and is prepared to lend it. Schemes with similar aims exist in other countries, the majority of which participate in international lending projects.

Photocopying and Microcopying

It is often advantageous, both for the requesting library and the lending library, for loan requests to be met by a photocopy of the work in question. Apart from reducing postage costs, this method also obviates the risk of loss of the originals or their damage in transit. Furthermore, the individual inquirer is usually able to borrow such a photocopy for a longer period, or may in some cases be made a gift of it. Photocopying in libraries is governed by the laws of copyright. In general, it is in order to make, for the purpose of private study or research, one copy of an article from a serial publication or of a limited extract from a book without infringement. Limited copying of this sort is known as "fair dealing", and an international agreement setting out the conditions of fair dealing in precise terms would be of great value to scholarship.

There are several ways in which documents are reproduced photographically. Apart from original-size copies there is a variety of microforms including continuous *microfilm*, opaque *microcards*, and their transparent counterparts called *microfiches*. A complete technical report can be reproduced on a single microcard, or a whole book on a roll of microfilm. It is, of course, necessary for this material to be used in a reading machine which throws up an enlarged image onto a screen. An increasing number of books and periodicals are being produced on microfilm to meet demands for out-of-print publications, and for the convenience of libraries wishing to conserve their limited shelf accommodation. A complete file of the *Journal of the Institute of Actuaries* for 1850–1959, for example, is available on microfilm from Micro Methods Ltd., England. Pioneers of this type of reproduction are University Microfilms of Ann Arbor, Michigan.

Microcard Editions (Washington, DC) publish an annual *Guide to Microforms in Print* and a *Subject Guide to Microforms in Print* in which about forty titles are listed under mathematics.

Library Catalogues

The function of the catalogue is to make the contents of the library known, to indicate the location of every item within the library. Catalogues take different forms. They may be printed books, sequences of filed index cards, loose-leaf sheets locked in binders, visible index strips, or even punched cards. Classic examples of printed catalogues are those of the British Museum

(Fig. 1) and the US Library of Congress. Most public, university, and college libraries have adopted index cards as being the most satisfactory from the point of view of updating. The Library of Congress printed catalogue was in fact produced from arrays of cards laid down in page form suitable for photo-mechanical reproduction. Catalogue cards produced by the Library of Congress may be purchased for use by other libraries. Each card bears full particulars of the book in question and the appropriate classification symbols both from the Library of Congress Classification and the Dewey Decimal Classification. Many new books indicate the relevant card numbers to facilitate ordering.

Whatever the form a catalogue takes, its entries will always follow a similar pattern. Minimum details are author, title, date and location symbol. The publisher and place of publication are also normally included. Generally the location symbol coincides with the classification notation since most libraries arrange their books on the shelves in a classified order. There are many rules governing the cataloguing of books which are designed to ensure consistency of treatment.

It is essential for the mathematician to familiarize himself with the arrangement of the library catalogues he intends to use. Much will be missed if too much reliance is placed on the shelf arrangement of the books themselves. This would be rather like consulting the alphabetic arrangement of articles in an encyclopedia and ignoring the analytical index. Most well-organized libraries issue a guide to the use of their catalogues.

Three aspects of a book may be used as the basis for arranging catalogue entries: its author, its title, and its subject. In consulting indexes and catalogues care should be taken to understand the filing rules that have been followed. For example, there is the difference between letter-by-letter and word-by-word arrangements. The former places *Logarithms* before *Log cabins*, whilst the latter places *New York* before *Newspapers*. Abbreviations, including numbers, are filed as if they were written out in full: M' and Mc equal Mac, St. equals Saint, 8 equals Eight.

When author, title, and subject entries are interfiled to form one

MATHEMATICS

See also AGGREGATES, *Theory of:* ALGEBRA: ANALYSIS: APPROXIMATION, *Theory of:* ARITHMETIC: CALCULUS: CONIC SECTIONS: DETERMINANTS: DYNAMICS: EQUATIONS: ERROR, *Theory of:* FUNCTIONS: GEOMETRY: GRAPHS: GROUPS: KINEMATICS: LOGARITHMS: MATRICES: MECHANICS: MENSURATION: NUMBERS: PHYSICS, *Mathematical:* POTENTIAL, *Theory of:* PROBABILITIES: QUATERNIONS: RELATIVITY: SERIES: STATICS: TIME AND SPACE: TRANSFORMATIONS: TRIGONOMETRY. For Mathematical Logic, *see* LOGIC. For the logic of mathematics, *see below, Philosophy, etc., of Mathematics.*

Bibliography

CRACOW.—*Polska Akademia Umiejętności.* [*Miscellaneous Publications.*]

—— Wykaz prac z działu nauk matematyczno-przyrodniczych, wykonanych w Polsce w okresie okupacji niemieckiej w latach 1939–1945. List of the works achieved in the field of mathematics and sciences in Poland during the German occupation, 1939–1945. [Edited by Jan Stach.] pp. 289. *Kraków,* 1947. 8°.   Ac. 750/188.

NEW YORK.—*New York Mathematical Society,* afterwards *American Mathematical Society.*

—— Catalogue of the Library. Edited by Arnold Dresden. pp. v. 173. *Menasha, Wis. & New York,* 1946. 8°. [*Bulletin of the American Mathematical Society.* vol. 52. no. 5. pt. 2.]   Ac. 4286.

FIG. 1. *British Museum Subject Index.* (By kind permission of the Trustees of the British Museum.)

alphabetical arrangement the result is known as a *dictionary catalogue*. Cross-references are included to guide the user. It can, of course, happen that the same filing word may be the name of an author, a place name, or a book title. The word "Bedford" is an example. In such cases the accepted filing order is:

Persons—authors	e.g.	Bedford, John Harold
—subjects		Bedford, John Harold (a biography)
Places —authors		Bedford Public Library
—subjects		Bedford (a town guide)
Titles		*Bedford Parks*

A *classified catalogue* is one in which the subject entries are filed separately in the order of classification notations, thereby paralleling the sequence of books on the shelves. This requires a subject index to enable the user to convert the name of his subject into the appropriate notation. The author index, including titles, is also in a separate run.

The library catalogues of professional societies and other organizations are often published in book form, for example the *Register of Mathematical Publications in NSA*, compiled by Jean Chandler Andrews (which lists the holdings of the US National Security Agency Library), and *Books and Periodicals in the Library of the Mathematical Association*—a catalogue of the Association's library maintained at the University of Leicester. Other examples are given in later chapters. Catalogues of this kind are particularly helpful in locating highly specialized material.

Classification of Mathematics

The object of classification in libraries is to bring together like material. Whilst printed records appear in different *forms* (books, periodicals, reports, etc.) they are classified by *subject*, since this is the approach most used by inquirers. Whilst new systems of classification are constantly being evolved, particularly to meet the requirements of those developing the use of electronic computers for information storage and retrieval, most libraries have adopted one or other of the established sets of published schedules. Undoubtedly the scheme most likely to be met with is the Dewey Decimal Classification, which was first published in the United States towards the end of the last century and is now in its sixteenth edition. It, and its descendant the Universal Decimal Classification, are the most widely used in libraries throughout the world.

Basically, the Dewey Decimal Classification consists of a distribution of the nine major divisions of human knowledge, plus a generalia class, through a three-figure array of main classes, as follows:

000: Generalia	500: Pure Science
100: Philosophy	600: Applied Science
200: Religion	700: Fine Arts
300: Social Sciences	800: Literature
400: Language	900: History and Geography

Each main class is subdivided into ten subclasses:

500: Pure Science	520: Astronomy
510: Mathematics	530: Physics, etc.

Further divisions are made in a similar way, each time decreasing the extention of the subject specified:

510: Mathematics	512: Algebra
511: Arithmetic	513: Geometry, etc.

The process continues after the decimal point:

517: Calculus

517.1: Infinitesimal calculus	517.3: Integral calculus
517.2: Differential calculus	517.4: Calculus of variations.

The extreme simplicity of this method of subdivision, plus the fact that Arabic numerals are almost universally understood, has made the Decimal Classification an important instrument in the documentation of recorded information at an international level.

Many articles in technical periodicals bear a Decimal Classification symbol; bibliographies and abstracting publications are arranged by the Decimal notation. Thus a mathematician knowing the symbols for his own field of specialization can readily exploit foreign sources which might otherwise remain inaccessible. He is immediately at home in a strange library.

Simple as the decimal notation is to memorize, additional mnemonic devices have been built into the scheme in the form of common subdivisions which are applicable throughout the schedules. Some examples are seen in the arrangement of the chapters which follow. 03 indicates dictionaries, 05 periodicals, and 09 history when added to subject numbers. When the basic notation ends in 0 this figure need not be repeated. Thus 510.3 equals mathematical dictionaries, 510.5 equals mathematical periodicals, and 510.9 equals history of mathematics.

This very brief outline of the scheme can be supplemented by the *Guide to the Use of Dewey Decimal Classification* (Forest Press, Lake Placid, NY) which is based on the practice of the Decimal Classification Office at the Library of Congress.

Philosophy and Foundations of Mathematics

In the preface to his book *Mathematics Manual: methods and principles of mathematics for reference, problem solving, and review* (McGraw-Hill, New York, 1962), Frederick S. Merritt talks of "The vine of mathematics, rooted in logic". A study of logic is indeed essential to a full understanding of the nature of mathematics. Quite naturally, the Dewey Decimal Classification makes provision for the subject in its philosophy class 100. Works on symbolic logic are specifically catered for in subclass 164. Whilst the primary function of classification is to bring like things together, it is inevitable that some subjects which in practice are *studied* together will be separated in the classification schedules. In Chapter 11 attention is drawn to other cases in which subjects of interest to the mathematician are classified outside the 510 heading.

An excellent "Bibliography of symbolic logic, 1666–1935" by Alonso Church appeared in vol. 1 of the *Journal of Symbolic Logic*, 1936. Additions and corrections were published in a later volume. This journal is the official organ of the Association for Symbolic Logic. It contains papers on symbolic logic and studies in the history of logic, and regularly features an extensive review section of relevant literature. Another important journal in the field is the German *Zeitschrift für mathematische Logik und Grundlagen der Mathematik*, Berlin, which commenced publication in 1955.

A Survey of Symbolic Logic by Clarence Irving Lewis (Dover, New York, 1960) also includes a very detailed bibliography. This edition is actually an adapted republication of the original work issued in 1918 by the University of California Press. In association with Cooper Harold Langford, the same author also produced *Symbolic Logic*, whose second edition was published by Dover in 1959.

An examination of mathematical books in a library will reveal that certain publishers tend to specialize in books on a particular subject. These are often produced as a uniform series. The North-Holland Publishing Company of Amsterdam is currently issuing such a series entitled *Studies in Logic and the Foundations of Mathematics*, edited by A. Heyting, A. Robinson, and P. Suppes. Some two dozen titles have already appeared. Representative of these are Beth, E. W., *The Foundations of Mathematics*, and Fraenkel, Abraham A., *Abstract Set Theory*, which again contains a useful bibliography. Van Nostrand's *University Series in Undergraduate Mathematics* contains additional works on set theory including Patrick Suppes' *Axiomatic Set Theory*. His *Introduction to Logic* is also in this series.

There are several great names which will always be linked with the development of the foundations of mathematics. Bertrand Russell is prominent among these. His *Principles of Mathematics* was originally published in 1903 by the University Press, Cambridge, and later published with a revised introduction. With A. N. Whitehead he produced the extremely important *Principia*

Mathematica (3 vols, Cambridge University Press, 1910–13). Alfred Tarski's contributions to mathematical logic and meta-mathematics were published by Oxford University Press in 1956 under the title *Logic, Semantics, Metamathematics.* The papers this volume comprises were all originally published in the years 1923–38 in various languages. Numerous references guide the reader to related writings. *Boolean algebra* and *Hilbert spaces* are terms well known to mathematicians. Both George Boole and David Hilbert made important contributions to the philosophy of mathematics. *Studies in Logic and Probability* (Waters, London, 1952) contains Boole's *Mathematical Analysis of Logic* and other writings. With P. Bernays, Hilbert wrote *Grundlagen der Mathematik* (2 vols., Springer, Berlin, 1934–9).

Recent works which may be noted are:

CURRY, H. B., *Foundations of Mathematical Logic*, McGraw-Hill, 1963.

EVES, HOWARD and NEWSOM, CARROLL V., *An Introduction to the Foundations and Fundamental Concepts of Mathematics*, Holt, Rinehart & Winston, revised edn., 1965.

FREUND, JOHN, *A Modern Introduction to Mathematics*, Prentice-Hall, New York, 1956.

KLEENE, STEPHEN COLE, *Mathematical Logic*, Wiley, New York, 1967. A sound undergraduate text with a good bibliography.

KNEEBONE, G. T., *Mathematical Logic and the Foundations of Mathematics: an introductory survey*, Van Nostrand, 1963. A valuable introduction to the modern terminology.

MATES, BENSON, *Elementary Logic*, Oxford University Press, 1965. An excellent introductory text in symbolic logic.

ROSSER, JOHN, *Logic for Mathematicians*, McGraw-Hill, New York, 1953.

WILDER, RAYMOND L., *Introduction to the Foundations of Mathematics*, Wiley, New York, 2nd edn., 1965.

WITTGENSTEIN, LUDWIG, *Remarks on the Foundations of Mathematics*, MIT Press, 1967. Has the German and English texts on facing pages.

Exercises

1. List some of the ways in which a librarian can help a student of mathematics.
2. Why is it more important for a mathematician to know how to use a library than to memorize a long list of books on his subject?
3. Distinguish between bibliographical and non-bibliographical sources of information and discuss their relative importance.
4. Write notes on the Dewey Decimal Classification with particular reference to mathematics.

CHAPTER 3

Mathematical Dictionaries, Encyclopedias and Theses

510.3: Dictionaries and encyclopedias
510.4: Essays and lectures

THE scope of this chapter heading has been widened a little to include something about translations of foreign mathematical literature. At the same time it has been restricted to the extent that Russian mathematical literature is dealt with separately in Appendix I.

Dictionaries

GENERAL DICTIONARIES

Though the concern here is primarily with dictionaries of mathematics, it should be noted that the better *general* dictionaries have a much wider use than the simple verification of spellings and definitions. Through their careful distinctions in shades of meaning, their explanations of historical derivations, and their cross references, they can be a considerable aid in initial orientation in unfamiliar and quite complex subjects. The following should be examined:

> *Oxford English Dictionary*, edited by Sir James Murray *et al.*
> *Webster's New International Dictionary*.
> FUNK and WAGNALL'S *New Standard Dictionary*.

MATHEMATICAL DICTIONARIES

An important work is the *Mathematics Dictionary*, edited by Glenn and Robert C. James (Fig. 2). It is described in the preface power series (2)]. Also called *Euler's method of summation*. See SUMMATION—summation of a divergent series.

Abel's problem. Suppose a particle is constrained (without friction) to move along a certain path in a vertical plane under the force of gravity. Abel's problem is to find the path for which the time of descent is a given function f of x, where the x-axis is the horizontal axis and the particle starts from rest. This reduces to the problem of finding a solution $s(x)$ of the *Volterra integral equation of the first kind* $f(x) = \int_0^x \frac{s(t)}{\sqrt{2g(x-t)}}\, dt$, where $s(x)$ is the length of the path. If f' is continuous, a solution is

$$s(x) = \frac{\sqrt{2g}}{\pi} \frac{d}{dx} \int_0^x \frac{f(t)}{(x-t)^{\frac{1}{2}}}\, dt.$$

Abel's tests for convergence. (1) If the series $\sum u_n$ converges and $\{a_n\}$ is a bounded monotonic sequence, then $\sum a_n u_n$ converges. (2) If $\left| \sum_{n=1}^k u_n \right|$ is equal to or less than a properly chosen constant for all k and $\{a_n\}$ is a positive, monotonic decreasing sequence which approaches zero as a limit, then $\sum a_n u_n$ converges. (3) If a series of complex numbers $\sum a_n$ is convergent, and the series $\sum (v_n - v_{n+1})$ is absolutely convergent, then $\sum a_n v_n$ is convergent. (4) If the series $\sum a_n(x)$ is uniformly convergent in an interval (a, b), $v_n(x)$ is positive and monotonic decreasing for any value of x in the interval, and there is a number k such that $v_0(x) < k$ for all x in the interval, then $\sum a_n(x)v_n(x)$ is uniformly convergent (this is Abel's test for uniform convergence).

Abel's theorem on power series. (1) If a power series, $a_0 + a_1 x + a_1 x^2 + \cdots + a_n x^n + \cdots$, converges for $x = c$, it converges absolutely for $|x| < |c|$. (2) If $\sum_0^\infty a_n$ is convergent, then $\lim_{t \to 1-} \sum_0^\infty a_n t^n = \sum_0^\infty a_n$, where the limit is the limit on the left at $+1$. An equivalent statement is that if $\sum_0^\infty a_n x^n$ converges when $x = R$, then S is continuous if $S(x)$ is defined as $\sum_0^\infty a_n x^n$ when x is in the closed interval with end points 0 and R. This theorem is designated in various ways, most explicitly by "Abel's theorem on continuity up to the circle of convergence."

A-BEL'IAN, *adj.* Abelian group. See GROUP.

A-BRIDGED', *adj.* abridged multiplication. See MULTIPLICATION.

A

A.E. An abbreviation for *almost everywhere*. See MEASURE—measure zero.

AB'A-CUS, *n.* [*pl.* abaci]. A counting frame to aid in arithmetic computation; an instructive plaything for children, used as an aid in teaching place value; a primitive predecessor of the modern computing machine. One form consists of a rectangular frame carrying as many parallel wires as there are digits in the largest number to be dealt with. Each wire contains nine beads free to slide on it. A bead on the lowest wire counts unity, on the next higher wire 10, on the next higher 100, etc. Two beads slid to the right on the lowest wire, three on the next higher, five on the next and four on the next denote 4532.

ABEL, Niels Henrik (1802–1829). Abel's identity. The identity

$$\sum_{i=1}^n a_i u_i = s_1(a_1 - a_2) + s_2(a_2 - a_3) + \cdots + s_{n-1}(a_{n-1} - a_n) + s_n a_n,$$

where

$$s_n = \sum_{i=1}^n u_i.$$

This is easily obtained from the evident identity:

$$\sum_{i=1}^n a_i u_i = a_1 s_1 + a_2(s_2 - s_1) + \cdots + a_n(s_n - s_{n-1}).$$

Abel's inequality. If $u_n \geq u_{n+1} > 0$ for all positive integers n, then $\left| \sum_{n=1}^p a_n u_n \right| \leq L u_1$, where L is the largest of the quantities: $|a_1|$, $|a_1 + a_2|$, $|a_1 + a_2 + a_3|, \cdots, |a_1 + a_2 + \cdots + a_p|$. This inequality can be easily deduced from Abel's identity.

Abel's method of summation. The method of *summation* for which a series $\sum_0^\infty a_n$ is *summable* and has *sum* S if $\lim_{x \to 1-} \sum_0^\infty a_n x^n = S$ exists. A convergent series is summable by this method [see below, Abel's theorem on

FIG. 2. James and James, *Mathematics Dictionary*. (By kind permission of D. Van Nostrand Company Inc.)

as "by no means a mere word dictionary, neither is it an encyclopedia. It is rather a correlated condensation of mathematical concepts, designed for time-saving reference work. Nevertheless the general reader can come to an understanding of concepts in which he has not been schooled by looking up the unfamiliar terms in the definition at hand and following his procedure to familiar concepts." This statement is worth re-reading, for this is the way a dictionary should be used. An appendix contains a selection of tables and a useful list of mathematical symbols, and there are indexes in French, German, Russian, and Spanish. A "New Multilingual Third Edition" was published by Van Nostrand in 1968.

There are three other recent dictionaries which may be noted here. Less expensive than James is C. C. T. Baker's *Dictionary of Mathematics* (Newnes, London, 1961) (Fig. 3). It is a concise work "suitable up to degree standard". A cheaper work still, and one which was "written specially for young people", is *A Dictionary of Mathematics* by C. H. McDowell (Cassell, London, 1961). The *Dictionary of Mathematics* by William and T. Alaric Millington (Cassell, London, 1966) is quite comprehensive, ranging from set theory and linear algebra to statistics and electricity, and is suitable for undergraduate-level work. Undoubtedly the value of the *Iwanami Mathematical Dictionary*, edited by the Mathematical Society of Japan (Iwanami Shoten, Tokyo, 1960), will be more fully appreciated when the planned parallel English text is also available.

Selection of a dictionary must be based on the requirements of the individual user. The only real way to test the suitability of a particular volume is to put it to practical use. Two or three should be available in most libraries. It is to be expected that the less expensive works will be more selective in their coverage. Baker, for example, does not include *topology*, but limitations of this sort must be matched against the price and the educational level for which the dictionary was compiled.

Other dictionaries of interest to the mathematician occur in more specialized fields, and are described in the appropriate

DICTIONARY OF MATHEMATICS

A

Abacus. A framework of wires carrying beads, and used for arithmetical calculations.

Abel. (1802–1829.) Abel proved the impossibility of solving the general quintic equation, $ax^5 + bx^4 + cx^3 + dx^2 + ex + f = 0$, by means of radicals. He wrote several papers on the convergence of series, "Abelian" integrals, and elliptic functions. One of the first to insist on rigor in Mathematics.

Abridged Notation, the Method of. The method of using a single letter, equated to zero, to represent a given locus. Thus, $l = 0$ may represent the straight line $Ax + By + C = 0$, and $S = 0$ may represent the point-conic

$$ax^2 + 2hxy + by^2 + 2gx + 2fy + c = 0.$$

This makes expressions more manageable. For example, let the line $x \cos \alpha + y \sin \alpha - p_1 = 0$ be represented by $l_1 = 0$, and the line $x \cos \beta + y \sin \beta - p_2 = 0$ be represented by $l_2 = 0$. Then any straight line through the point of intersection of the two lines is represented by $l_1 - kl_2 = 0$, where k is an arbitrary constant. Similarly, the equation of the conic passing through the points of intersection of the conics $S_1 = 0$ and $S_2 = 0$ is $S_1 = kS_2$.

Abscissa. Using rectilinear coordinates, it is the perpendicular distance of a point from the y-axis, with the proper sign attached. Using inclined axes, it is the distance from the point to the y-axis along a line parallel to the x-axis. In Fig. A-1, PN is the abscissa of P.

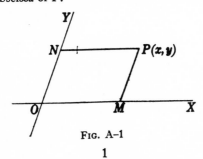

Fig. A–1

1

FIG. 3. Baker, *Dictionary of Mathematics*. (By kind permission of Butterworths.)

chapters. Representative of these is the *Dictionary of Statistical Terms* by M. G. Kendall and W. R. Buckland (Oliver & Boyd, Edinburgh, 2nd edn., 1960).

For those with a knowledge of foreign mathematical terminology, interlingual dictionaries can also be used to advantage in clarifying the meanings of terms in one's native tongue. Some German terms, for example, will be found to be more expressive than their English counterparts.

SIGNS AND SYMBOLS

A uniquely useful compilation is David D. Polon's *Dictionary of Physics and Mathematics, Signs and Symbols* (Odyssey, New York, 1965); and serving a special need is Karel Wick's *Rules for Type-setting Mathematics*, a 100-page manual for editors and others, published by Mouton, The Hague, 1965.

TRANSLATIONS

Mathematics is international. New developments are reported in the scientific press of many countries and in many different languages. Abstracting and translating services endeavour to facilitate the use of foreign-language literature, but it is frequently necessary for the mathematician to be able to read material in its original form. It is for this reason that many universities have a language requirement in their degree programmes. The three most useful languages are French, German, and Russian. A reading knowledge of *two* foreign languages is often specified for doctoral degrees.

It is only natural that English-speaking mathematicians will first try to discover whether a particular foreign work has appeared in translation. In this connection the American Mathematical Society's *Notices* gives details of Russian books currently being translated into English by the Society, under grants from the National Science Foundation, under the heading "Mathematics in translation". This feature also reports translations, both from

Russian and from other languages, being prepared by other organizations. An *Index to AMS Selected Translations* is also available.

A standard reference work on translations is the *Index Translationum: international bibliography of translations* published in Paris by UNESCO. The stated object of UNESCO in publishing the index is to provide an annual listing of all translated books published throughout the world on any subject. Some idea of its proportions may be gained from the fact that the 1965 annual volume contains more than 36,000 items representing 70 different countries. The section on the United States is compiled by the General Reference and Bibliography Division of the Library of Congress, and that on the United Kingdom by the British Museum. Translations are first arranged according to the country in which they were published, and are further subdivided by the 10 major headings of the Universal Decimal Classification. Thus mathematics comes under the heading "Natural and Exact Sciences". A complete alphabetical index of authors is provided. The system of the International Organization for Standardization for transliterating Cyrillic alphabets (e.g. Russian) is used. Each entry follows the plan: author; title of the translation; place of publication; name of publisher; number of pages, etc.; price (in the country of translation); the language in which the work was written, or from which it was translated; title of the original work; place, publisher, and date of the original.

A recent *Survey of Translation Activities in Universities, Societies and Industry in the Fields of Science and Technology*, prepared by the Special Libraries Association, New York, indicates that the US sources most frequently checked for existing *unpublished* translations are the SLA Translations Center in Chicago, and the Office of Technical Services. In Britain, Aslib and the National Lending Library for Science and Technology are the principal sources. *The Foreign Language Barrier*, edited by C. W. Hanson (Aslib, 1962), deals with the learning of languages by scientists, cover-to-cover translations, and pools and services.

INTERLINGUAL MATHEMATICAL DICTIONARIES

It will be remembered that the *Mathematics Dictionary* of Glenn and Robert C. James has indexes in French, German, Russian, and Spanish. Indexes in the same four languages are given in the large *International Dictionary of Applied Mathematics* (Fig. 4), which was compiled by experts in different fields under the general editorship of W. F. Freiberger and published by Van Nostrand, Princeton, NJ, in 1960. Dictionaries which cover more than two languages are called *polyglot*.

The majority of interlingual dictionaries are restricted to two languages. These are termed *bilingual*. They are usually more detailed than polyglot dictionaries, often giving individual terms in the context of phrases. In mathematics, German and Russian have received most attention. German–English compilations include:

> HERLAND, LEO., *Dictionary of Mathematical Sciences: German–English and English–German*, 2 vols., Harrap, London; Ungar, New York, 2nd edn., 1965 (Fig. 5). The statistical entries are by Gregor Sebba and the commercial entries by Robert Grossbard. Numerous cross-references increase the value of the work.
>
> HYMAN, C., *German–English Mathematics Dictionary*, Interlanguage Dictionaries, New York, 1960 (Fig. 6). Whilst lacking some of Herland's helpful features, it is very useful when dealing with the newer terminology.
>
> KLAFTEN, E. B., *Mathematical Vocabulary, English–German/ German–English*, Wila Verlag, Munich, 1961. Has an unusual arrangement in that each part is divided into 8 sections: general terms, arithmetic, algebra, plane geometry, solid geometry, trigonometry, calculus, and coordinates. Of handy format, it gives examples of usage as well as individual terms.

A compact and inexpensive work published in 1959 by the VEB Deutscher Verlag der Wissenschaften, Berlin, has two braic equation of degree $n + 1$; that they are real; and that when ordered they satisfy $a < x_0 < x_1 < \cdots < x_n < b$. They are, in fact, zeros of one of a set of orthogonal polynomials.

If a and b are both finite, one can make a substitution of variable, if necessary, and suppose $b = -a = 1$. Then if $w(x) = 1$, the x_i are zeros of the **Legendre polynomial** of degree $n + 1$, and one has for

$n = 1$: $\mu_0 = \mu_1 = 1$;

$n = 2$: $\mu_0 = \mu_2 = \frac{5}{9}$, $\mu_1 = \frac{8}{9}$;

$n = 3$: $\mu_0 = \mu_3 = (18 - \sqrt{30})/36$, μ_1

$= \mu_2 = (18 + \sqrt{30})/36$.

By choosing $w(x)$ appropriately either a or b or both can be infinite. (For tabulation of x_i and μ_i, and for a discussion of the theory, see Zdenek Kopal, *Numerical Analysis*, John Wiley & Sons, Inc., 1955.)

GAUSSIAN REPRESENTATION. See **spherical representation of a surface.**

GAUSSIAN UNITS. A nonrationalized mixed system of units in which electrical quantities, such as electric charge and electric potential, are carried in the esu system, while currents and magnetic quantities are carried in the emu system. The constant c is then carried in the equations where it is needed, as a conversion factor. (See **electromagnetic units.**)

GAUSSIAN WAVE GROUP. A **wave group** for which

$$q(\kappa) = ae^{-\alpha(\kappa - \kappa_0)^2}.$$

Here κ is the "wavelength constant" but actually the independent variable, while α and κ_0 are arbitrary constants.

GAUSS LAW. The total **electric flux** passing out from a closed surface (c.s.) is (in rationalized units) equal to the total charge, q, enclosed within the surface. In unrationalized units the flux is equal to 4π times the enclosed charge q.

In integral form, the Gauss law may be written (in rationalized MKS units)

$$\int_{c.s.} \mathbf{D} \cdot d\mathbf{A} = \int_{c.s.} \epsilon_0 \mathbf{E} \cdot d\mathbf{A} = q = \int \rho dV,$$

where $\mathbf{D}$ is the electric flux density, $\mathbf{E}$ is the electric field strength, ρ is the electric charge density, and ϵ_0 is the permittivity of free space.

Applying Gauss theorem, the law may be put in the differential form,

$$\nabla \cdot \mathbf{D} = \rho, \quad \text{or} \quad \nabla \cdot \mathbf{E} = \rho/\epsilon_0.$$

Using $\mathbf{E} = -\nabla V = -\rho/\epsilon_0$, the law may be put into still another form, the **Poisson equation:** $\nabla^2 V = -\rho/\epsilon_0$.

GAUSS LAW FOR A MAGNETIC MEDIUM. Applying the **Gauss law** to the magnetic case:

$$\int \mathbf{B} \cdot d\sigma = 4\pi \int \rho dT$$

$$\nabla \cdot \mathbf{B} = 4\pi\rho$$

where $\mathbf{B}$ is the **magnetic induction,** and ρ the density of magnetic charge. Since free magnetic charge does not exist in nature, $\rho = 0$, and the Gauss law states:

$$\int \mathbf{B} \cdot d\sigma = 0$$

$$\nabla \cdot \mathbf{B} = 0.$$

GAUSS LAW OF NORMAL GRAVITATIONAL FORCE. The surface integral of the normal component of the gravitational force on a particle of unit mass, taken over any closed surface is equal to $-4\pi G$ times the total mass enclosed by the surface.

GAUSS-MARKOV THEOREM. A theorem to the effect that an unbiased linear estimator of a parameter has minimal **variance** when obtained by the method of **least squares.**

GAUSS PLANE. See **Argand plane.**

GAUSS PRINCIPLE OF LEAST CONSTRAINT. See **least constraint, Gauss principle of.**

GAUSS-SEIDEL METHOD. A term sometimes applied to the total step iteration for solving linear equations (see **matrix inversion**), although both Gauss and Seidel used a **relaxation method.**

GAUSS-STOKES THEOREMS, FOUR-DIMENSIONAL. See **Minkowski world.**

GAUSS THEOREM. See **integral theorems of vector analysis.**

FIG. 4. *International Dictionary of Applied Mathematics* (By kind permission of D. Van Nostrand Company Inc.)

maß·gleich *adj.* isometric [=ISO-METRISCH]

Maß·gleichheit *f* isometry [=ISOMETRIE]   ·

Maß·größe *f* **1.** (*messende Größe*) measuring quantity, measurement: *die Maßgrößen auf der y-Achse* the measurements along the *y*-axis [*cf.* MASS 1.]. **2.** (*meßbare Größe*) dimensional (or measurable) quantity

Maß·stab *m* **1.** (*Meßgerät*) measuring device; (*Stab*) measuring stick (or rod), yardstick; (*Band*) measuring tape; *mit demselben M. meßbar* commensurable [*cf.* STAB 2.]. **2.** (*Vergleichsmaß*) yardstick (of comparison) [*cf.* VERHÄLTNISMÄSSIG]. **3.** (*relative Größe* relative size) scale (of measurement): *in großem* (*kleinem*) *Maßstab* on a large (small) scale; *in vergrößertem* (*verkleinertem*) *M.* at an enlarged (a reduced) scale; *im M.* 100 : 1 in scale of 100 : 1; *eine Zeichnung in großem M.* a large-scale drawing [*cf.* VERJÜNGEN; VERJÜNGUNG; VERKLEINERN; VERKLEINERUNG]

Maßstab·fehler *m* scale error

maßstab·getreu *adj.* (in) scale, (true) to scale: *maßstabgetreue Zeichnung* scale diagram [*cf.* ZEICHNEN]

Maßstab·karte *f* scale map (or chart)

maßstäblich *adj.* (in) scale, (true) to scale; *m. verkleinern* scale down

Maßstabs·verhältnis *n* scale ratio

Maß·system *n* system of measurement (or: of units): *absolutes M.* absolute (system of) units [=ZENTIMETER-GRAMM-SEKUNDEN-SYSTEM]

Maß- und Gewichtskunde *f* metrology [=METROLOGIE]

Maß·zahl *f* coefficient of measure; (*statist.*) (*Indexziffer*) relative, index number; (*Stichprobenmaßzahl*) statistic [*cf.* CHI-MASSZAHL]

Material *n* material [*cf.* AUFGLIEDERN; ERHEBUNG]

Materie *f* (*phys.*) matter [=STOFF]

materie·frei *adj.* (*phys.*) free of matter

materiell *adj.* (*phys.*) material: *materieller Punkt* material point, particle; *Linie* (*Fläche*) *von materiellen Punkten* material line (surface) [*cf.* MASSENMITTELPUNKT]

Materie·welle *f* matter (or electron, or phase, or: de Broglie) wave [=ELEKTRONENWELLE, PHASENWELLE]

Mathematik *f* mathematics: *angewandte* (*reine*) *M.* applied (pure) mathematics [*cf.* GRÖSSENLEHRE]

Mathematiker *m* mathematician

mathematisch *adj.* mathematical: *mathematische Erwartung* (*oder Hoffnung*) mathematical expectation [*cf.* GLIED; LOGIK; PENDEL; RECHNUNG 1.; WAHRSCHEINLICHKEIT]

Matrix *f* (*pl. Matrizes oder Matrizen*) (*alg.*) matrix: *M. der Koeffizienten eines Systems simultaner linearer Gleichungen* matrix of the coefficients of a system of simultaneous linear equations; *quadratische M.* square matrix; *symmetrische M.* symmetric

169

FIG. 5. Herland, *Dictionary of Mathematical Sciences.* (By kind permission of George G. Harrap & Co. Ltd. and Frederick Ungar Publishing Co. Inc.)

A

Abakus (m) abacus

abändern (vt) alter, change, modify, vary

Abänderung (f) change, modification, variation

Abbild (n) image; map

abbilden (vt) map, make an image of

Abbildung (f) map, mapping, image, projection, representation.
flächentreue A: equal-area mapping or projection;
konforme oder winkeltreue A: conformal or angle-preserving or isogonal mapping;
längentreue A: length-preserving or equidistant mapping

Abbildungsprinzip (n) mapping principle

Abbildungssatz (m) mapping theorem or principle

Abbildungsverfahren (n) mapping method or process

abböschen (vt, vi) slant, slope

Abböschung (f) slant, slanting, sloping

abbrechen (vt, vi) break off.
die Reihe nach dem ersten Glied abbrechen: break off the series after the first term

Abel (N) Abelscher Hilfssatz: Abel's lemma; Abelsche Asymptotik: Abelian asymptotic behavior; Abelsches Integral: Abelian integral

Aberration (f) aberration

Aberrationswinkel (m) angle of aberration

Abfall (m) decrease, drop

abfallen (vi) decline, decrease, slope down

abflachen (vt, vr) flatten; level off; become flat or level

Abflachung (f) flattening

abgeflacht (a) flattened

abgekehrt (a) turning away. (cf. abgewandt)

abgekürzt (a) abbreviated, abridged, short, shortened

abgeleitet (a) derived

Abgeleitete (f) derivative, derived function

abgeplattet (a) flattened, oblate

abgeschlossen (a) closed

Abgeschlossenheit (f) closure.
A. einer Funktionenfolge: closure of a sequence of functions

Abgeschlossenheitskriterium (n) closure criterion

abgestumpft (a) truncated

abgestutzt (a) truncated

abgewandt (a) turning away.
abgewandte Spitze: ceratoid cusp

abgrenzen (vt) bound, delimit, demarcate

Abgrenzung (f) delimitation, demarcation

abhängig (a) dependent

Abhängigkeit (f) dependence

Abhängigkeitskoeffizient (m) coefficient of contingency

abkürzen (vt) abridge, shorten

Ablauf (m) course, passage

ablaufen (vi) pass; run down; wane

ableitbar (a) deducible

Ableitbarkeit (f) deducibility

ableiten (vt) deduce, derive; differentiate (a function)

Ableitung (f) derivation; derivate; derivative, derived function, differential coefficient, differential quotient, differentiation

Ableitungsformel (f) formula for derivatives

Ableitungsgleichung (f) equation for derivatives

Ableitungsindex (m) derivative subscript

Ableitungsordnung (f) order of derivative

Ableitungsprozess (m) process of differentiation, derivative process

Ableitungsvektor (m) derivative vector

ablenken (vt) deflect, divert

Ablenkung (f) deflection, deviation, diversion

Ablenkungsfeld (n) deflecting field

Ablenkungskraft (f) deflecting force

Ablenkungsprisma (n) deviation prism

FIG. 6. Hyman, *German–English Mathematics Dictionary*. (By kind permission of Plenum Publishing Corp.)

complementary title pages to explain its purpose: *Mathematical Dictionary—Russian–English, with a Short Grammar*, and *Mathematisches Wörterbuch—Russisch–Deutsch, mit einer kurzen Grammatik*. An outline Russian grammar in both English and German precedes the trilingual vocabulary. The Russian words determine the alphabetical order, and each is followed by German and English equivalents. Some names of mathematicians are given, together with very brief biographical notes. These appear in the German column but not in the English.

Yet another function is filled by the *German–English Mathematical Vocabulary*, by Sheila Macintyre and Edith Witte (Oliver & Boyd, Edinburgh, 2nd edn., 1966). An inexpensive work of less than 100 small pages, it comprises a vocabulary and grammatical sketch. It is essentially a practical tool for those who need an introduction to the language with a mathematical bias.

Although Russian is separately dealt with in Appendix I, it is appropriate to mention here an excellent dictionary which was prepared under the auspices of the American Mathematical Society, the National Academy of Sciences of the USA, and the Academy of Sciences of the USSR. It is the *Russian–English Dictionary of the Mathematical Sciences*, compiled and edited by A. J. Lohwater with the collaboration of S. H. Gould, published by the Society in 1961 (Fig. 19). It contains over 15,000 terms preceded by a short grammar of the Russian language. The English–Russian companion volume is the *Anglo-Russkiĭ Slovar' Matematicheskikh Terminov*, edited by P. S. Aleksandrov (Izdatel'stvo Inostrannoĭ Literatury, 1962).

Specialized dictionaries are also available for other languages, among them the following:

GARCIA, MARIANO, *Mathematics Dictionary, Spanish–English/ English–Spanish*, Hobbs Dorman, New York, 1965. Contains about 3000 expressions.

GOULD, S. H. and OBREANU, P. E., *Romanian–English Dictionary and Grammar for the Mathematical Sciences*, American Mathematical Society, 1967.

TONIAN, A. H. and TONIAN, V. A., *Mathematikakan Ter-minner Barbaran . . . Dictionary of Mathematical Terms: English, Russian, American, German and French.* Academy of Sciences of the Armenian SSR, Erevan, 1965. The basic numbered order is in English, with references from terms in the other languages.

GUIDES TO FOREIGN SCIENTIFIC AND TECHNICAL DICTIONARIES

Many libraries and most individuals content themselves with acquiring a selection of general scientific and technical dictionaries in the more popular languages, and these are often sufficient in translating mathematical texts. The *German–English and English–German Dictionary for Scientists* by O. W. and Irma S. Leibiger (Edwards, Ann Arbor, Mich., 1950), for example, certainly includes a large number of mathematical terms.

It is, however, necessary, on occasions, to be able to identify more specialized works, particularly in the less common languages. Two compilations will prove of value here:

MARTON, TIBOR W., *Foreign-language and English Dictionaries in the Physical Sciences and Engineering: a selected biblio-graphy, 1952 to 1963*, US National Bureau of Standards, USGPO, Washington, DC, 1964.

UNESCO, *Bibliography of Interlingual Scientific and Tech-nical Dictionaries*, Paris, 4th edn., 1961. *Supplement*, 1965. Entries are arranged according to the Universal Decimal Classification, and there are English, French, and Spanish indexes of authors and subjects. (A 5th edn. is in prepara-tion).

Encyclopedias

Encyclopedias are often either overlooked or spurned by those seeking scientific and technical information. This is due to two main reasons. Firstly, general encyclopedias cover the entire range of human knowledge, and are therefore sometimes considered to treat their material superficially. Secondly, they take so long to compile, edit and print that they are out of date long before they are published.

To put these arguments into proper perspective it is necessary to examine the functions of the general encyclopedia. These are twofold. By the very wealth of its factual content it acts as a comprehensive fact-finding tool which yields an immediate end-product. But it is not necessarily intended as an end in itself. An equally important function is that of orientating the inquirer.

The articles which comprise a first-rate encyclopedia are prepared by specialists. They write according to a preconceived plan which is designed to relate the contributions on individual topics to the context of knowledge in general. Cross-references between articles lead the inquirer to related topics. Each article contains a bibliographical guide to other more specialized or detailed writings on the subject with which it deals. It also gives leads to further research in the form of terminology which can be noted and used as keywords; in establishing date limits; in providing the names of those who have made important contributions to the subject in question; and in specifying the countries and institutions in which research has been carried out. These features make the general encyclopedia a very valuable starting point for information searches which may eventually end with the examination of a highly abstruse monograph, or even the writing of a letter to a research institute.

On the question of up-to-dateness, two points should be noted. The first concerns their method of revision. Broadly speaking, there are two ways in which general encyclopedias are revised. They are either completely revised in every section and then published as new editions, or else they undergo what is known as *continuous revision*. This means that editorial staff are constantly working at the task of keeping the encyclopedia up-to-date, introducing new facts, recording historical events, replacing outdated illustrations. Whilst in the case of an encyclopedia having completely revised editions such editions will only be published at intervals of 5 years or more, printings of the continuous revision encyclopedia may be made annually or even oftener. At each successive printing amendments made since the last printing are incorporated. This results in a certain unevenness of revision, but is a method adopted by several of the more important works. Updating is also achieved in several cases by the publication of yearbooks.

The second consideration in the question of up-to-dateness is that mere age by no means invalidates all information. The history of so-called invention is full of cases in which the development of a discovery made many years previously had to await advances in another field to make it practicable. The principle of jet propulsion was not discovered by Whittle, though it was he who developed its practical application.

The publishers of important encyclopedias usually operate a research service, and draw upon wide specialist resources. They are often able to put inquirers into touch with sources of information. Reprints of encyclopedia articles are sometimes produced, and they make useful monographs. An example of such services is the series of *Home Study Guides* produced by the Library Research Service of the *Encyclopaedia Britannica*. One volume in the series is entitled *Mathematics*. It is intended to guide the reader through the mathematical contents of the multi-volume encyclopedia.

The evaluation of encyclopedias is adequately dealt with in Constance M. Winchell's *Guide to Reference Books* (American Library Association, Chicago).

General Encyclopedias

It is generally advisable to consult several different encyclopedias as the treatment of individual subjects varies, sometimes quite considerably, from one to another.

The following are four of the most important English-language general encyclopedias:

Chambers's Encyclopaedia, Pergamon, new edn., 1967. This appears in 15 volumes, and does not follow the policy of continuous revision. It is anticipated that completely new editions will appear every 4 or 5 years.

Collier's Encyclopedia, Collier, New York, new edn., 1967. This completely revised edition is in 24 volumes, 4 more than the 1st edn. The last volume contains 11,500 bibliography entries and the index.

Encyclopedia Americana, New York and Chicago, Annual printings. Uses continuous revision and the *Americana Annual* (1923–) to keep up to date. Most of the articles are short. Vol. 30 contains the index, which has an alphabetical, dictionary arrangement and is revised with each printing.

Encyclopaedia Britannica. Chicago, 14th edn., annual printings. Still of great value where up-to-dateness is not essential is the 9th edn. (1875–89) in 25 volumes. In the current edition the previous monograph-type articles have been replaced by shorter articles. The index, which should always be consulted when using the *Encyclopaedia*, is in vol. 24. Continuous revision and the *Britannica Book of the Year* (1938–) are its means of keeping up to date.

Mathematical Encyclopedias

Encyclopedias are produced which cater for more specialized needs. Scientific and technical encyclopedias vary in the degree of their specialization, some covering a number of fields, others restricted to an individual discipline. In their respective areas they offer more detailed information than the corresponding sections of general encyclopedias, and their bibliographies may be more ample.

An example of a multi-volume special encyclopedia is the *McGraw-Hill Encyclopedia of Science and Technology* (New York, revised edn., 1966). The basic 15 volumes are supplemented by the *McGraw-Hill Yearbook of Science and Technology*. Most of the longer articles are followed by bibliographies. *Van Nostrand's Scientific Encyclopedia* is shorter, but also of great value.

A mathematical encyclopedia which contains bibliographies is the large *Encyklopädie der mathematischen Wissenschaften mit Einschluss ihrer Anwendungen* (Teubner, Leipzig, 1898/1904–1904/35). Despite its age it remains a valuable source of reference. A new edition did in fact commence publication in 1950 and is progressing slowly.

Other important German works are the *Encyklopädie der Elementarmathematik* (Deutscher Verlag der Wissenschaften, Berlin, 1954–) and the *Mathematisches Wörterbuch*, edited by Joseph Naas and H. L. Schmid (Akademie Verlag, Berlin; Teubner, Stuttgart, 3rd edn., 1965), a 2-volume encyclopedic dictionary which includes over 400 biographies.

The *Universal Encyclopedia of Mathematics*, translated and adapted from *Meyers Rechenduden* and published by Allen & Unwin in 1964, has a remarkably wide range and makes a generally useful reference work for the student. Not an encyclopedia, but a comprehensive collection of definitions, theorems, and formulae which also includes a large number of references is the *Mathematical Handbook for Scientists and Engineers* by Granino A. Korn and Theresa M. Korn (McGraw-Hill, 2nd edn., 1968).

Theses

Great strides have been taken in recent years both in making degree theses more readily available and in bringing their existence to the notice of those who can best make use of them. A considerable amount of valuable research work has for long remained unexploited in the files of university archives. Sometimes theses see the light of day in the form of published books, but more often they are shelved in their original form, having once fulfilled their primary academic function. Several organizations, however, are now tackling the problem of thesis documentation. The sources of information for three countries will be described.

In the United States, University Microfilms Inc. of Ann Arbor, Michigan, have for several years published a bibliography of theses entitled *Dissertation Abstracts*. This is "a monthly compilation of abstracts of doctoral dissertations submitted to

University Microfilms Inc. by more than 135 co-operating institutions". Naturally, the listing is only as comprehensive as these institutions make it. The abstracts are quite full, and are grouped under subject headings. Subject and author indexes are included in each issue. Copies of the dissertations are available either as photocopies or as microfilms from the publishers.

While *Dissertation Abstracts* is the principal source of reference, other compilations of more restricted scope have been published from time to time. An example in the mathematics field is Summers, E. G. and Stochl, J. E., "Bibliography of doctoral dissertations completed in elementary and secondary mathematics from 1918–1952", *School Science and Mathematics*, **61,** 323–35, 431–9, 1961.

The standard British work is the *Index to Theses Accepted for Higher Degrees in the Universities of Great Britain and Ireland*, published by Aslib, London (Fig. 7). The first volume, covering the 1950–1 academic year, was published in 1953. Volume 11, for 1965–6, records 5417 entries arranged in subject groups. There is an author index and an index to subject headings. In this particular volume the Mathematics heading is further subdivided into Theory of Numbers, Algebra, Geometry, Topology, Analysis, Numerical Methods, and Probability and Statistics. A tabulated statement of the availability of theses from each university is provided.

French theses are deposited with the Bibliothèque Nationale and the Bibliothèque de la Sorbonne in Paris. Exchange systems exist between the latter and universities in other countries. Photocopies and microcopies of French theses are usually available from the Service de Documentation of the Centre National de la Recherche Scientifique in Paris. Mathematical listings have been published in *French Doctoral Theses—Sciences* by French Cultural Services of New York. A more comprehensive listing is the *Catalogue des Thèses et Ecrits Académiques*, which is issued in the form of a supplement to *Bibliographie de la France* under the authority of the Ministère de l'Education Nationale.

Duplication of research work is an obvious waste of any

Topology

1331. REID, G. A. (C). Studies in the theory of topological groups. PH.D.
1332. TIPPLE, D. A. (M). On the homotopy groups of some Stiefel manifolds. M.Sc.
1333. GARSIDE, F. A. (O). The theory of knots, and associated problems. D.PHIL.
1334. HODGKIN, L. H. (O). Some problems in algebraic topology. D.PHIL.
1335. SWINFEN, R. A. (O). Some problems in topological algebra. B.Sc.
1336. WOOD, R. M. W. (O). Some problems in algebraic topology. D.PHIL.
1337. McCRUDDEN, M. (Q). Topological semigroups. M.Sc.
1338. ARMSTRONG, M. A. (Wk). Some theorems in combinatorial topology. PH.D.

Analysis

1339. WARD, D. J. (Br). Some problems arising from the Eilenberg inequality. PH.D.
1340. DE BRUYN, G. F. C. (C). Asymptotic properties of linear operators. PH.D.
1341. WATSON, G. A. (E). Chebyshev methods for the solution of linear ordinary differential equations. M.Sc.
1342. CAREFOOT, W. C. (Li). Some inequalities of harmonic analysis. PH.D.
1343. ABD-EL-A, L. M. F. (LKC). Conformal mapping of convex and starlike domains. [Diss.] M.Sc.
1344. BARNSBY, R. M. (LCCST). A survey of the Stokes phenomenon. [Diss.] M.Sc.
1345. COLMAN, W. J. A. (LWoolP). The approximate integration of functions of one variable. [Diss.] M.Sc.
1346. CRAZE, W. C. (LNernP). Approximate methods for the evaluation of multiple integrals. [Diss.] M.Sc.
1347. EKE, B. G. (LIC). Asymptotic behaviour and valency of functions regular in the unit disk. PH.D.
1348. EKE, V. R. (LIC). On entire functions of small order. M.Sc.
1349. FISHEL, B. (LBkC). An order-free structure related to the theory of integration. PH.D.
1350. HIRST, K. E. (LUC). The Hausdorff dimension of a certain packing of circles and other problems involving Hausdorff measures. PH.D.
1351. LARMAN, D. G. (LUC). Properties of Hausdorff measure. PH.D.
1352. LOCKER, J. W. (LExt). Application of Lyapunov's direct method to problems of stability of non-linear differential equations, including control systems. [Diss.] M.Sc.
1353. MURTHY, T. K. S. (LCassC). Approximations to the solutions of non-linear differential equations. [Diss.] M.Sc.
1354. OFFEI, D. N. (LExt). Vector methods for second-order linear differential operations. [Diss.] M.Sc.

FIG. 7. *Index to Theses Accepted for Higher Degrees in the Universities of Great Britain and Ireland.* (By kind permission of Aslib.)

country's resources of scientific manpower, a subject which has received a great deal of government attention. Vast sums of money are invested in university research activities, and one way in which the dividends may be increased is by more efficient use of doctoral theses.

Exercises

1. List the features you would look for in assessing the value of a mathematical dictionary for a particular class of user. First define the class of user you have selected.
2. As a mathematician, for what kinds of information would you consult a multi-volume general encyclopedia?
3. Why do you consider that a knowledge of sources of information on theses is important?
4. Write notes on the question of up-to-dateness in encyclopedias.

CHAPTER 4

Mathematical Periodicals and Abstracts

510.5: Mathematical periodicals

Importance of Periodicals

Every scientific book is out of date even on the day it is published. During the time-lag between the author putting his first words on paper and the appearance of the finished product new opinions, new theories, new facts will have come to light which it is virtually impossible to incorporate. This applies particularly in the case of works treating subjects of current importance and in which active research is being carried out. Books are not, therefore, always the ideal medium for the dissemination of scientific information. The research worker demands a rapid service, and this is provided to a more satisfactory degree by periodicals. It must, however, be confessed that periodicals themselves are being overwhelmed by the pace of today's scientific progress. The American Mathematical Society regularly publishes details of the backlogs which have accumulated in such important publications as the *American Journal of Mathematics, Annals of Mathematics, Canadian Journal of Mathematics*, and the *Proceedings* and *Transactions of the American Mathematical Society*. At given times journals of the highest quality have up to 2000 pages of text waiting to be published. The American Mathematical Society itself has taken urgent steps to eliminate its backlog. In particular, the *Proceedings* changed (1969) from bi-monthly to monthly, and whereas the number of pages in 1968 was some 1500, the number for 1969 is some 4000. Similar increases have also occurred in the *Transactions*.

Mathematicians have for many years engaged in the practice of exchanging research papers as a means of keeping informed of the advances made by their colleagues. In an endeavour to widen the range of communication, the American Mathematical Society began in 1968 a new Mathematical Offprint Service (MOS) through which individuals are supplied, on subscription, with offprints and title listings according to their interests.

Guides to Periodicals

The history of scientific periodicals goes back to the 1660's, to the inception of the world-renowned *Philosophical Transactions* and the *Journal des Sçavans*. At first, periodical publications were the product of large academies and for many years remained few and select. Today, some estimates of the number of scientific and technical periodicals reach 100,000. These are published in over 60 languages. An account of the early history of this form of literature is given in David Kronick, *A History of Scientific and Technical Periodicals: the origins and development of the scientific and technological press, 1665–1790* (Scarecrow Press, New York, 1962).

There are some 1500 journals which contain mathematical contributions. The list of periodical titles given in the latest index issue of *Mathematical Reviews* constitutes a comprehensive, international list of periodicals of mathematical interest. Another list of international scope is that in the mathematics section of Malclès, Louise Noëlle, *Les Sources du Travail Bibliographique* (see pp. 110–11).

Ulrich's Periodicals Directory (Fig. 8) is revised every few years by Bowker, New York. The latest edition is the 12th, 1967; it is in 2 volumes, of which the first, covering science and technology, lists over 12,000 titles from all over the world under broad subject headings: Abstracting and Indexing Services, Automation—Computers, Mathematics, etc. Information on each title includes price; frequency of issue; name and address of publisher; whether the periodical carries abstracts, reviews, or bibliographies;

ACTA LOGICA.* (Text in English and French) 1958. irreg.
Price not given. Biblioteca Centrala Universitara, Str.
Oneşti nr.1, Bucharest, Rumania. cum.index: 1958-1963?

ACTA MATHEMATICA. 1882. 2 vols.per yr. Kr.80 per vol.
Ed. N.E. Norlund. Institut Mittag-Leffler, Auravagen 17,
Djursholm, Sweden. bibl. index. cum.index: v.1-100.
Indexed: Math.R.

ACTA MATHEMATICA. (Academiae Scientiarum Hungaricae)
(Text in English, French, German or Russian) 1950. q.
$12. Ed. Gy. Hajós. Akadémiai Kiadó, Publishing House of
the Hungarian Academy of Sciences, Alkotmány u.21, Buda-
pest V, Hungary. adv. bibl. bk.rev. index.
Indexed: Math.R.

ACTA POLYTECHNICA SCANDINAVICA. MATHEMATICS
AND COMPUTING MACHINERY SERIES. 1958. irreg.
Price not given. Scandinavian Council for Applied Research,
Box 5073, Stockholm 5, Sweden. bibl. charts. illus. index.
Indexed: Biol.Abstr. Chem.Abstr. Eng.Ind. Math.R.
Met.Abstr. Sci.Abstr.

ACTA SCIENTIARUM MATHEMATICARUM. (Text in English,
French, German, Italian and Russian) 1922. s-a. $8.50.
Ed. Prof. Béla Sz.-Nagy. Bolyai Institute of Attila Fózsef,
Univ. of Szegad, Szeged, Hungary. adv. bk.rev. charts.
index. Indexed: Math.R.

AKADEMIE DER WISSENSCHAFTEN.UND DER LITERATUR.
MATHEMATISCH-NATURWISSENSCHAFTLICHE KLASSE.
ABHANDLUNGEN. (Text in English, French and German)
1950. irreg. Price not given. Franz Steiner Verlag GmbH.,
Bahnhofstr. 39, Postfach 743, Wiesbaden, Germany. abstr.
charts. illus. index.

AKADEMIYA NAUK ARMYANSKOI SSR. IZVESTIYA.
SERIYA: MATEMATIKA. (Text in Russian; summaries in
Armenian) 1965. bi-m. $3.50. Mezhdunarodnaya Kniga,
Moscow G-200, U.S.S.R. index.

AKADEMIYA NAUK AZERBAIDZANSKOI SSR. DOKLADY.
(Text in Azerbaijani and Russian) 1945. m. $9. Ed. Z.I.
Khalilov. Mezhdunarodnaya Kniga, Moscow G-200, U.S.S.R.
charts. illus. index. circ. 820. Indexed: Chem.Abstr.

AKADEMIYA NAUK AZERBAIDZHANSKOI SSR. IZVESTIYA
SERIYA FIZIKO-TEKHNICHESKIKH I MATEMATICHESKIKH
NAUK. See PHYSICS—General

FIG. 8. *Ulrich's Periodicals Directory.* (By kind permission of
R. R. Bowker Company.)

whether it is abstracted or indexed in any other publication; and its date of origin. A companion work is *Irregular Serials and Annuals*, of which the first edition was published by Bowker in 1967.

Another directory, published by the Library Association, London, in 1962, is a *Guide to Current British Periodicals*, edited by Mary Toase. Approximately 3800 titles published in the United Kingdom are arranged according to the Dewey Decimal Classification, and there is a comprehensive index.

Different Kinds of Periodicals

In general terms, mathematical periodicals are issued by three different kinds of publishers: universities, societies, and commercial houses. There are a few periodicals issued by government departments which are of value to the mathematician, but they are not so numerous as in other fields. The *Journal of Research of the National Bureau of Standards;* B: *Mathematics and Mathematical Physics* is a publication in this category.

UNIVERSITIES

In mathematics, the universities of many countries continue to make valuable contributions by the publication of the results of basic research. The following are examples from the United States:

Duke Mathematical Journal, Duke University Press, Durham, North Carolina, 1935– .

Illinois Journal of Mathematics, University of Illinois Press, Urbana, Ill., 1957– .

Journal of Mathematics and Physics, MIT, Cambridge, Mass., 1921– .

Michigan Mathematical Journal, University of Michigan Press, Ann Arbor, Mich., 1952– .

Notre Dame Journal of Formal Logic, University of Notre Dame Press, Notre Dame, Indiana, 1960– .

Pacific Journal of Mathematics, University of California Press, Berkeley, Calif., 1951– .

Annals of Mathematics is edited with the co-operation of Princeton University and the Institute for Advanced Study, and published by Princeton University Press.

SOCIETIES

The periodical publications of mathematical societies (proceedings, transactions, etc.) may be classified, with other works relating to societies, under 510.6, which is the subject of Chapter 5. This is reasonable in view of the fact that they report society business, give notice of meetings, and include other matters of interest to members. They are, however, considered here for their subject content. When a particular society has a large publishing programme it may issue several periodicals, each devoted to a different aspect of the society's activities. The American Mathematical Society is a good example. It publishes the results of original research in both its *Proceedings* and its *Transactions*, with longer papers usually in the latter. The *Bulletin* is the Society's official organ, containing reports of meetings, book reviews, the full texts of invited addresses, and other matter of interest to professional mathematicians. Its fourth publication, the *Notices*, contains details of meetings, personal notes and other news items.

The London Mathematical Society was established in 1865, and in the same year commenced publication of its *Proceedings* which now contains more lengthy papers. Shorter papers are printed in the society's *Journal* (1926–) which also contains the records of proceedings at meetings, obituary notices, and other matter of interest to members. Other important British periodicals are the Royal Society of London's *Philosophical Transactions;* Series A: *Mathematical and Physical Sciences* (1665–), and *Proceedings;* Series A: *Mathematical and Physical Sciences* (1800–); and the Mathematical Association's *Mathematical Gazette* (1894–), published by Bell, London.

M—C

It is not possible to give a full list of mathematical society publications here, but it may be noted that the organizations listed in the next chapter between them issue an important body of periodical literature.

COMMERCIAL PUBLISHERS

In many subjects, whilst the publication of research is largely the province of institutional periodicals, the practical applications of that research are reported by the commercial press. This is not true to the same extent in mathematics, except perhaps in such specialized cases as electronic computers.

Two important titles of recent origin are:

> *Advances in Mathematics*, Academic Press, 1961– . Each volume is issued in paper-bound parts at irregular intervals. These are later available as hard-bound books.
>
> *Topology, an International Journal of Mathematics*, Pergamon Press, 1962– . A research journal with special emphasis on topology, topological techniques, Lie groups, differential geometry and algebraic geometry.

There are several instances where mathematical societies place the publication of their periodicals in the hands of commercial houses. Bell of London, for example, is the publisher for the Mathematical Association. Such, however, are purely arrangements to suit administrative convenience.

Foreign-language Periodicals

Most countries with an academic tradition produce at least one mathematical periodical which is internationally recognized. English-language abstracts of their contents regularly appear in *Mathematical Reviews* (see p. 57). In France, for example, Gauthier-Villars of Paris publishes the *Journal de Mathématiques Pures et Appliquées* (1836–). In Germany, Springer of Berlin publishes *Mathematische Annalen* (1869–), and *Mathematische*

Zeitschrift (1918–). Scandinavia produces the important *Acta Mathematica*, which was begun by Mittage-Leffler in Stockholm in 1882.

Other foreign titles may be traced in *Ulrich's Periodicals Directory*, described above. Some idea of their relative importance may be gained by observing the frequency with which they are cited as references in research papers and analysed in abstracting publications.

Locating and Using Periodicals

Periodicals may be used for current perusal, to keep abreast of new developments, and for retrospective searching, to locate specific pieces of information in past issues. Articles which may have little relevance to one's present interests may assume a greater importance at a later date. By considering some of the work involved in searching back files of periodicals, the value of systematic current perusal, or *scanning*, may be better appreciated.

SEARCHING PERIODICAL LITERATURE

It is the normal practice in libraries to preserve all issues of those periodicals which report information of permanent value. They are usually bound into volumes to facilitate shelving and handling, and to prevent individual issues from becoming damaged or even lost. Each volume will contain an index of authors and subjects produced by the publisher. It may be included in one of the issues or sent separately at the end of the year. The value of periodicals for which no index is produced is considerably reduced. Some publishers, on the other hand, issue cumulative indexes covering perhaps 10 years or more.

One's choice of which files of periodicals to examine will be conditioned by various factors. Naturally, those are examined first which are most readily available and which are known to be likely sources. Most people who are accustomed to periodical searching have their favourites which they always try first.

Abstracting and indexing publications, considered in the following sections, provide short cuts and obviate the tedious job of checking annual indexes of individual titles.

It is essential when undertaking a lengthy search to keep a careful record of the sources which have been examined, and the key words which have been used when consulting indexes. Information discovered while the search is in progress may necessitate a re-examination of the indexes already consulted, this time under one or more new subject headings. It may also happen that a second person is brought in to assist with the search, and he will need to know what ground has already been covered. Search records are conveniently made on index cards and filed in alphabetical order of the titles checked.

When a potentially useful reference is located it should be fully transcribed, preferably in one of the accepted bibliographical forms. A typical reference might read:

> BOOT, C. G., "On trivial and binding constraints in programming problems", *Management Science*, **8** (4), 419–41, July 1962.

This represents an article with the stated title by C. G. Boot which appeared on pp. 419–41 in issue number 4 of vol. 8, dated July 1962, of the periodical *Management Science*.

Odd as it might at first seem, information on a particular *subject* can often be traced by consulting *author* indexes. The author of a useful paper which has already been traced may well have written others of equal or even greater value. Furthermore, it is as a rule easier to check author indexes than it is to check those arranged by subject.

When a reference has been located in the volume-index of the periodical in question, then consulting the full article is a simple matter. It is merely a question of turning to the appropriate pages in the (bound) volume. If, on the other hand, the reference has been noted in an abstracting or indexing publication, or found in a bibliography, then it will be necessary first to locate a copy of the original document. This is usually a case of discovering a library which possesses a file of the periodical concerned.

Directories of the combined periodical holdings of a large number of libraries have been compiled to facilitate the location of individual titles. The *Union List of Serials in Libraries of the United States and Canada*, edited by Edna Brown Titus (commonly known as "Gregory" after Winifred Gregory, editor of the first two editions), has an alphabetical arrangement of titles and indicates the holdings of each library represented by means of symbols, to which there is a key. The five basic volumes, published by H. W. Wilson, New York, are updated by a Library of Congress publication called *New Serial Titles*. A similar service is provided by the *British Union Catalogue of Periodicals* (4 vols., Butterworths, London, 1955–8). Its subtitle is "A record of the periodicals of the world from the seventeenth century to the present day, in British libraries". Five volumes cover the period to 1960, and the work is updated by quarterly lists and annual cumulations of the *British Union Catalogue of Periodicals: new periodical titles*.

A great deal of time spent in literature searching can be saved by the systematic perusal of current periodicals coupled with the creation and maintenance of one's own information index. It is both frustrating and wasteful of time to have to search for the particulars of a paper which one has already read but omitted to index.

Abstracting and Indexing Publications

The days have long since passed when a mathematician could reasonably hope to be acquainted with most if not all of the important literature of his subject. With the increasing applications of mathematics has come more intensive specialization by its practitioners. Probably Poincaré was the last of the mathematicians who could fairly be considered to be abreast of the advancing front along its whole length. Greater specialization in one field has led to further remoteness from the others. As new subfields are developed to cope with the requirements of our technological age so the amount of literature grows in total volume.

Each topic of specialization acquires its own bibliographical resources which in turn divide and grow in amoeba-like fashion as the process develops. The problem confronting the mathematician is twofold: how to keep track of the progress made both in his own and other branches of the subject, and how to trace back through the welter of literature and pinpoint the particular facts or theories which he may need from time to time in this work.

This situation is prevalent throughout the twin fields of science and technology. In fact the mathematician is looked to for a solution to the problem of what is now termed *information storage and retrieval*. Computers are being specially designed to digest bibliographical information and to reproduce it on demand according to individual requirements and specifications. It will be some years yet, however, before the computer completely replaces the manual methods now in use. The printed abstracting publication has not yet in fact reached its full development.

Abstracts of literature are generally considered to fall into two classes: indicative and informative. The former simply sets out the bibliographical details of the original (author, title, publisher, date, etc.) and gives a broad indication of the subject matter. The informative abstract, on the other hand, includes the essential "meat" of the original, reproducing factual data, the steps of experiments, and the main lines of theories. It can sometimes obviate the need to consult the original, and is especially valuable in the case of material which is not readily accessible or is perhaps written in a foreign language. As a rule, however, it is unwise to rely solely on abstracts as they are naturally subject to bias on the part of the abstractor. When an abstract includes an evaluation of the original it is properly called a *review*.

Abstracting publications usually appear in the form of periodicals, though some are issued on index cards which can be selectively filed to suit the individual requirements of the subscriber. Their major use is in documenting periodical articles. The number of abstracting services has grown with the number of new periodicals. The more periodicals there are the more abstracting services there will be; and the more services there are the more those periodicals will be used.

Abstracts are produced by different kinds of organizations. In *A Guide to the World's Indexing and Abstracting Services in Science and Technology* (National Federation of Science Abstracting and Indexing Services, 1963) nearly 2000 publications are listed.

The literature of mathematics from the second half of the nineteenth century to the present day is covered by the following publications:

> *Bulletin de Bibliographie, d'Histoire et de Biographie Mathématiques* (8 vols.), 1855–62.
> *Bullettino di Bibliografia e di Storia delle Scienze Matematiche e Fisiche*, 1868–87. Now available from the Johnson Reprint Corporation, New York.
> *Jahrbuch über die Fortschritte der Mathematik*, 1868–1934.
> *Revue Semestrielle des Publications Mathématiques*, 1893–1934.
> *Zentralblatt für Mathematik*, 1931– .
> *Mathematical Reviews*, 1940– .
> *Referativyni Zhurnal—Matematika*, 1953– .

A useful guide to earlier work is the Royal Society of London's *Catalogue of Scientific Papers, 1800–1900*, published by Cambridge University Press. Volume 1 of its subject index, 1908, covers pure mathematics.

Mathematical Reviews (Fig. 9), published by the American Mathematical Society, is undoubtedly the most valuable current English-language publication in the field. It is used monthly. Its subject coverage is extremely wide, and it is international in scope. Some entries are reprinted from other services whose titles should be noted:

Applied Mechanics Reviews,	*Referativyni Zhurnal,*
Computing Reviews,	*Science Abstracts,*
Mathematics of Computation,	*Zentralblatt für Mathematik.*
Operations Research,	

Mathematical Reviews

Vol. 29, No. 3　　　　　　　　March 1965　　　　　　　　Reviews 2140–3328

GENERAL

See also 2157, 2265, 2686, 2823, 2881, 3028, 3267.

Pemberton, John E.　　　　　　　　　　　2140
★**How to find out in mathematics.**
Libraries and Technical Information Division, Vol. 2.
Pergamon Press, Oxford-Paris-Frankfurt; The Macmillan Co., New York, 1963. x + 158 pp. $2.45.
This is an up-to-date comprehensive guide to the use of libraries, journals, reference works, and so on for the purpose of obtaining information on mathematical topics at all levels.　　　　　　　*S. S. Cairns* (Urbana, Ill.)

★**American Mathematical Society Translations.**　　2141
Series 2, Vol. 35: 12 papers on analysis and applied mathematics.
American Mathematical Society, Providence, R.I., 1964.
iv + 363 pp. $5.20
This volume contains translations of two papers by I. P. Mysovskih, and one each by V. A. Zalgaller, M. I. Višik, S. N. Mergeljan and A. P. Tamadjan, L. A. Ljusternik, E. A. Volkov, V. K. Saul'ev, Ju. M. Berezanskiĭ, I. C. Gohberg and M. G. Kreĭn, K. K. Golovkin, and S. V. Iordanskiĭ.

★**American Mathematical Society Translations.**　　2142
Series 2, Vol. 36: 14 papers on groups and semigroups.
American Mathematical Society, Providence, R.I., 1964.
iv + 395 pp. $5.40.
This volume contains translations of four papers by M. A. Naĭmark, three papers by V. V. Vagner, two by P. A. Gol'berg, one each by I. M. Gel'fand and D. A. Raĭkov, V. A. Kurbatov, A. I. Kostrikin, S. N. Černikov, and L. M. Gluskin.

★**American Mathematical Society Translations.**　　2143
Series 2, Vol. 40: 9 papers on functional analysis and numerical analysis.
American Mathematical Society, Providence, R.I., 1964.
iii + 290 pp. $4.40.
This volume contains translations of two papers by V. N. Kublanovskaja, and one each by V. N. Kublanovskaja and V. N. Fadeev, A. V. Štraus, Ju. I. Ljubič, O. V. Besov, M. M. Džrbašjan, Ju. L. Šmul'jan and V. B. Lidskiĭ.

★**American Mathematical Society Translations.**　　2144
Series 2, Vol. 41: 4 papers on partial differential equations.

1—M.R.

American Mathematical Society, Providence, R.I., 1964.
iii + 317 pp. $4.40.
This volume contains five papers on differential equations: two by S. D. Ėĭdel'man [Mat. Sb. (N.S.) **38** (80) (1956), 51–92; MR **17**, 857; ibid. **53** (95) (1961), 73–136; MR **24** #A925], and one each by L. N. Slobodeckiĭ [ibid. **46** (88) (1958), 229–258; MR **22** #830], Yuh-lin Jou [ibid. **47** (89) (1959), 431–484; MR **22** #1751], and A. A. Dezin [Trudy Mat. Inst. Steklov. **68** (1962), 88 pp.; MR **28** #2344].

Courant, Richard　　　　　　　　　　　2145
Mathematics in the modern world.
Sci. Amer. **211** (1964), 40–49.
An expository article on the nature of mathematics.

Dyson, Freeman J.　　　　　　　　　　　2146
Mathematics in the physical sciences.
Sci. Amer. **211** (1964), 128–146.
An expository article on the role of mathematics (in particular, group theory) in the physical sciences.

Davis, Philip J.　　　　　　　　　　　2147
Number.
Sci. Amer. **211** (1964), 50–59.
An expository article on the nature of numbers.

Bouligand, G.　　　　　　　　　　　2148
Sur les caractères heuristiques, latents ou tangibles, d'œuvres mathématiques récentes.
Math. Notae **19** (1964), 121–131.

Hawkins, David　　　　　　　　　　　2149
★**The language of nature.· An essay in the philosophy of science.**
Drawings by Evan L. Gillespie.
W. H. Freeman and Co., San Francisco, Calif.-London, 1964. xii + 372 pp. $7.50.
This text is primarily intended for students preparing for a career in science whose general intellectual interests have been aroused by their professional studies and who are seeking an outlet for these interests in the general philosophical questions created by specific scientific topics. The book does, therefore, assume some familiarity with scientific concepts and modes of argument, a fact which enables it to proceed at a somewhat more advanced level than is normal in general treatments of this kind. Of particular interest is the attempt of the author to pay more attention than usual to thermodynamics and

413

FIG. 9. *Mathematical Reviews*. (Reprinted with permission of the publisher, the American Mathematical Society, Copyright © 1965, vol. 29, no. 3, p. 1.)

Each of these is an important abstracting publication in its own right. In addition to the American Mathematical Society, over a dozen other national mathematically oriented societies sponsor the publication which also has financial aid from the National Science Foundation. This is such an important tool for the mathematician that the majority of the headings used in the arrangement of the reviews are reproduced here: General, history and biography; logic and foundations, set theory, combinatorial analysis, order, lattices, general mathematical systems; theory of numbers, fields and polynomials, algebraic geometry, linear algebra, associative rings and algebras, non-associative algebra, homological algebra; group theory and generalizations, topological groups and Lie theory; functions of real variables, measure and integration, functions of a complex variable, potential theory, several complex variables, special functions, ordinary differential equations, partial differential equations, finite differences, and functional equations; sequences, series, summability, approximations and expansions, Fourier analysis, integral transforms, operational calculus, integral equations, functional analysis, operator theory, calculus of variations; geometry, convex sets and geometric inequalities, differential geometry, general topology; algebraic topology, topology and geometry of differential manifolds; probability, statistics, numerical methods; economics, operations research, programming, games; biology and behavioral sciences; systems, control; information and communication.

A separate index issue is published for each 6-month volume. Some idea of the exhaustiveness of this work is given by the fact that in the first 6 months of 1968, 4060 reviews were published.

In 1969 the American Mathematical Society commenced bi-weekly publication of *Contents of Contemporary Mathematical Journals* which reproduces the contents pages of a worldwide range of mathematical periodicals. Each issue also includes the names and addresses of the authors of individual papers and articles.

The *Bulletin Signalétique* of the Centre de Documentation du Centre National de la Recherche Scientifique in Paris is published in a number of separate subject sections of which no. 110 is entitled *Mathématiques* (Fig. 10). It appears monthly. The abstracts are arranged in a classified sequence, and an author index is included in each issue.

The purely mathematical abstracting publications may be supplemented, especially for subjects of marginal interest, by compilations of a more general character. *Engineering Index*, for example, contains several thousand brief abstracts every year selected from about 2000 different periodicals. After 1957 the *Industrial Arts Index* (Wilson, New York) split into the *Applied Science and Technology Index* and *Business Periodicals Index*. Both index articles in a large number of periodicals, the titles covered being in fact decided by a vote of subscribers. 1962 saw the inception of *British Technology Index*, published monthly by the Library Association, London.

Cover-to-Cover Translations

Over the past few years several societies and commercial organizations, usually with strong government backing, have undertaken the complete translation of important foreign periodicals on a current basis. Although annual subscription rates tend to be rather high in some cases, one of the objects of such publications is to reduce the total amount of money spent on individually commissioned translations. But it is more than a simple cost-sharing process. Complete translations of periodicals make personal selection of articles unnecessary. If items for translation were selected according to the probable demand for them it is quite likely that material which at present would appear to have little use would remain untranslated, and therefore possibly overlooked by later workers to whom it might be of value. A relevant statement by Aslib is quoted on pp. 161–2.

Most cover-to-cover translations concern Russian-language periodicals and are therefore dealt with in Appendix I. The American Mathematical Society has however recently commenced publications of *Chinese Mathematics—Acta*, which is a

Espaces vectoriels topologiques et espaces de Banach.

Voir aussi Analyse (espaces fonctionnels).

Voir : 30–*110*–1013,   30–*110*–1141,   30–*110*–1356, 30–*110*–1768,   30–*110*–1870.

30–*110*–776. *BROWDER (F. E.)*. **Semicontractive and semiaccretive nonlinear mappings in Banach spaces.** *Bull. amer. math. Soc.* (1968), *74*, n° 4, 660-5, bibl. (15 réf.). — Amélioration de résultats antérieurs (*ibid*, 1967, *73*, 867-82).

30–*110*–777. *BROWDER (F. E.), PETRYSHYN (W. V.).* **The topological degree and Galerkin approximations for noncompact operators in Banach spaces.** *Bull. amer. math. Soc.* (1968), *74*, n° 4, 641-6, bibl. (20 réf.). — Notion de degré topologique généralisé et propriétés analogue à celle du degré de Leray-Schauder.

30–*110*–778. *BLUMENTHAL (L. M.).* **Note on normed linear spaces.** *Rev. r. Acad. Ci. exact. fis. nat. Madrid* (1968), *62*, n° 2, 307-10, bibl. (4 réf.). — Critère pour qu'un espace normé admette un produit scalaire.

30–*110*–779. *BUMCROT (R. J.).* **Algebraic versus metric concepts in a normed linear space.** *Simon Stevin, Belg.* (1967-68), *41*, n° 3, 252-5, bibl. (1 réf.). — Comparaison de définitions algébriques et de définitions métriques dans un espace normé.

30–*110*–780. *BUCHER (W.).* **Différentiabilité de la composition et complétitude de certains espaces fonctionnels.** *Comment. math. helv.* (1968), *43*, n° 3, 256-88, bibl. (6 réf.).

30–*110*–781. *CROFTS (G. W.).* **Echelon spaces, co-echelon spaces, and steps.** *Dissert. Abstr., B, U. S. A.* (1968), *28*, n° 10, 4194. — Rés. thèse.

30–*110*–782. *SENECHALLE (D. A.).* **A characterization of inner product spaces.** *Dissert. Abstr., B, U. S. A.* (1968), *28*, n° 10, 4207. — Rés. thèse.

30–*110*–783. *BROWDER (F. E.), BUI AN TON.* **Nonlinear functional equations in Banach spaces and elliptic superregularization.** *Math. Z., Dtsch.* (1968), *105*, n° 3, 177-95, bibl. (20 réf.).

30–*110*–784. *KOMATSU (H.).* **An ergodic theorem.** *Proc. Jap. Acad.* (1968), *44*, n° 2, 46-8, bibl. (4 réf.). — Pour les opérateurs bornés d'un espace de Banach.

30–*110*–785. *VAJNIKKO (G. M.)* [Univ. Etat, Voronej]. **En russe.** ((Sur les opérateurs semblables). *Dokl. Akad. Nauk S. S. S. R.* (1968), *179*, n° 5, 1029-31, bibl. (8 réf.). — Théorème sur les opérateurs semblables en rapport avec la convergence des méthodes projectives. Estimations de la convergence de la méthode de Boubnov-Galerkine.

30–*110*–786. *WASHIHARA (M.), YOSHIDA (Y.).* **Remarks on bounded sets in linear ranked spaces.** *Proc. Jap. Acad.* (1968), *44*, n° 2, 73-6, bibl. (3 réf.).

FIG. 10. *Bulletin Signalétique—Mathématiques Pures et Appliquées.* (By kind permission of the Centre National de la Recherche Scientifique.)

complete translation of *Acta Mathematica Sinica*, the official journal of the Institute of Mathematics of the Academy of Sciences, Peking. It contains papers on research in pure and applied mathematics. The first translated volume corresponds with vol. 10, 1960, of the original.

Exercises

1. List the uses to which an abstracting publication can be put. Assume that both the current issue and a complete back file are available.
2. Why is it important for a mathematician to peruse a number of periodicals as a regular part of his work?
3. What is meant by the term *cover-to-cover translations*, and why are they considered necessary?
4. If you were given a grant to cover subscriptions to three mathematical periodicals, how would you choose the titles?

CHAPTER 5

Mathematical Societies

510.6: Mathematical societies

ANY differences which may exist between societies, associations, and institutions need not be considered here. All are treated together as organizations comprising memberships of persons interested or actively engaged in and devoted to the furtherance of a particular subject. For convenience, they will be referred to as societies.

An outline of the history and rise of mathematical societies will be found in most large general encyclopedias. As the rate of scientific and technical progress has increased and research has become more specialized, so the related societies have grown in number and become more restricted in their subject scope. Early societies, such as the Royal Society of London, covered a wide range of scientific disciplines. Some of those which have survived have divided their activities into sections in order to enable them to deal more efficiently with each individual subject. Newer societies are almost invariably devoted to quite narrow subject fields.

Functions

All societies publish a statement of their aims and purpose, and indicate the nature of their membership. Sometimes, conditions for membership are very stringent and demand high academic attainment. Membership, especially in the case of those bodies which bestow qualifications based on examinations or other criteria, may be at various levels according to the individual member's state of advancement. This arrangement is clearly shown in the Institute of Actuaries, London, where provision is made for membership in the grades of student, associate, and fellow. Generally speaking, specialized societies are anxious to encourage students, and are pleased to answer inquiries, within their capabilities, from any *bona fide* inquirer.

An analysis of the publications of the major mathematical societies reveals the following as their main areas of activity:

Meetings—local, national, international.
Publications—periodicals (including proceedings and transactions), books, monographs, etc.
Library and information services.
Educational programmes.
Vocational guidance services.
Co-operation with other organized bodies.

Societies are often called upon to advise on problems concerning mathematics which arise in a wide variety of situations. Their influence can be quite considerable, for example, in determining the course of official inquiries into matters affecting mathematical education. Referring to mathematics in industry, J. T. Combridge, President of the Mathematical Association, London, recently said in a letter to fellow-members: "There are large companies who have the vision to realise that they need a body like our Association to advise them. . . ." It will be found that true scholars who dedicate themselves to the advancement of their subject are invariably willing to give others the benefit of their knowledge and experience.

Advantages of Membership

No society can function without members, and the advantages of membership should therefore be seen as the product of a co-operative effort. This does not simply mean the payment of dues, but the active support of as many of the society's activities as possible. The society, by virtue of the nature of its constitution, may attract other sources of income such as government grants, contributions from industry, and institutional membership. Reciprocal arrangements may be made with societies in other countries whereby certain privileges, such as reduced prices of publications, may be extended in both directions. The American Mathematical Society, for instance, has entered into reciprocity agreements with nearly thirty overseas mathematical organizations.

It cannot be too strongly emphasized that an important complement to bibliographical sources of information is the fund of knowledge possessed by subject specialists. The advantage of society membership in this direction is immeasurable, for it provides a ready means of contacting such specialists and becoming aware of their respective fields of interest. Most societies issue membership directories, which are supplemented by biographical notes in their periodical publications.

International Organizations

The Union of International Associations (1 rue aux Laines, Brussels 1, Belgium) publishes a directory entitled *Yearbook of International Organizations*. Improved communications have led to an increasing number of international meetings in a wide range of subjects. Much valuable work is done, but due to inadequate documentation of resultant publications this has remained a largely untapped source of information. In an endeavour to improve the situation the Union commenced publication, in January 1961, of *Bibliographie Courante des Documents, Comptes Rendus et Actes des Réunions Internationales*. As from January 1968 this has been published monthly in the review *International Associations*. Another of its publications is *International Congress Calendar*. The 1967 edition gives details of scheduled international meetings in all fields for the years 1968–85.

The *World List of Future International Meetings* is published by the Library of Congress, Washington, DC.

International mathematical conferences are held every 4 years. Two valuable prizes are awarded at each conference for the best work done by a mathematician under 40 years of age. In 1962 the International Congress of Mathematicians was held in Stockholm, with the Swedish National Committee for Mathematicians and the Swedish Mathematical Society acting as hosts. The first International Congress was held in 1897. The principal international mathematical body is the International Mathematical Union, Djursholm, Sweden. The object of the sixty-four member countries is to foster international co-operation in mathematics. The Union assists the International Congress of Mathematicians and encourages other activities which are intended to stimulate the development of mathematics.

National Organizations

National mathematical organizations cover the three major areas of research, education, and applications. In some countries all these areas are gathered under the wing of one society, in others different societies cater for each separately. They normally issue, free of charge, a brochure detailing their aims, activities and conditions of membership. Other brochures will deal with such topics as careers, education, and publications. Regular publishing activities vary considerably between societies, but most national bodies issue, as a minimum, a journal in which their transactions and proceedings appear. The contributions to such periodicals and the content of any other scientific literature published by national societies can usually be relied upon to be authoritative. When information is required on any branch or aspect of mathematics in a foreign country, the national society will be the first source to approach.

The Moscow Mathematical Society (Moskovskoe Matematicheskoe Obshchestvo) is one of the oldest of the major national mathematical societies. Founded in 1864, it publishes *Matematicheskiĭ Sbornik* and its transactions, *Trudy*. In the following year came the London Mathematical Society, and in 1894 the American Mathematical Society which actually began in 1888 as the New York Mathematical Society. Two articles in the

Fiftieth Anniversary Issue of the *American Mathematical Monthly* (1967)—one by A. A. Bennett the other by R. A. Rosenbaum—recount the history of the Mathematical Association of America.

The following are all national bodies and are arranged alphabetically by country:

Argentina:	Unión Matemática Argentina Casilla 3588 Buenos Aires
Australia:	Australian Mathematical Society Department of Mathematics Secondary Teacher's College Parkville, Victoria
Austria:	Österreichische Mathematische Gesellschaft Karlsplatz 13 Vienna IV
Belgium:	Société Mathématique de Belgique 317 Avenue Ch. Woeste Brussels 9
China:	Chinese Mathematical Society Peking
France:	Association des Professeurs de Mathématiques de l'Enseignement Public Musée Pédagogique 29 rue d'Ulm, Paris 5e Comité National Français de Mathématiciens 11 rue Pierre et Marie Curie Paris 5e Société Mathématique de France 11 rue Pierre et Marie Curie Paris 5e
Germany:	Berliner Mathematische Gesellschaft eV 1 Berlin 12 Strasse des 17 Juni 135

Germany: (cont.)	Deutsche Mathematiker-Vereinigung eV
	Kiel
	Mathematisches Seminar der Universität
	Olshausenstrasse
	Gesellschaft für Angewandte Mathematik und
	Mechanik
	51 Aachen,
	Templergraben 55
Hungary:	Bolyai János Matematikai Társulat
	Budapest, V
	Szabadság-tér 17
India:	Indian Mathematical Society
	Hans Raj College
	Delhi 7
Israel:	Israel Mathematical Union
	Faculty of Mathematics
	Technion
	Israel Institute of Technology
	Haifa
Italy:	Istituto Nazionale di Alta Matematica
	Città Universitaria
	Rome
Japan:	Nippon Sugaku Kai
	Faculty of Science
	University of Tokyo
Mexico:	Sociedad Matemática Mexicana
	Tacuba 5
	Mexico 1, DF
Netherlands:	Wiskundig Genootschap
	Amsterdam C
	Singel 421
Norway:	Norsk Matematisk Forening
	Universitetet
	Blindern
	Oslo

Poland:	Polskie Towarzystwo Matematyczne Warsaw ul. Śniadeckich 8
Scotland:	Edinburgh Mathematical Society 20 Chambers Street, Edinburgh 1
	Glasgow Mathematical Association Department of Mathematics University of Glasgow Glasgow W2
Spain:	Real Sociedad Matemática Española Serrano 123 Madrid
Sweden:	Makarne Mittag-Lefflers Matematiska Stiftelse Auravägen Djursholm
	Svenska Matematikersamfundet Näckvosgatan 6 Molndal
Switzerland:	Stiftung zur Förderung der Mathematischen Wissenschaften in der Schweiz Genferstrasse 3 Zürich
USSR:	Academy of Sciences of the USSR Section of Physics and Mathematical Sciences Lenin Prospect 14, Moscow
United Kingdom:	Institute of Mathematics and its Applications Maitland House, Warrior Square, Southend-on-Sea, Essex
	London Mathematical Society Burlington House London, W1

United Kingdom: Mathematical Association
    (cont.)   22 Bloomsbury Square
                  London, WC 1

United States: American Mathematical Society
                  P.O. Box 6248
                  Providence
                  Rhode Island 02904

                  Industrial Mathematics Society
                  100 Farnsworth Street
                  Detroit 2
                  Michigan

                  Mathematical Association of America, Inc.,
                  1225 Connecticut Avenue, NW
                  Washington, DC 20036

                  Society for Industrial and Applied
                  Mathematics
                  33 South 17th St.
                  Philadelphia
                  Pennsylvania 19103

Tracing Societies

Tracing societies is facilitated by the existence of directories which specialize in this kind of information. Difficulties may sometimes be experienced where a society has no fixed headquarters, and the only address that can be located is that of an elected secretary or president whose term of office may have expired. When directory information is inadequate, recourse has to be made to the national society. This may also be the case with newly created societies of which the particulars have not yet appeared in directories. Guides to societies vary in the amount of detail they provide. Some go as far as indicating the total membership and the names of the principal officers. Those which list the societies' publications are useful. Others, however, content themselves with names and addresses. Usually the entries are grouped under broad headings with indexes giving a complete alphabetical list of the societies and a subject guide to their activities. Directories which are frequently revised, either as completely new editions or by the publication of supplements, are obviously to be preferred, but it should be remembered that any directory is out of date even on the first day of publication. When writing to a society it is therefore usually best to address either the society or the secretary impersonally rather than an individual officer.

Some useful directories are as follows:

World of Learning, Europa Publications, London, Annual (18th edn., 1968). Information useful to the mathematician includes societies, research institutes, libraries, and universities.

Encyclopedia of Associations. Gale, Detroit, 5th edn., 1968. Volume 1 covers national organizations of the United States.

Scientific and Learned Societies of Great Britain: a handbook compiled from official sources, London, Allen & Unwin, 61st edn., 1964. Part I covers government and public bodies conducting scientific research in Great Britain. Part II is a list of scientific and learned societies. It is divided into 16 sections of which Section 1: *General Science*, and Section 11: *Mathematical and Physical* are of most interest to the mathematician.

Scientific and Technical Societies of the United States, Washington, National Academy of Sciences—National Research Council, 8th edn., 1968. The societies are arranged alphabetically, and there is an analytical index to them. Particulars include history, purpose, membership, meetings, library, research funds, and publications.

The names, addresses, and activities of societies in related fields are given in later chapters together with other information in the respective subjects.

Society Library and Information Services

Many societies maintain libraries for the use of members. Those belonging to old-established bodies may be considerable in size and very comprehensive. They form unique special collections in their subject fields, and their catalogues constitute excellent subject bibliographies. Most societies are prepared to allow inquirers who are not members to use their libraries. This function does indeed make an important contribution to the increasing volume of inter-library loans in which all types of libraries participate.

Those societies which publish journals having book review sections are in a very strong position with regard to book accessions, since they receive many review copies of new publications. Societies also frequently enter into exchange agreements with other learned bodies. Usually periodical publications are the subject of such agreements. In January 1962 a US–USSR exchange agreement came into effect, under which the American Mathematical Society and the Library of the Academy of Sciences of the USSR undertook to ensure the supply of each other's mathematical journals and book series. Initially, some 1000 subscription exchanges were involved. UNESCO has published a third edition of its *Handbook on the International Exchange of Publications* which lists libraries with publications available for exchange. The list is updated by a regular section in the *UNESCO Bulletin for Libraries*. Among recent listings has been the Society of Mathematical and Physical Sciences of the Rumanian People's Republic, offering its *Buletinul Matematic* and other publications in exchange for publications on the same subject. Not all the participants, of course, are societies. Another example is the Mathematical Institute of the Serbian Academy of Sciences and Arts which offers its periodicals in exchange for other publications on mathematics. Donations of personal collections from members are yet another valuable source of stock acquisition for society libraries.

The library of the Mathematical Association, London, may be

described by way of example. It is housed in the Library building of the University of Leicester. Members of the Association may borrow books and journals for periods not exceeding three months at a time, on payment of the cost of postage. A catalogue of *Books and Periodicals in the Library of the Mathematical Association* was published by the Association in 1962.

Another example is that of the Royal Statistical Society, whose library comprises some 80,000 volumes. About 350 different periodicals are currently received. The library covers the entire range of statistics, including the employment of statistical methods in industry.

Other British examples will be found in vol. 1 of the new edition of the *Aslib Directory* (Aslib, London, 1968). American society libraries are included in the *American Library Directory* and *Subject Collections*, both described in Chapter 2, and in *Scientific and Technical Societies of the United States* mentioned above. Most industrial countries of the world have active special libraries (society, industry, government, etc.) and these are the subject of directories. Australian special libraries, for example, are covered by *Directory of Special Libraries in Australia*, published by the Library Association of Australia, Sydney. Up-to-date information on the society libraries of other individual countries may be sought from the respective national library associations.

Locating Society Publications

Locating society publications in catalogues and bibliographies can present certain problems of which it is important to be aware. It is not the present purpose to deal with cataloguing procedures, but their implications can often hamper the layman. The difficulty revolves around the form of author heading which is used. A society publication may be produced either by the society itself, or by an individual or individuals on behalf of the society. In the latter case, although recognized cataloguing codes give specific rulings on the point, some compilers of bibliographies adopt the method of using the society's name as the author entry while others use the name of the individual as the entry word. Confusion also arises in the alternative treatment given to the form of entry used in listing the periodical publications of a society. The most widely adopted rule here is that when the title of the periodical is distinctive and does not include the name of the society, then the entry simply consists of the title of the publication itself. An example is the *Journal of Symbolic Logic*, which is the official organ of the Association for Symbolic Logic. Where, however, the name of the society does figure in the title, this comes first. Thus the *Proceedings of the American Mathematical Society* would be listed as *American Mathematical Society, Proceedings*. One of the advantages of this method is that the exact title of a particular society's publication need not necessarily be known in order to locate it. One would otherwise have to proceed by trial and error using the numerous possible alternatives: Journal, Proceedings, Transactions, Annals, and so forth. This is even more hazardous when a foreign title is in question. Nevertheless, practice is by no means uniform in the major bibliographical works. When using such tools it is essential to read the editor's instructions. These will also deal with the way in which title changes have been catered for. Care must always be taken to try all possible alternatives before abandoning a search.

Exercises

1. Write notes on any four activities of mathematical societies, quoting actual examples where possible.
2. Discuss the advantages and responsibilities attaching to membership of a professional society.
3. In what respects are the library resources of a mathematical society complementary to those of a large public library?
4. Name some of the difficulties of locating society publications in catalogues and bibliographies.

CHAPTER 6

Mathematical Education

510.7: Mathematics, study and teaching

EVERY professional mathematician has a responsibility towards the advancement of his subject and the spread of mathematical knowledge. He is concerned with one aspect of education or another throughout his career. Many qualified men and women who are primarily occupied in non-educational work undertake part-time lecturing. Others contribute articles to periodicals, write books, or simply take part in discussions at society meetings. It is essential to the advancement of science that the fruits of the latest research should be digested into the main body of mathematical education.

The principal responsibility in this process clearly lies with those actively engaged in teaching, from the university professor down to the primary school teacher. They must be aware of the most efficient pedagogical techniques and at the same time absorb as much as they can of the increasing bulk of mathematical literature. If they are to succeed they need to be thoroughly conversant with the appropriate sources of information in both fields.

Purely educational matters lie outside the scope of the present work, but attention should be drawn to the fact that in the Dewey Decimal Classification the literature of education is catered for in the 370 class. Often, in classifying educational literature, its level is a deciding factor. This may be exemplified by reference to textbooks. In the sixteenth edition of Dewey, provision is made for textbooks on a specific subject *at elementary school level* at 372.3–372.8. Textbooks on a specific subject *at higher levels*, however, are placed with that subject, for example, algebra 512.

It is for this reason that textbooks are here dealt with in Chapter 9. Secondly, it should be observed that many of the works referred to in the following sections would, in a general library, normally be classified under education.

Mathematics Courses and Financial Aid

Grants are made to students by a number of different bodies, among which universities, colleges and governments figure most prominently. The American Mathematical Society publishes an annual survey of *Assistantships and Fellowships in Mathematics* as a special issue of its *Notices*. It lists institutions in the United States and Canada offering degree courses in mathematics and closely allied subjects with details of assistantships and fellowships. There is a complete alphabetical index of departments of mathematics, statistics and applied mathematics. A *Guidebook to Departments in the Mathematical Sciences in the United States and Canada*, published by the Mathematical Association of America (3rd edn., 1968), comprises 1200 entries which provide information on the location, size, staff and library facilities for both undergraduate and graduate courses. The *New American Guide to Colleges* by Gene R. Hawes (Columbia University Press, 3rd edn., 1966) gives information on 2000 US colleges and universities, including admission policy and financial aid. A useful feature is its "College Discovery Index". For Britain there is W. J. Langford's *Honours Courses in the Universities and Colleges of Advanced Technology in Great Britain*, originally published in 1963 and since supplemented and updated.

More and more students are undertaking studies in overseas countries. Officially sponsored programmes aim to facilitate this, particularly in the field of student exchanges. Every year UNESCO publishes a directory entitled *Study Abroad*. The 1968–70 edition contains information on some 215,000 individual opportunities for international study and travel offered by 1773 awarding agencies throughout the world. The proportional distribution of awards is given as about 45 per cent for post-graduate study and research, some 30 per cent for study towards university degrees, and the remaining 25 per cent covering other forms of educational activity. The number of awards in mathematics is high. Particulars of organizations offering advisory services and practical help to persons wishing to study abroad are given for each county represented.

Fulbright Travel Grants are well known, and have assisted a great many students. Every two years the Association of Commonwealth Universities, London, publishes *United Kingdom Postgraduate Awards*.

Every university, college, or other educational institution issues its own prospectus, often called a *calendar*, in which particulars of courses offered, admission requirements, and financial assistance are among the information given. Still greater detail is available in the publications of individual departments.

There are several comprehensive guides to universities and colleges covering different areas. Among these are:

> *American Universities and Colleges*, American Council on Education, Washington, DC, 10th edn., 1968. Published every 4 years.
>
> *Commonwealth Universities Yearbook*, Association of Commonwealth Universities, London.
>
> *World of Learning*, Europa Publications, London, 18th edn., 1968.

Need for Teachers

In countries such as the United States and Great Britain, out of the total number of qualified mathematicians approximately one half find employment in academic institutions. In Britain, the majority of university trained mathematicians are employed in teaching of one kind or another. The actual number is over 10,000. To this figure must be added many more who have not passed through university. The new Diploma in Mathematics awarded by the Mathematical Association is primarily intended for non-graduate teachers. Recent figures for the United States show a tendency for younger mathematicians to seek employment outside teaching. In fact, while the proportion of all mathematicians in educational institutions is 42 per cent, the figure for those under 40 years of age is only 36 per cent.

There is an undoubted shortage of mathematics teachers, and the position is being aggravated by the increasing opportunities for less qualified mathematicians in business and industry. These two fields combined are showing only a slightly lower intake of qualified people than teaching. The situation has now arisen when mathematics enrolments at all levels of education are increasing, while the proportion of qualified mathematicians entering the teaching profession is diminishing. More mathematics teachers are required not only to train potential mathematicians but also to provide adequate mathematical training for the other scientists and technologists whose services are in such great demand, especially physicists, chemists, and engineers.

Teacher Training

Teacher qualifications vary according to the level at which the subject is to be taught. They also differ in certain respects from one country to another.

British conditions are covered by the following three publications which are all available free of charge from the Department of Education and Science, Curzon Street, London, W 1:

Becoming a Teacher.
A Career in Education for University Graduates.
Courses of Professional Training for Teachers and Intending Teachers in Technical Colleges and Similar Establishments of Further Education.

Another useful publication, which deals with conditions in Scotland, is entitled *Prospects in Teaching* and is available upon application to the Scottish Education Department, St. Andrew's House, Edinburgh 1.

For America there is the *Occupational Outlook Handbook* prepared by the US Department of Labor, and published in a new edition every other year. Separate reprints of sections dealing with particular occupations may be obtained. *Mathematics Teaching as a Career* is available free from the National Council of Teachers of Mathematics.

The American Mathematical Society's annual survey of *Assistantships and Fellowships in Mathematics* gives details of master's and doctoral degrees designed for the teaching of mathematics. These are offered at nearly 200 institutions and branches of institutions. A new degree, Master of Arts in Teaching (MAT), is offered at some two to three dozen institutions. Almost three-quarters of the institutions make it possible to earn a degree through summer study alone. In order to stimulate much-needed research into mathematics education the Graduate School of Education in the University of Pennsylvania has instituted a new programme in mathematics education research leading to the EdD and PhD degrees; and this programme has ranked for support from the US Office of Education.

To assist in the adoption of new changes in curricula, the US National Science Foundation operates a programme of institutes for high school mathematics teachers. A relevant publication is Andrée, R. V., *The Road for Modern Mathematics* (NSF Summer Institute for Teachers of Secondary School Mathematics, Oklahoma Agricultural and Mechanical College, Stillwater, 1955).

The problem of adapting to new techniques particularly affects experienced teachers and has been treated in a report by the Mathematical Association, London, entitled *The Supply and Training of Teachers of Mathematics* (1963); in the Joint Mathematical Council of the United Kingdom's report on *In-Service Training for Teachers of Mathematics* (1965); and in the Royal Society's pamphlet on *In-service Training for Teachers of Science and Mathematics in England* (1966).

Mathematical Education Periodicals

Periodicals concerned with mathematics teaching should be regularly perused in order to keep abreast of the latest techniques. Full publication details are given here of some English-language journals.

> *Arithmetic Teacher* and *Mathematics Teacher*
>> National Council of Teachers of Mathematics
>> 1201 Sixteenth Street, NW
>> Washington 6, DC, USA.
>> (A *Cumulative Index: the mathematics Teacher, 1908–1965* was published in 1967).
>
> *Journal of Teacher Education*
>> National Education Association
>> Washington 6, DC, USA
>
> *Mathematical Gazette* (Journal of the Mathematical Association)
>> G. Bell & Sons, Ltd.
>> Portugal Street, Kingsway
>> London, WC 2, England
>
> *Mathematics Teaching*
>> Vine St. Chambers
>> Nelson, Lancs., England
>
> *School Science and Mathematics*
>> Central Association of Science and Mathematics Teachers, Inc., Box 246, Bloomington, Indiana 47401, USA.

Periodicals written from the viewpoint of the student are also published, such as *Mathematics Student Journal*, National Council of Teachers of Mathematics, Washington, DC.

Curricular Changes

Until quite recently the method of teaching mathematics had barely changed in over 200 years. During that time, however, considerable advances had been made in the subject with the result that a serious dislocation developed between secondary school and university instruction. This is the centre of the problem, which is being tackled quite vigorously in certain countries. What is required is an adjustment of the type of mathematical thinking among students. Such adjustment requires new pedagogical methods. Mathematics has developed a new language, and it is commonly felt that there is a need to teach modern symbolism in secondary schools as early as possible. Teachers themselves need to be fully conversant with the concepts of what is commonly referred to as *the new mathematics*. An analysis of the present situation and the need to match new methods to contemporary requirements are the subject of *The Changing Curriculum: mathematics*, prepared for the Association for Supervision and Curriculum Development by Robert B. Davis, and published by the National Education Association (Washington, DC) in 1967.

Evidence of the introduction of revised programmes in the United States is reflected, for example, in the examinations of the College Entrance Examination Board and the Educational Testing Service. The bodies mainly responsible for the changes are the School Mathematics Study Group (see *SMSG: the making of a curriculum* by William Wooton, Yale University Press, 1966), the University of Illinois Committee on School Mathematics, the Ball State College Project, and the Commission on Mathematics. In their *Analysis of Research in the Teaching of Mathematics* (USGPO, 1965), Kenneth E. Brown and Theodore L. Abell review research projects and comment on a number of special topics, for example televised instruction.

In Great Britain, a significant event was a conference convened at Southampton in 1961. The findings of the delegates were reported in the Southampton Mathematical Conference, 1961, *On Teaching Mathematics: a Report on Some Present-Day Problems in the Teaching of Mathematics* (Pergamon Press, Oxford, 1961).

The response of mathematics teachers has been impressive: a number of continuing experiments have been initiated and school texts to suit the changing curriculum are being published. Among the principal projects are the following:

Midland Mathematical Experiment. (School texts published by Harrap.)

Nuffield Mathematics Teaching Project (5–13). (Early guides published by John Murray and Chambers; new guides under trial.)

School Mathematics Project. (School texts published by Cambridge University Press.)

Schools Council Secondary School Mathematics Project. (No school texts.)

Scottish Mathematics Group. (School texts published by Blackie/Chambers.)

Shropshire Mathematics Experiment. (School texts published by Penguin.)

Modern Mathematics

There is no lack of literature on the nature of the new mathematics, and the following is a selection of some of the more useful writings:

ADLER, IRVING, *The New Mathematics*, Dobson, London, 1959.

BANKS, JOHN HOUSTON, *Elements of Mathematics*, Allyn & Bacon, Boston, 2nd edn., 1961.

COURANT, RICHARD and ROBBINS, HERBERT, *What is Mathematics? An Elementary Approach to Ideas and Methods*, Oxford University Press, New York, 1941.

FEHR, HOWARD F. and HILL, THOMAS J., *Contemporary Mathematics for Elementary Teachers*, Heath, Boston, 1966.

MCFARLAND, DORA and LEWIS, EUNICE M., *Introduction to Modern Mathematics for Elementary Teachers*, Heath, Boston, 1966.

NATIONAL COUNCIL OF TEACHERS OF MATHEMATICS, *Insight into Modern Mathematics*, The Council, Washington, DC, 1957.

SAWYER, WILLIAM WARWICK, *Prelude to Mathematics*, Penguin Books, 1955. This gives the ideas behind mathematics, and deals especially with matrices and determinants.

SAWYER, WILLIAM WARWICK, *A Path to Modern Mathematics*, Penguin Books, 1966.

STABLER, E. R., *Introduction to Mathematical Thought*, Addison-Wesley, Cambridge, Mass., 1953.

On the international level, a conference was organized by the Office for Scientific and Technical Personnel of the Organization for European Economic Co-operation. An edited version of the papers, discussions, and recommendations may be found in OEEC, *New Thinking in School Mathematics*, The Organization, Paris, 1961.

Also, in 1967, UNESCO published the first volume of a biennial series entitled *New Trends in Mathematics Teaching*; covering the work of the previous 2 years it includes papers given at conferences, articles and reprints, and particulars of periodicals on mathematical education.

Exercises

1. This chapter has many links with the four preceding chapters. Write brief notes on some of these, and quote the titles of publications given in the previous chapters to illustrate your answer.
2. What is the main reason for the revised mathematical curricula which are now being introduced?
3. What steps can the mathematics teacher take to keep abreast of the new developments in teaching methods in his subject?

CHAPTER 7

Computers and Mathematical Tables

510.78: Computation instruments and machines
510.83: Mathematical tables

Computers

ORIENTATION

Electronic computers have brought about an industrial revolution. In many industries they have completely changed the character of research, being applied to the solution of complex problems which might not otherwise have seemed capable of solution. This is the case, for example, in today's vitally important research in aerodynamics, where factors governing the design of the latest supersonic aircraft and space vehicles are calculated with the aid of computers. Indeed, new applications are being devised to meet the requirements of an ever widening range of scientific and technological endeavour. The design, development, and application of computers have stimulated a vast amount of basic research in mathematics, demanding a constant supply of fully qualified mathematicians.

In its turn, this research has generated a steady flow of new literature of all forms: handbooks, textbooks, periodicals, technical reports, tables, conference proceedings, standards and patents. Every aspect of the subject has received its share of attention, from mathematical theory to the most detailed refinements of the applications of individual machines to individual problems.

The difficulty for the student or layman is to know where to start. A distinction must first be drawn between the literature dealing with the manufacture of computers and that covering theoretical design and ultimate applications. Manufacture is largely the province of the instrument engineer, whilst the other aspects are of varying degrees of interest to the mathematician. Standard reference works on computers must necessarily cover every facet of the subject. This is true of Harry H. Husky and Granino A. Korn's *Computer Handbook* published by McGraw-Hill in 1962. The essential mathematical theories are included, and extensive lists of references are appended to each of the sections into which the work is divided. Another recent handbook is Francis J. Murray's *Mathematical Machines* (Columbia University Press, 1961). It is in 2 volumes, the first entitled *Digital Computers*, and the second *Analog Devices*. Again, references follow each chapter.

Marvin L. Minky's *Computation: finite and infinite machines* (Prentice-Hall, 1967) is an imaginative work requiring a minimum of algebra; it includes a bibliography and glossary index. An unusual book is *The Arithmetic of Computers* by Norman A. Crowder (English Universities Press, London, 3rd edn., 1962; first published in the USA by Doubleday). It is issued under the trademark *Tutor Text*, and the presentation is intended to simulate a conversation between teacher and student. The pages are not read consecutively. The reader's response to questions determines which page he should turn to next. If a question is not answered correctly he cannot move on to the next step.

Mathematics and Logic for Digital Devices by James T. Culbertson (Van Nostrand, 1958) is a textbook dealing with permutations and combinations, probability, number systems, traditional logic, Boolean algebra of classes, Boolean algebra of proportions, and applications to switching circuits. Ivan Flores' *Computer Logic: the Functional Design of Digital Computers* (Prentice-Hall, 1960), contains a useful annotated bibliography. Two additional books which contain bibliographies are *Arithmetic Operations in Digital Computers* by R. K. Richards (Van Nostrand, 1955), and *Mathematics and Computers* by George R. Stibitz and Jules A. Larrivee (McGraw-Hill, 1957).

Books on numerical anaylsis are grouped in the Dewey class 517.6. Here will be found such works as the 2-volume *Mathematical Methods for Digital Computers*, edited by Anthony Ralston and Herbert S. Wilf (Wiley, New York, 1960–7). In its preface it is described as "a reference text for many of the more commonly used mathematical methods for digital computers", and it covers the generation of elementary functions; matrices and linear equations; ordinary differential equations; partial differential equations; statistics; and miscellaneous methods. Somewhat similar ground is covered in *Numerical Methods for High Speed Computers* by G. N. Lance (Iliffe, London, 1960). An excellent, annotated bibliography of over 200 items is included in the National Physical Laboratory's *Modern Computing Methods* (Notes on Applied Science, No. 16), (HMSO, London, 2nd edn., 1961).

For those requiring a textbook for secondary school and early college level, there is R. Wooldridge's *An Introduction to Computing* (Oxford University Press, London, 1962). Other titles to note are:

> CALINGAERT, PETER, *Principles of Computation*, Addison-Wesley, Reading, Mass., 1965.
>
> HULL, T. E., *Introduction to Computing*, Prentice-Hall, 1966.

COMPUTER PROGRAMMING

The above are, of course, only representative of a very extensive literature. Some of the works quoted include guides to further reading, and new contributions are reviewed in the periodicals described in the following section. Before leaving books, however, computer programming should be mentioned. A short list must suffice to illustrate the literature of this subject:

> BOOTH, KATHLEEN H. W., *Programming for an Automatic Digital Calculator*. Butterworth, London, 1958.
>
> DORN, WILLIAM S. and GREENBERG, HERBERT J., *Mathematics and Computing; with FORTRAN Programming*, Wiley, New York, 1967.

EVANS, G. W. and PERRY, C. L., *Programming and Coding for Automatic Digital Computers*, McGraw-Hill, New York, 1961.

NATHAN, ROBERT and HANES, ELIZABETH, *Computer Programming Handbook: a guide for beginners*, Prentice-Hall, Princeton, NJ, 1961.

SHARPE, WILLIAM F., *BASIC, an Introduction to Computer Programming*, Free Press, New York, 1967.

WRUBEL, MARSHALL H., *A Primer of Programming for Digital Computers*, McGraw-Hill, New York, 1959.

THE ABACUS

Despite the existence of the high-speed computer the Japan Chamber of Commerce and Industry remains firmly convinced of the continued value of the abacus, and has accordingly prepared a volume entitled *Soroban, the Japanese Abacus, its Use and Practice* (Charles E. Tuttle, Rutland, Vermont, 1967).

BIBLIOGRAPHIES, PERIODICALS, ABSTRACTS

The latest results of research in the field of computers are published in periodicals and technical reports. *United States Government Research and Development Reports* are particularly valuable sources of information. To quote an example, UCRL–6581, 1961, by Rose Kraft and Carl J. Wensrich, is entitled *Monte Carlo Methods: a bibliography covering the period 1949 to June 1961.* Another US Government publication is the *Computer Literature Bibliography* of W. W. Youden (US National Bureau of Standards Miscellaneous Publication 266, USGPO 1965). It contains over 6100 references to which there is an author index and a subject index arranged on the keyword-in-context principle.

A substantial compilation of references has resulted from the joint efforts of researchers in different countries. Entitled *International Computer Bibliography: a guide to books on the use, application and effect of computers in scientific, commercial, industrial and social environment*, it was published in 1968 in Manchester by the National Computing Centre in co-operation with Stichting het Nederlands Studiecentrum Administratieve Automatisering.

Mathematics of Computation is published quarterly by the American Mathematical Society for the National Academy of Sciences—National Research Council. The topics which it covers include advances in numerical analysis, the application of numerical methods and high-speed calculator devices, the computation of mathematical tables, and the theory of high-speed calculating devices and other aids to computation. It also contains reviews and notes.

Another valuable American publication is *Communications of the Association for Computing Machinery* which contains papers dealing with both fundamental research and with computer applications.

In Britain, the *Computer Journal*, published by the British Computer Society, has been established for several years and is an acknowledged source of information of high quality.

There are several other titles which contain a significant amount of fundamental information. These include: *Avtomatika i Telemekhanika* (Russian, available in English translation), *Chiffres* (French), *IEEE Transactions on Computers*, *Journal of Mathematics and Physics*, *Journal of the Association for Computing Machinery*, *Numerische Mathematik* (German), and *Quarterly of Applied Mathematics*. *Computers and Automation* is particularly noteworthy for its reviews of equipment and services and for its *Computer Directory*.

A convenient way of discovering which periodicals include relevant information is to note the titles which regularly appear in the references contained in the two complementary abstracting publications which cover the field. These are *Computer Abstracts*, Technical Information Company, St. Helier, Jersey, and *Computing Reviews* published bi-monthly by the Association for Computing Machinery. The former is issued monthly and includes relatively brief abstracts, while the latter's entries are more selective and evaluative. Although these two are specifically computer-orientated, the relevant sections of the invaluable *Mathematical Reviews* should not be left out of account.

DICTIONARIES

A useful reference work is David D. Polon's *Dictionary of Computer and Control Systems Abbreviations, Signs and Symbols* (Odyssey, New York, 1965).

COMPUTER SOCIETIES

Evidence of the recognition of the necessity for an efficient system of recording and disseminating newly acquired information is further provided by the rapid growth of computer societies. Right from the start, national organizations have been keen to co-operate, to pool their human resources. This spirit can be seen in the foundation of the International Computation Centre, in the co-operative effort behind *Computing Reviews*, and in various other reciprocal arrangements between individual societies. Exchanges of information between East and West have undoubtedly led to a more rapid advancement of the *state of the art* than would otherwise have been the case.

It is essential that information seekers recognize the value of scientific and technical societies. Their interests and activities embrace every aspect of the fields they cover. One has only to peruse a year's back file of a computer society's journal to appreciate this. From its book reviews and advertisements one can compile a bibliography; from an examination of the titles of its principal articles one can grasp the range of computer applications; from its details of meetings one can assess the respective importance of areas of current research.

The International Computation Centre (CP 10053, Viale della Civilita 23, EUR, Rome 100144, Italy) has stated that one of its aims is to undertake mathematical research. The *ICC Newsletter* reports on the activities of members as well as the Centre itself. Particulars of national computer societies are available from the Centre. Activities in Great Britain and the United States are based on:

> The British Computer Society
> Finsbury Court, Finsbury Pavement
> London, EC 2

and

> The Association for Computing Machinery
> 211 East 43rd Street
> New York, NY 10017.

MATHEMATICAL LINGUISTICS AND MACHINE TRANSLATION

Computers provide one of the links between the mathematician and the linguist. Although the subject of machine translation may more properly be considered a part of linguistics, and therefore belonging to the Dewey 400 class, the subject is of growing interest to mathematicians. A brief guide to some sources of information is therefore appropriate here.

One of the barriers to the free interchange of information has been the language problem. English, French and German have traditionally been the principal modern languages in which scientists have recorded their discoveries and technologists their practical accomplishments. Who can say which will be the most used languages of the future? Russian, Chinese and Japanese are certainly claiming an increasingly large measure of the attention of research workers. The launching of the first sputnik by the Soviet Union signalled the start of a gigantic campaign in Western countries to make Russian literature more readily available to English-speaking scientists. The governments of Britain and the United States have added weighty support to programmes of abstracting and translation, and professional societies have undertaken an unprecedented amount of work of this nature. Scientists are being encouraged to learn foreign languages and special courses have been designed by educational institutions to cater for them. There has been a spate of textbooks tailored to the needs of the busy scientist.

The language problem still remains, and efforts are being made to harness machines to the task of solving it. Machine translation is very much a co-operative venture. In an introduction to the *Proceedings of the (1960) National Symposium on Machine Translation* (Prentice-Hall, 1961), H. P. Edmunson remarked that "These interdisciplinary investigations have required the researches of linguists, mathematicians, and computer scientists".

That mathematicians and linguists are equally concerned in the application of mathematics to language problems in general is evidenced by the origin of relevant literature. The *Proceedings of the Eighth International Congress of Linguists* (Oslo University Press, 1958), for example, contains one paper by Joshua Whatmough entitled: "Mathematical linguistics", and another by Paul L. Garvin on machine translation. On the other hand, vol. 12 of the American Mathematical Society's *Proceedings of Symposia in Applied Mathematics* is concerned with the "Structure of language and its mathematical aspects".

It would be wrong to infer that it is only Western countries which have a stake in this field. The first volume in a series of *Prague Studies in Mathematical Linguistics* was published in 1966 by the University of Alabama Press, and a great deal of work in mathematical linguistics is being carried out in Russia, particularly at the University of Leningrad. An article by N. D. Andreyev of that institution, entitled "Models as a tool in the development of linguistic theory", appeared in the April–August issue of *Word* (**18**, 186–97). Pioneer work in machine translation has been done in Russia by D. Yu Panov, whose book *Automatic Translation* was published in English in 1960 by Pergamon Press. Several useful bibliographies have been produced, including:

DELAVENAY, EMILE, *An Introduction to Machine Translation*, Thames & Hudson, London, 1960. A comprehensive study dealing with all aspects of the subject. It includes a bibliography and glossary.

DELAVENAY, E. and K., *Bibliography of Mechanical Translation*, Mouton, The Hague, 1960.

OETTINGER, A. G., *Bibliography of Mathematical Linguistics*, Harvard Computation Laboratory, 1957 (mimeographed).

The Office of Technical Services has also published a selected bibliography of some fifty book reviews and reports on machine translation that have appeared in *Technical Translations* and *US Government Research and Development Reports*. In 1954 the Massachusetts Institute of Technology commenced publication of a periodical called *Machine Translation* to report current information on the subject.

Mathematical Tables

GENERAL

Mathematical tables are tools designed to save the time and labour of those engaged in computing work. They are by no means new. The Greek, Claudius Ptolemy of Alexandria, gave a table of values of the chords of a circle at intervals of one half degree in his *Almagest*. Indeed, tables of a more elementary kind date from much further back in history. The origins and development of tables are dealt with in a convenient manner in the *Encyclopaedia Britannica*.

With the onset of the electronic computer era the rate at which new tables are produced has been enormously accelerated, particularly in Great Britain, the United States, and the USSR. The problem now is to keep track of them so as to ensure their maximum use, and to avoid duplication of effort. To this end the National Research Council in the United States commenced publication of a quarterly journal entitled *Mathematical Tables and other Aids to Computation*. Its aim was to help scientists keep abreast of new tabular computations. This is now issued by the American Mathematical Society under the title *Mathematics of Computation*.

UNITED STATES

In the United States much valuable work was done by the Mathematical Tables Project which was financed by the Works Projects Administration of the City of New York. The Project was eventually absorbed into the Applied Mathematics Division of the National Bureau of Standards, which had been organized in Washington DC in July 1947.

The Computation Laboratory of the National Bureau of Standards is the centre of government activity today. Its *MT* series of publications are important contributions to the subject. They have appeared in a number of publications. Many, for example, have been published in the *Journal of Mathematical Physics* and the *Bulletin of the American Mathematical Society*.

UNITED KINGDOM

An important series are the *British Association Mathematical Tables*. The series began publication in 1931, and ten volumes were published. In 1948 the Royal Society undertook responsibility for the work on mathematical tabulation and has published revised editions of some of the BA volumes. Any subsequent revisions and new volumes will be issued by the Royal Society. The Royal Society also publishes its own *Royal Society Mathematical Tables* and the *Royal Society Shorter Mathematical Tables*.

The object of the *National Physical Laboratory Mathematical Tables Series* is to make available tables of mathematical functions prepared by the Mathematics Division of the NPL which are likely to have a wide application but may not come within the range of the more fundamental tables published in the *Royal Society Mathematical Tables* series. One of the duties of the Mathematics Division is the construction of new mathematical tables and the revision of existing ones. Examples of this work are the tables contributed by F. W. J. Olver to the *Journal of Applied Physics* (1946) and the assistance given in the calculation and arrangement of the mathematical tables in Kaye and Laby's

Tables of Physical and Chemical Constants. The Division constructs new tables in response to requests from clients, and some of these have been published in scientific journals. Others have been contributed to the depository for unpublished tables (from which photocopies may be obtained) which is maintained by the Royal Society Mathematical Tables Committee.

L. Fox's *The Use and Construction of Mathematical Tables*, 1956, is the first volume of the NPL series. The subject is arranged under three headings: the use of mathematical tables, derivation of formulae and analysis of error, and the construction of mathematical tables. A list of references at the end of the volume constitutes a useful bibliography.

CONVERSION TABLES

Cassell's Concise Conversion Tables, compiled by S. Naft and R. de Sola (Cassell, revised by P. H. Bigg, 1966), have been specially adapted for schools from the larger *International Conversion Tables*. Headings and explanations are given in French, German, Italian, and Spanish, as well as in English.

GUIDES TO TABLES

Indexes of tables help to avoid duplication and reduce the work involved in compiling new ones. They also show where gaps exist. Some of the more important, in chronological order of publication, are:

LEHMER, D. H., *Guide to Tables in the Theory of Numbers*, National Research Council, Washington, DC, 1940.

BATEMAN, H. and ARCHIBALD, R. C., Guide to tables of Bessel functions, *Mathematical Tables and Other Aids to Computation*, no. 7, 1944.

FLETCHER, A. *et al.*, *Index of Mathematical Tables*, Blackwell Scientific Publications, 2nd edn., 1962 (Fig. 11). This was first published by Scientific Computing Service Ltd.,

NATURAL AND LOGARITHMIC VALUES OF EXPONENTIAL AND HYPERBOLIC FUNCTIONS

10·0. Introduction

The arrangement of this section will be evident from the short summary given below. It is convenient in many cases to consider separately the ascending exponential e^x and the descending exponential e^{-x}, where x is positive.

The tabulation of exponential and hyperbolic functions is more modern than that of, for instance, trigonometrical functions, but fairly adequate tables are now available.

For $e^{\pm x^2}$ and similar functions, see Section 15. For $e^x/\sqrt{2\pi x}$, etc., see Art. 18·59. For the Poisson distribution, see Art. 23·86.

10·1. Natural Values of e^x

We consider first fundamental tables, mostly to 9 or more figures, and then other tables, mostly to 8 or fewer figures. Radix tables of e^x are considered in Art. 10·15.

On the value of e, see Art. 5·2, where a few integral powers of e to very many figures are mentioned.

10·11. Fundamental Tables of e^x

Extended tables of e^x are somewhat difficult to arrange in order of precision, as often neither the number of decimals nor the number of figures is fixed. In this article we therefore adopt chronological order. The N.B.S. tables are now the chief source of information. On errors in some of the earlier tables, see Van Orstrand 1921, Holtappel 1938 and Glaisher 1883a.

28–29; 32–33 f.	1(1)24; 25, 30 and 60		Schulze 1778

Reproduced in Glaisher 1883a. Last few figures uncertain.

20 dec.	1(1)10	No ⊿	**Bretschneider 1843**
9 fig.	0(·001)·1(·01)2(·1)10(1)500	No ⊿	**Glaisher 1883a**
24 dec.	10(1)20		
4–15; 10–15 dec.	5(·2)6·8; 7(·2)20 ⎫	No ⊿	**Gram 1884**
1 dec.	0(·1)15 ⎭		
12; 16 dec.	0(·001)2; 0(·1)3	No ⊿	**Newman 1889**
20–22 fig.	0(·1)32	No ⊿	Van Orstrand 1913
33; 42 fig.	0(·1)50; 1(1)100	No ⊿	**Van Orstrand 1921**
32 fig.	1(1)32	No ⊿	**Peters & Stein 1922**

230

FIG. 11. Fletcher, *Index of Mathematical Tables*. (By kind permission of Blackwell Scientific Publications.)

London, in 1946. The new edition (in two volumes) is divided into four parts. Part 1 is arranged according to mathematical function and contains an extensive index. Part 2 lists several thousand references to tables. Part 3 lists known errors in published tables. Part 4 is an index to the Introduction and to Part 1.

SCHÜTTE, KARL, *Index of Mathematical Tables*, Oldenburg, Munich, 1955. Throughout this German work all textual matter, including headings and subheadings, is given in both German and English. The work is divided into 16 chapters and over 130 subdivisions. Within these the tables are arranged chronologically. Some 1200 tables are listed. There is an author index and an alphabetical list of institutions.

LEBEDEV, A. V. and FEDOROVA, R. M., *Guide to Mathematical Tables*, Pergamon Press, 1960. This is a translation of a work compiled for the Computing Centre of the Academy of Sciences of the USSR. It is in two parts, of which the first, comprising 15 chapters, is arranged by mathematical function or type of operation. The items in this part bear reference numbers to the second part which contains bibliographical descriptions of the tables. Coverage is international. Supplement No. 1, by N. M. Burunova, Pergamon Press, 1960, contains information on tables published since the compilation of the main work, and other tables which had previously escaped the notice of the compilers. The arrangement is substantially the same as that of the parent work.

COLLECTIONS OF TABLES

There are many useful collections of mathematical tables designed for different classes of users. The following are given as examples of recent publications of this type:

ABRAMOWITZ, MILTON and STEGUN, IRENE A. (Eds.), *Handbook of Mathematical Functions*, Dover, New York, 1965.

Arranged in some 30 sections, each with a full introduction, a bibliography and a set of tables; this is an extremely comprehensive and authoritative work.

Barlow's Tables of Squares, Cubes, Square Roots, Cube Roots and Reciprocals of all Integers up to 12,500, 4th edn. by L. J. Comrie, Chemical Publishing Co., New York, 1956.

BURINGTON, RICHARD STEVENS, *Handbook of Mathematical Tables and Formulas*, McGraw-Hill, 4th ed., 1965.

DWIGHT, HERBERT BRISTOL, *Tables of Integrals and Other Mathematical Data*, Macmillan, New York, 4th edn., 1961.

Handbook of Mathematical Tables, Chemical Rubber Publishing Co., Cleveland, Ohio, 2nd edn., 1964. A substantial volume of nearly 700 pages.

JAHNKE, E. *et al.*, *Tables of Higher Functions*, McGraw-Hill, New York, 6th edn., 1960.

RYSHIK, I. M. and GRADSTEIN, I. S., *Tables of Series, Products and Integrals*, VEB Deutscher Verlag der Wissenschaften, Berlin, 1957. Translation of a standard Russian work containing indefinite and definite integrals as well as Fourier and Laplace transformations. These are supplemented by some numerical tables. There is also a very good bibliography.

Logarithmic tables occur in a variety of forms and prices. One of the variants is the number of decimal places to which they have been worked. Chambers, for example, have catered for different requirements in this respect. A recent, compact volume issued by this publisher is *Chambers's Seven-Figure Mathematical Tables: full edition*, compiled by James Pryde (London, 1961).

The recently inaugurated series by C. Attwood (Pergamon Press) also offers volumes of convenient size; an example is C. Attwood, *Practical Five-Figure Mathematical Tables* (Pergamon, 2nd end., 1967).

Mention of two further works must suffice as illustrations. They are, firstly, *Logarithmetica Britannica: being a Standard Table of Logarithms to Twenty Decimal Places* by Alexander John Thompson. This is a publication of the Department of Statistics of University College, London, issued by Cambridge University Press. Nine separate sections have been published, and the complete work, consisting of logarithms of numbers 10,000–100,000, is available in two bound volumes. Cambridge University Press also issues the Department's series of *Tracts for Computers*; and, secondly, *Smithsonian Logarithmic Tables to Base e and Base* 10, Smithsonian Institution, Washington, DC, 1952.

Exercises

1. How are the various aspects of computers and computation treated in the Dewey Decimal Classification?
2. How would you decide which periodicals would be particularly useful to a mathematician in keeping abreast of new developments in the computer field?
3. Describe, with examples, the different types of sources of information on mathematical tables.
4. Write notes on the value of mathematical tables.

CHAPTER 8

Mathematical History and Biography

510.9: History of mathematics

Introduction

It is impossible to separate the history and biography of mathematics. The history of the subject is largely an account of the work of individuals. It should, however, be noted that in most libraries, biographies of mathematicians will be classified in the Biography class at the number 925.1. For libraries specializing in mathematics, provision is made in the schedules of the Dewey Decimal Classification for placing biography at 510.92.

Poggendorff

Deserving special mention is the monumental *Biographisch–literarisches Handwörterbuch der exakten Naturwissenschaften* of J. C. Poggendorff. It is a systematic bio-bibliographical reference work covering scientists of many nations. Following the biography of each individual is a full listing of his writings. The current volume, which is being issued in parts, began publication by Akademie Verlag of Berlin in 1955. Earlier volumes were issued by Barth of Leipzig in the following order:

1 }	Origins–1858 (1864)	4	1883–1904 (1904)
2 }		5	1904–22 (1925)
3	1858–83 (1898)	9	1923–31 (1936–40)

The years in brackets indicate the dates of publication. Poggendorff will not, of course, be found in all libraries, but all serious students of mathematics should be aware of its existence.

General Histories

Articles in the better encyclopedias are very useful for scene-setting and orientation. Articles on mathematical history may be supplemented by others dealing with the lives and works of individual mathematicians. Usually they also provide guides to further reading.

Similarly, the well-known histories of science may be profitably used, particularly for seeing the development of mathematics against a wider scientific background. Three such works are Charles Singer's *A Short History of Scientific Ideas* (Oxford University Press, 1957), George Sarton's *A History of Science* in two volumes (Harvard University Press, 1952–9), and Salomon Bochner's *The Role of Mathematics in the Rise of Science* (Princeton University Press, 1966).

General histories of mathematics vary in their treatment of the subject. Some give a straightforward chronological picture from the earliest times to the present, others concentrate on a particular period, and a few offer selections of original writings by eminent mathematicians.

Excellent especially for class use is Carl B. Boyer's *History of Mathematics* (Wiley, New York, 1968) which has been well received.

High on one's reading list also should be *The Development of Mathematics* by Eric Temple Bell (McGraw-Hill, New York, 2nd edn., 1945), and the same author's *Mathematics, Queen and Servant of Science* (McGraw-Hill, 1951). Lancelot Hogben's *Mathematics in the Making* (Macdonald, London, 1960) is a beautifully illustrated account of the development of mathematics which breaks away from the style of the more formal history text. Hogben is perhaps best known for his popular *Mathematics for the Million*.

There are several other comprehensive histories of note which are well represented in libraries. They include:

BALL, W. W. ROUSE, *Short Account of the History of Mathe-*

matics, Dover, New York, 1960. This was previously published by Macmillan, 3rd edn., 1901.

CAJORI, FLORIAN A., *History of Mathematics*, Macmillan, New York, 2nd edn., 1919 (repr. 1953).

EVES, HOWARD, *Introduction to the History of Mathematics*, Holt, Rinehart & Winston, New York, 3rd edn., 1964.

FREEBURY, H. A., *History of Mathematics for Secondary Schools*, Cassell, London, 1958.

SMITH, DAVID EUGENE, *History of Mathematics*, 2 vols., Dover, New York, 1958 (originally published by Ginn, Boston, 1923–5). Vol. 1: *General Survey of the History of Elementary Mathematics*. The chapters are arranged chronologically and are preceded by a selective bibliography. Vol. 2: *Special Topics of Elementary Mathematics*.

STRUIK, DIRK JAN, *A Concise History of Mathematics*, Dover, New York, 3rd edn., 1967. Now clearly the best-known general history of mathematics.

Morris Kline has been concerned to show the relationship of mathematics to general cultural development. His *Mathematics: a cultural approach*, was published in 1962 by Addison-Wesley, Reading, Mass., and London. Two other contributions by Kline are *Mathematics and the Physical World* (Crowell, New York, 1959) and *Mathematics in Western Culture* (Oxford University Press, New York, 1953).

Selections from 125 different, important mathematical writings are presented in David Eugene Smith's *A Source Book in Mathematics*, 2 vols., Dover, New York, 1959. This Dover edition is stated to be an unabridged and unaltered republication of the first edition originally published as a one-volume work in 1929. The extracts, translated into English where necessary, are grouped under broad headings: the field of number, algebra, geometry, probability, calculus, functions, and quaternions. Usefully supplementing Smith's selection is *The Treasury of Mathematics*, edited by Henrietta O. Midonick (Peter Owen, London, 1965).

Special Periods

Whilst the general histories usually recount the story from the early beginnings of mathematics up to the present, other works deal selectively with particular periods of development.

Otto Neugebauer's *The Exact Sciences in Antiquity* (Brown University Press, Providence, RI, 2nd edn., 1957) includes Babylonian, Egyptian and Greek manuscripts. With A. J. Sachs, Neugebauer also wrote *Mathematical Cuneiform Texts* (American Oriental Society, New Haven, Conn., 1945) relating to mathematics in Mesopotamia.

Greek mathematics is covered by Sir Thomas Little Heath's *A History of Greek Mathematics* (2 vols., Oxford University Press, 1921). This is a detailed work for the advanced student. A more recent volume is that by Tobias Dantzig entitled *The Bequest of the Greeks* (Scribner, New York, 1955).

The seventeenth and eighteenth centuries figure in *Classical Mathematics: a concise history of the classical era in mathematics* by Joseph Ehrenfried Hofmann (Philosophical Library, New York, 1959), and E. G. R. Taylor has written two lively and stimulating books, *Mathematical Practitioners of Hanoverian England, 1714–1840*, and *Mathematical Practitioners of Tudor and Stuart England*, both published by Cambridge University Press in 1966.

Biography

The outstanding single work under this heading is Poggendorff's handbook which was described at the beginning of this chapter.

Detailed studies of individual mathematicians have also appeared, of which J. F. Scott's *The Mathematical Work of John Wallis, DD, FRS (1616–1703)* may be taken as an example. Originally presented as a doctoral thesis, this was published by Taylor and Francis, London, in 1938. One appendix gives brief biographies of Wallis's contemporaries and immediate predecessors, and another lists the subject's mathematical writings, including his contributions to the *Transactions of the Royal Society.*

Two series of memoirs may be noted as examples of another useful type of source material. They are the annual *Biographical Memoirs of Fellows of the Royal Society,* in which each entry is followed by an exhaustive bibliography of the subject's writings arranged chronologically, and the *Biographical Memoirs* of the National Academy of Sciences of the USA which is published annually by Columbia University Press. An example from the latter is an article on Eliakim Hastings Moore, 1862–1932, contributed by G. A. Bliss and L. E. Dickson (**17,** 83–102, 1937). The piece is followed by a bibliography of 75 references to Moore's writings.

Whilst Smith's source book reproduces extracts from the writings of many eminent mathematicians, the *Collected Works of John Von Neumann,* edited by A. H. Taub (Pergamon Press, 1961), covers the output of a single man in great detail.

In order to keep abreast of new publications on mathematical history and biography, the appropriate section of *Mathematical Reviews* should be regularly perused. Articles on historical research in mathematics are included in such periodicals as the new *Archive for History of Exact Sciences* (1960–) published by Lange and Springer, Berlin–Wilmersdorf, Germany.

Reprints of Classics

The recent increase of interest in the writings of the pioneers in mathematics is manifested in the number of classics being reprinted by major publishers in this field. Frank Cass of London, for instance, have reprinted A. de Moivre's *The Doctrine of Chances or a Method of Calculating the Probabilities of Events in Play.* Cambridge University Press have undertaken to reproduce *The Mathematical Papers of Sir William Rowan Hamilton* in 4 volumes, and G. H. Hardy's *A Mathematician's Apology*; and the Johnson Reprint Corporation have collected *The Mathematical Works of Isaac Newton* in 2 volumes. Also available from

Johnson are the *Lexicon Technicum or an Universal English Dictionary of Arts and Sciences* by John Harris (1704), Guglielmo Libri's *Histoire des sciences mathématiques en Italie depuis la Renaissance des lettres jusqu'à la fin du dix-septième siècle* (1865), and Karl Immanuel Gerhardt's *Geschichte der Mathematik in Deutschland* (1877).

Mathematicians of Today

Those actively engaged in mathematics usually belong to one or more of the national societies, and the simplest way of identifying them is through society membership directories. The libraries of most mathematical societies usually have a collection of membership lists of corresponding overseas organizations, whilst the larger public libraries are also able to produce a number of them, particularly those which appear in issues of a society's periodical publication. Individual membership lists which may be specifically required are available direct from the societies concerned. Names and addresses of societies will be found both in Chapter 5 and in chapters dealing with special branches and aspects of mathematics. An excellent example is the *Combined Membership List* of the Mathematical Association of America, the Society for Industrial and Applied Mathematics, and the American Mathematical Society. It is regularly revised. Published annually by the American Mathematical Society, the *Mathematical Sciences Administrative Directory* lists the names and addresses of chairmen of departments of mathematical sciences in the United States and Canada, heads of industrial mathematics groups, and key personnel in government agencies involved in the mathematical sciences.

The most up-to-date biographical notes of all are to be found in the news or personalia sections of mathematical periodicals.

A service offered by some societies is their membership mailing list which can be supplied, for purposes in the interest of members, either in a form convenient for affixing to envelopes or already printed on a set of envelopes. The American Mathematical Society's list comprises some 8500 names.

A *World Directory of Mathematicians* has been published in Bombay by the Tata Institute of Fundamental Research.

The H. W. Wilson Company of New York publishes *Current Biography* each month (except August). It covers all fields and has a yearly cumulation known as *Current Biography Yearbook*. Each entry is followed by a list of references to the sources of biographical information used. A criterion for inclusion in this publication is that the person must be "prominent in the news". A recent example is the Russian mathematician Mstislav Keldysh (Feb. 1962, pp. 19–21). Obituary notices are published for persons whose biographies have previously appeared.

Eminent mathematicians are represented in the various who's who publications, and certainly appear in *American Men of Science* (Bowker, New York, 11th edn., 1965–7). This is a 6-volume work which gives biographical data on living Americans, including Canadians. Supplementary volumes are issued. French mathematicians are listed in the *Répertoire des scientifiques français;* Tome I, *Mathématiques pures*, published by the Office National des Universités et Écoles Françaises. The *Directory of British Scientists* in 2 volumes (Benn, London, 1966–7 edn., 1966) is an alphabetical directory followed by a classified directory in which names are listed under subject headings: over 4000 under mathematics.

Mathematicians working in academic institutions may be identified in prospectuses and calendars, together with details of any research projects they are engaged upon. *Scientific Research in British Universities and Colleges*, issued annually by the Department of Education and Science, lists researchers and their subjects.

Another way of locating individuals working on particular problems is through the publishers of journals in which their papers have appeared. Occasionally, publishers give the addresses of contributors specifically for this purpose.

Exercises

1. Discuss the question of whether biographical works should be classified with the appropriate subjects or grouped together under the general heading, Biography.
2. How would you set about tracing the mathematical contributions of any living British or American mathematician?
3. Write notes on the relative importance to the student of mathematical history, of histories of mathematics and the original writings of eminent mathematicians.

CHAPTER 9

Mathematical Books. Part 1: Bibliographies

511: Arithmetic
512: Algebra
513–516: Geometry
517: Calculus

Introduction

Opinions about books as sources of information vary greatly. They are largely based on the type of book used during the individual's period of most intensive study, or in the practice of his chosen profession. The textbook naturally figures prominently in many minds. In fact, one's respect for books as working tools can be conditioned by the quality of the particular volumes used during organized courses of study. Those who use books in their occupations consider them in the light of their own experience. The travelling salesman might see them as directories, the research worker as technical reports, and the actuary as statistical handbooks. There are many different kinds of books, each filling a specific role which is primarily determined by the type of information required. It should, however, be noted that different people may well use the same book for different purposes. It is important to be aware of the variety of books, to discover the relationships that exist between them, and to know which to use in particular circumstances.

Finding the books which meet one's requirements at any given time involves three processes: *identification* (discovering appropriate lists of authors and titles), *acquisition* (locating copies of the actual books), and *evaluation* (assessing the ability of the individual books to yield the required information). Identification is a pre-requisite to the other processes, and normally presents the greatest difficulties. Its various aspects therefore occupy the whole of this chapter. Acquisition and evaluation are dealt with in Chapter 10.

Bibliographies

PROBLEMS OF IDENTIFICATION

The need for books may be said to arise through the desire or necessity to find information on a particular subject, and it depends on the purpose for which the information is required what *kind* of books will be sought. For general orientation in a subject an encyclopedia article may suffice. Factual data may be found in handbooks and standard reference works. Even general engineering handbooks, for example, usually contain useful chapters on mathematical fundamentals. Monographs deal exhaustively with narrowly defined subjects, and normally contain guides to further reading. For consecutive study, the textbook is the ideal medium. The advanced textbook having a systematic layout and detailed index may also, of course, be used as a reference tool.

It will be appreciated that the title of a book does not necessarily indicate what kind of book it is. Whilst the most satisfactory way of selecting books is to examine them physically, this is clearly not always possible. One frequently needs to have some means of discovering the authors and titles of books which *may* be suitable. This requirement is met by bibliographies.

FUNCTIONS OF BIBLIOGRAPHIES

Bibliographies are compilations of references to published (and sometimes unpublished) literature on a particular place, person or subject. They are produced as aids to book selection and literature searching, and include sufficient information about each item to enable the inquirer to specify his requirements accurately either to a bookseller or to a librarian. This information consists of author, title, publisher, date and place of publication. Additional particulars may be supplied such as size, a note on the subject content, whether a list of references is included, and whether the text is supplemented by illustrative material such as maps, photographs, diagrams, facsimiles, tables, and so forth. A bibliography may be issued serially on a continuing basis, or as a complete work. In the latter case, the date limits of material quoted are usually indicated.

Bibliographies are used as means of identification of required material. They indicate the existence of this material without necessarily indicating where it may be obtained. In this way they differ from library catalogues and booksellers' lists. It should, however, be noted that both the latter can constitute bibliographies, especially when they have a subject limitation. Thus, for example, the printed catalogue of a mathematical society's library would form a valuable bibliography of mathematics, as would a bookseller's list of his mathematical stock.

The nature of the information provided in each entry determines whether the bibliography is *descriptive* or *evaluative*. The *List of Books Suitable for Training Colleges* mentioned in the section on textbooks is an evaluative bibliography as it is graded according to user requirements and contains comments on the suitability of individual titles for particular purposes.

Whilst it is essential to know the books dealing with one's own subject, this is not sufficient. Problems of finding information frequently occur in fields bordering one's own. This is especially so in mathematics. The mathematician often finds himself working as a specialist member of a team which may be engaged on a problem outside the range of his subject experience. It then becomes imperative for him to be able to find his way through the literature of the new subject.

The following sections contain descriptions of the principal mathematical bibliographies, and to that extent the present work is itself a guide to bibliographies. The titles listed below have been selected both for their value as guides to mathematical bibliographies and as aids to orientating the inquirer in related fields.

GENERAL MATHEMATICAL BIBLIOGRAPHIES

The first bibliography of mathematics was probably Cornelius à Beughem's *Bibliographia Mathematica*, comprising some 3000 entries, and published in Amsterdam in 1685 and 1688. Other early bibliographies, many of them German, are listed in *Scientific Books, Libraries and Collectors* by John L. Thornton and R. I. J. Tully (Library Association, London, 2nd edn., 1962). Nathan Grier Parke's *Guide to the Literature of Mathematics and Physics Including Related Works on Engineering Science* (Dover, New York, 2nd edn., 1958), is divided into 2 parts: 1, General Considerations; and 2, The Literature. The first deals with the principles of reading and study, self-directed education, literature search, and periodicals. Part 2 is the bibliography proper. It contains over 5000 entries grouped under subject headings arranged alphabetically. The number of entries is in fact more than double that of the first edition (McGraw-Hill, 1947), and not surprisingly the emphasis is on applied mathematics. New headings include Actuarial Mathematics (7 entries), Games of Strategy (mathematics) (12 entries), Operations Research (3 entries), and Sets, Theory of (8 entries). Introductory remarks, sometimes including additional references, precede the bibliographical entries under each subject heading. Definitions of the subjects are accompanied by notes on some of the titles quoted. Full author and subject indexes are included. A large proportion of the work naturally deals with the literature of physics. Peripheral topics of a wide range also appear, among them: chemistry, mechanical and general engineering, and technical writing. Sylvia Goldman's *Guide to the Literature of Engineering, Mathematics, and the Physical Sciences* (Johns Hopkins University, Applied Physics Laboratory, 2nd edn., 1964) is selective but not annotated. Northeastern University's *A Selective Bibliography in Science and Engineering* (G. K. Hall, Boston, 1964) is, in effect, a photographic copy of the library's classified catalogue, and mathematical works therefore appear under the classifications given in the present guide.

Volume 3 of *Les Sources du Travail Bibliographique* by L. N.

Malclès (Minard, Paris, 1958), is entitled *Bibliographies Spécialisées*. Mathematical sciences are dealt with in detail on pp. 30–55, arranged in two parts as follows:

Partie générale

Histoire: mathématiques générales; recueils de tables et de formules; bibliographies retrospectives; bibliographies internationales courantes; principaux périodiques mathématiques.

Partie spéciale

Arithmétique et algèbre; théorie des nombres, des groupes, des ensembles; calcul numérique. Analyse: calcul infinitésimal; théorie des fonctions; calcul des variations. Géométrie. Calcul des probabilités; statistique mathématique; logique mathématique.

SPECIAL MATHEMATICAL BIBLIOGRAPHIES

Many bibliographies of value to mathematicians have been produced to cover more specialized needs. They may deal with a particular branch of the subject, the mathematical publications of a particular country, the literature of a specified period, or the requirements of an identified class of user, textbooks for example. Those which have a geographical specification include:

Bibliografia Brasileira de Matemátika e Física, Instituto Brasileiro de Bibliografia e Documentação, 1956– , vol. 1, 1950–4.

CHIA KUEI TSAO, *Bibliography of Mathematics Published in Communist China during the Period 1949–1960*, American Mathematical Society, 1961. 1335 entries with over 1160 titles.

FORSYTHE, GEORGE ELMER, *Bibliography of Russian Mathematics Books*, Chelsea Pub. Co., New York, 1956.

French Bibliographical Digest, Series II, No. 14: *Science and Mathematics*, Part 1: *Pure Mathematics*, French Cultural Services of New York, 1955.

KARPINSKI, L. C., *Bibliography of Mathematical Works Printed in America through 1850*, University of Michigan Press, 1940.

LA SALLE, J. and LEFSCHETZ, S., *Recent Soviet Contributions to Mathematics*. Macmillan, New York, 1962. A state-of-the-art review, with references.

Matematika v SSSR za Sorok Let 1917–1957, 2 vols., Fizmatgiz, Moscow, 1959. Volume 2 contains an exhaustive bibliography for the period.

UNIONE MATEMATICA ITALIANA, *Bibliografia Matematica Italiana*, Tip. Ed. Cremonese, 1950– .

Individual aspects and branches of mathematics—history, teaching, statistics, and so forth—are the subject of other chapters of the present work, each containing details of appropriate bibliographies. It is a useful exercise to create a personal file of references to bibliographies in special mathematical fields. This can be done on standard index cards, and the exercise is doubly effective if these are arranged according to their Dewey classification.

Special bibliographies are frequently met with in periodicals. Examples are:

GOODSTEIN, R. L., Pure mathematics, *British Book News*, no. 221, 1–5, 1959. A brief survey of books on pure mathematics published in Great Britain since 1948. It deals with textbooks and monographs.

HOUSEHOLDER, ALSTON S., Bibliography on numerical analysis, *Association for Computing Machinery Journal*, **3**, 85–100, 1956.

It will be apparent that potentially useful bibliographies occur in publications that will not normally be scanned by mathematicians. A check can, however, be kept on them by regularly consulting the *Bibliographic Index: a cumulative bibliography of bibliographies*. This notes bibliographies appearing in about 1500 different periodicals. It is published semi-annually, with annual and larger cumulations, by Wilson, New York.

The fact should not be overlooked that many books on special aspects of mathematics contain excellent select bibliographies. Monographs are a particular case. They are books which aim to treat a topic exhaustively, setting down as far as possible all that is known about it. When dealing with monographs of mathematical topics it is important to consider their publication dates, and to consult the abstracting publications for possible new contributions. Clearly, the topic under consideration must be very specific. Early examples in scientific literature concerned the life cycles of individual insects. An important mathematical series are the *Carus Mathematical Monographs* of the Mathematical Association of America which may be exemplified by *Non-Commutative Rings* by I. N. Herstein (no. 15, 1968). The Association's *MAA Studies* may be represented by *Studies in Modern Topology*, edited by P. J. Hilton (no. 5, 1968). The American Mathematical Society publishes two series of monographs: *Mathematical Surveys* and *Colloquium Publications*. The latter deal with advanced mathematical subjects such as orthogonal polynomials, algebraic topology, and point set theory. Other such series include the *Princeton Mathematical Series* (1939–), and the *Cambridge Tracts in Mathematics and Mathematical Physics* (1905–).

A unique kind of bibliography is Eleanora A. Baer's *Titles in Series*, published in 2 volumes by Scarecrow Press, Metuchen, NJ, 2nd edn., 1964). Individual volumes are listed under the title of their parent series. The series themselves are arranged alphabetically. Thus under *Carus Mathematical Monographs* appear the authors, titles and dates of publication of the single works. An alphabetical author and title index is provided.

Bibliographies of Textbooks

The textbook hardly needs defining. With the general shortage of mathematics teachers, textbooks are assuming an increasing importance. Whenever mathematicians meet to discuss the modification of mathematical curricula, they invariably stress the urgent need for new and more appropriate textbooks. They see the short-comings of many of the works at present in use as twofold: firstly, they do not adequately reflect the significant advances in the subject over the past decades; and secondly, they need to be designed as more complete units, since greater reliance has to be placed on them in the face of the teacher crisis. "The Textbook as a Teaching Aid", *Education Abstracts*, **7,** 1955, is the title of a UNESCO contribution to the subject. Pages 1517–24 of the *Encyclopedia of Educational Research* (Macmillan, New York, 3rd edn., 1960), are devoted to textbooks and include a bibliography of fifty-five items.

There are several services available to educational authorities which may assist in the identification and selection of suitable material. *Current College-level Book-selection Service* is the title of a bi-weekly joint publication of the American Library Association and the Council on Library Resources. It reviews books as soon as possible after publication, and draws upon a panel of experts for evaluations. A standard work in the United States is Jane Clapp, *College Textbooks* (Scarecrow Press, New York, 1960). This is a classified listing of 16,598 textbooks used in sixty colleges and universities in the United States. There are full subject and author indexes, and a directory of publishers and distributors. Additional titles are listed in a supplement published in 1965. *Textbooks in Print* is issued annually by the R. R. Bowker Company of New York. It indexes the textbooks of some 200 United States publishers. Publishers of this kind of literature have, incidentally, formed an association called the American Textbook Publishers Institute. An example of a specialized listing is the "Guide for the mathematics books of the travelling high school science library of the American Association for the Advancement of Science and the National Science Foundation" by M. F. Willerding, published in *School Science and Mathematics*, **61,** 1961. The bibliography is on pages 101–13. An excellent, annotated list of works published since 1945 is E. P. Vance's "College textbooks", *American Mathematical Monthly*, **62,** 265–88, 1955.

More up to date is the *Basic Library List*, comprising some 300 recommended titles in the area of undergraduate mathematics, published in 1965 by the Mathematical Association of America's Committee on the Undergraduate Program in Mathematics. A third edition of *High School Mathematics Library* was published by the National Council of Teachers of Mathematics in 1967, and in 1968 the Council published *Mathematics Library: Elementary and Junior High School.*

In Britain, school-level textbooks are listed in the annual publication, *Education Book Guide* (National Book League, London). This is "designed to bring together the titles of all books suitable for use in schools . . . published in the United Kingdom in a given year". The entries are based on information supplied by the book publishers, a list of which is given in each volume. The section on mathematics is divided into Primary and Secondary.

The Mathematics Section of the Association of Teachers in Colleges and Departments of Education (151 Gower Street, London, WC 1) has produced a very useful working tool under the title *A List of Books Suitable for Training Colleges.* Apart from bibliographical details, including prices, the value of this publication lies in its annotations contributed by people closely associated with mathematics teaching. In 1963 the Library of the Institute of Education of Southampton University published *Aspects of Modern Mathematics: a select list of books for teachers*, comprising some 200 annotated entries. A. P. Rollett's *School Library Mathematics: a list of books from which a selection could be made for school libraries* was published by Bell, London, in 1966.

Exercises

1. What features would you look for in evaluating a bibliography of mathematics?
2. What are the main differences between a library catalogue and a bibliography?
3. If you were trying to sell a bibliography of mathematics to a mathematician, what would be the main points of your sales talk?

CHAPTER 10

Mathematical Books.
Part 2: Evaluation and Acquisition

511: Arithmetic
512: Algebra
513–516: Geometry
517: Calculus

THE identification of authors and titles of mathematical books was discussed in the previous chapter. The next step is to assess the appropriateness of individual works according to the needs of the inquirer. As there are certain ways in which a preliminary selection can be made without consulting the actual volumes, the question of evaluation will be considered first.

Evaluation

BOOK REVIEWS

Some of the bibliographies already described are evaluative in character. That is, by virtue of their arrangement according to user requirements and their annotations they assist making a preliminary selection. A few in fact quote from reviews which have appeared. Book reviews are a useful aid, but great care has to be taken to maintain a sense of proportion when consulting them. It is not unusual to find two reviews at variance in their assessment of the same book. Most reviewers find something to criticize, and one should not therefore always attempt to discover a book which all the reviewers wholeheartedly and unanimously praise. The distinction must also be drawn between reviews proper and publishers' announcements.

Three guides to reviews of mathematical books are:

AMERICAN MATHEMATICAL SOCIETY, *Twenty-Volume Author Index to Mathematical Reviews*, 1940–59, 2 vols. Covers the first 20 volumes of *Mathematical Reviews*, and includes cross-references to joint authors. Where there is no personal author the entry is under the editor or title.

AMERICAN MATHEMATICAL SOCIETY, *Author Index of Mathematical Reviews, 1960–1964, Vols. 21–28*, 2 vols., 1966.

Book Review Digest, Wilson, New York, monthly (except February and July). The entries are arranged in alphabetical order of authors, and include extracts of reviews taken from a wide range of publications. Plus and minus signs are included, when needed and possible, to indicate the verdict of the complete review. Each issue contains a subject and title index, and cumulative volumes are published.

Technical Book Review Index. Special Libraries Association, New York, monthly (September–June). Quotes reviews from some 1200 scientific, technical and trade journals.

Some Useful Criteria

There is no such thing as the *best* book on a subject. The value of any book to a person seeking information is rigidly determined by its ability to yield that information accurately and in an understandable manner. So much depends on the user's own ability to interpret the information presented in books that it is virtually impossible to name a book "the best". Certain criteria do none the less exist for the evaluation of books in general terms. If applied as a matter of habit when handling books they can frequently save frustration and minimize the risk of acquiring erroneous data.

The date of publication is probably the first thing to look for. Here it is necessary to know the difference between a new edition, a reprint and a re-impression. A new edition really constitutes a new book, earlier editions having been superseded by the inclusion of new material, corrections and re-shaping. A re-impression, on the other hand, is exactly the same as the work previously published. It is in fact printed from the same type. A reprint is produced from new type and contains very few, if any, textual amendments. Unfortunately, many books are called new editions by their publishers when only the slightest textual differences exist. A detailed comparison of the actual volumes is the only sure test. Some indication of the qualifications of the author are frequently given either on the title-page of the book or on its paper jacket. Depending on the type of book will be the desirability of a subject and/or author index. The value of many potentially useful books is seriously reduced by the lack of an adequate index. The inclusion of a bibliography or guide to further reading is another desirable feature, though this again depends on the nature of the individual work. Publication dates of the items comprising such a bibliography can also serve as a guide to the currency of the textual matter.

Acquisition

BOOKS IN PRINT

Acquisition of books will naturally be influenced to some extent by their cost. Whilst library services are constantly improving, it is not, however, their intention to obviate the need for personal collections. Efforts are being made to encourage people to buy their own copies. Special paperback editions are produced at reasonable prices, and several good mathematical series are available in such editions. There is even a catalogue of paperbacks in print. Most professional societies offer specially reduced rates for their publications to members. Unfortunately, this question of cost often adversely influences the selection of books for purchase, though it is emphatically *not* true to say that cheaper books are necessarily inferior.

Guides are published to books which are currently available. *Books in Print: an author–title–series index to the Publishers' Trade List Annual* is issued by Bowker, New York. For those using a subject approach, Bowker also publish a *Subject Guide to Books in Print*, which again is an index to their *Publishers' Trade List Annual*. Their annual *American Scientific Books*, edited by Phyllis B. Steckler, is a cumulation from the *American Book Publishing Record* and is arranged by the Dewey classification. Author and title indexes are included.

Covering the British book trade is *British Books in Print* (Whitaker, London) which replaces the *Reference Catalogue of Current Literature*. It is published annually; the 1968 edition recorded 200,000 British books in print and on sale in the United Kingdom at April of that year. The list can be updated by reference to the quarterly and annual issues of *Whitaker's Cumulative Book List* and the weekly *Bookseller*; also published by Whitaker are *Technical Books in Print* and *Paperbacks in Print*; in 1968 the former listed nearly 15,000 books classified into 37 main classes, and the latter listed 33,000 titles under 53 classifications. Both have an author and title index. The *British National Bibliography* is described on p. 123 and illustrated in Fig. 13.

After drawing up a list of titles selected from mathematical bibliographies, therefore, one can easily discover which of them may be purchased without difficulty.

Great care has to be exercised in purchasing second-hand books, particularly mathematical textbooks. Existing works are rapidly being superseded by those which match the revised curricula. Many older mathematical works still do have current value, but as far as textbooks are concerned early volumes are now usually only of academic interest.

OUT-OF-PRINT BOOKS

Obtaining out-of-print works can be a protracted affair, and the inquirer is advised to put the matter in the hands of a good bookseller. There does exist a unique catalogue of older mathematical texts. This is the *Bibliotheca Chemico-Mathematica: catalogue of works in many tongues on exact and applied science, with a subject index*, compiled and annotated by H. Z. (Heinrich Zeitlinger) and

H. C. S. (Henry Cecil Sotheran) (2 vols., Henry Sotheran & Co., London, 1921). Entries are arranged in an author sequence. The preface states that "As the Catalogue is one of actual books for sale it is of course not complete, but it is believed that few of the great books will be found lacking". The *First Supplement*, published in 1932, has a classified arrangement with mathematics occupying 45 pages. The *Second Supplement*, in 2 volumes, was published in 1937, and comprises 22,895 items. Again the arrangement is classified, with mathematics this time having 133 pages. The annotations are very valuable and the entire catalogue is a work of great scholarship. The *Mathematical Gazette* is quoted as saying that "Sotheran's lists are almost a liberal education in mathematical history". Seperate mathematics lists contain additional titles.

Keeping Abreast of New Mathematical Books

CURRENT BIBLIOGRAPHIES

Mathematical Reviews is of course the main guide to new publications in all branches of mathematics, and from all countries. The American Mathematical Society also produces a list of *New Publications* as a separate item which is available from the Society's Special Projects Department. It was previously published as a regular feature of the *Bulletin*. The Society's own new publications are still given in the *Bulletin*. In 1961 a new service was commenced by the American Bibliographic Service, Darien, Connecticut. It is a *Quarterly Checklist of Mathematica: an international index of current books, monographs, brochures, and separates.*

As a matter of course, the mathematician will read the book reviews in the periodicals which he has selected to peruse. Sometimes overlooked are the lists of new publications which the editor has received for reviewing purposes. Space limitations alone will prevent the review of all the material he receives, and whilst he will attempt to cover the books of widest interest, he is bound to pass over many others. The more specialized the nature of the

Barnard, D. S. Adventures in mathematics. $4.50 '67 Hawthorn bks; 21s '65 Pelham bks.

Dudeney, H. E. 536 puzzles & curious problems. $7.95 '67 Scribner

Emmet, E. R. 101 brain-puzzlers for the young of all ages. 30s '67 Macmillan (London)

Trigg, C. W. Mathematical quickies. $7.95 '67 McGraw

Wadler, L. L. comp. Brain burnishers and rib ticklers. $5 '67 Exposition

See also
Chess

Mathematical statistics
Backhouse, J. K. Statistics. pa 12s 6d '67 Longmans, ltd.

Bancroft, H. Introduction to biostatistics, reprint 21s '65 Harper

Beneš, J. Statistical dynamics of automatic control systems. 70s '67 Iliffe

Berkeley symposium on mathematical statistics and probability. Proceedings of the fifth Symposium, 1965-1966; v 1-3, 5 in 5pts v 1 $20; v2 pt 1, v5 ea $14; v2 pt 2 $15; v3 $10 '67 Univ. of Calif. press

Chakravarti, I. M. and others. Handbook of methods of applied statistics. 2v v 1 $12.95 (98s); v2 $9 (68s) '67 Wiley

Cuming, H. G. and Anson, C. J. Mathematics and statistics for technologists. $12.50 '67 Chemical pub. co.

Fisz, M. Probability and mathematical statistics. $7.50 '66 Verry

Lambe, C. G. Statistical methods and formulae. 27s 6d '67 English univs.

Morrison, D. F. Multivariate statistical methods. $9.95 (80s) '67 McGraw

Neyman, J. and Pearson, E. S. Joint statistical papers. $7 '67 Univ. of Calif. press; 45s Cambridge

Noether, G. E. Elements of nonparametric statistics. $7.95 '67 Wiley

Rickmers, A. D. and Todd, H. N. Statistics. $7.95 (64s) '67 McGraw

Saxena, H. C. and Surendran, P. U. Statistical inference. Rs 12.50 '67 Chand

Shapiro, J. M. Elementary analysis and statistics. $10.95 '67 Merrill

Shaw, L. W. Management information and statistical method. 10s '67 Inst. of chartered accountants in England & Wales. City house, 56-66 Goswell rd, London, E.C. 1

Sverdrup, E. Laws and chance variations. v2 90fl '67 North-Holland pub. co.

See also
Least squares
Mathematical linguistics
Probabilities
Sampling (statistics)
Statistics
Time series analysis

Mathematical statistics: a decision theoretic approach. $14.50 (£5 16s) Academic press

Mathematical symbols. See Abbreviations

Mathematical techniques in electronics and engineering analysis. Head, J. W. pa 18s Iliffe

Mathematical theory of control. Conference on mathematical theory of control, University of Southern California, 1967. $14 Academic press

A mathematician's apology. Hardy, G. H. $2.95 (15s) Cambridge

FIG. 12. *Cumulative Book Index.* (By kind permission of H. W. Wilson Company.)

periodical, the more complete the reviewing coverage is likely to be for the field in question.

Most of the larger libraries, including those of industrial organizations, issue their own accessions lists, often in a classified arrangement. In using them, care should be taken to consult all possible headings under which relevant titles could appear. Titles on linear programming, for example, might be placed under the management heading, or those on econometrics under economics. Every subject specialist should consider it a professional responsibility to acquaint the head of the library service which he regularly uses with details of publications he feels should be purchased. Book selection policies, though always subject to financial considerations, are fundamentally geared to the known and potential requirements of library users. Clearly, the best way in which a librarian can become aware of his clients' needs is from the clients themselves. When a librarian is not sure from the bare bibliographical details of a book whether or not it is worthy of inclusion in his stock he may order an examination copy from the publisher. Close liaison with the librarian in this connection will prove to be of mutual benefit.

There are a number of comprehensive bibliographies, issued on a continuing basis, which are used by librarians as book selection tools, and which are therefore readily accessible to the mathematician.

The *Cumulative Book Index* (*CBI*) (Fig. 12) is issued monthly by the H. W. Wilson Company, New York, and later appears in cumulated volumes designed to reduce the work of consultation. The entries are arranged in one dictionary-type sequence of authors, titles and subjects. Cross-references between subject headings are provided. On looking up geometry, for example, one finds the following headings used:

Geometry	Geometry, Analytic
see also	*see also*
Curves	Conic sections
Topology	Curves
Trigonometry	Geometry, Algebraic

Geometry, Algebraic	Geometry, Descriptive
see also	*see also*
Geometry, Analytic	Perspective
Topology	Etc.

The *British National Bibliography* (*BNB*) (Fig. 13) is arranged according to the Dewey Decimal Classification, and is issued weekly with subsequent cumulations. Whilst the *CBI* covers all English-language publications, the entries in the *BNB* are based on copies received at the British Museum Copyright Office. The publication comprises two sections. In the first the entries are arranged according to the Dewey Decimal Classification. The second section contains entries and references under authors, titles, editors, series and subjects in one alphabetical sequence. First issues of new periodicals and those which have changed their titles are also noted. *British Book News* is largely devoted to reviews of books and new periodicals. Every review bears a Dewey class number, and entries are arranged in a classified order.

Since 1956 the Library of Congress in Washington, DC, has published the *National Union Catalog: a cumulative author list representing Library of Congress printed cards and titles reported by other American libraries*. This extremely valuable work reporting the current intake of the largest libraries in the United States is published monthly with quarterly and annual cumulations. It was previously known as the *Library of Congress Catalog —Books: authors*, and constitutes an updating service to the multi-volume printed catalogue of the Library of Congress. Its quarterly counterpart which began publication in 1950 is the *Library of Congress Catalog—Books: subjects* (Fig. 14). This is arranged in an alphabetical sequence of subject headings and a location in at least one library in the United States is given for each title, additional locations being supplied in the *National Union Catalog*. An idea of the type of headings used can be gained from examining what one finds on looking up algebra in a sample issue:

MATHEMATICS

510—MATHEMATICS. GENERAL WORKS—*cont.*

KORN, Granino Arthur, *and* KORN, Theresa Marie
Manual of mathematics. New York [etc.],
London, McGraw-Hill, 40/-. Jun[1967]. [18],391p.
tables, diagrs. 21cm.

(B67-17334)

LIVING mathematics. London. Cassell.
Part 1, by T. Alaric Millington; designed and illustrated by
Patricia and T. Alaric Millington. 6/6. Oct 1967. [3],62p.
illus., diagrs. 25cm. Lp.

(B67-25278)

MACDONALD, Peter
**Mathematics and statistics for scientists and
engineers.** Windsor House, 46 Victoria St., London, S.W.1, Van Nostrand, 65/-(32/6 Sd.) Oct
[1966]. xii,299p. tables, diagrs., bibliog. 23½cm.

(B67-1558)

MARJORAM, Derek Thomas Elliott
Further exercises in modern mathematics. Oxford,
London, Pergamon, 17/6. 1966. ix,267p. tables,
diagrs. 20cm. Lp. (Commonwealth and inter-
national library. Mathematics division, edited by
W. J. Langford and E. A. Maxwell)
With answers.

(B67-19056)

MATHEMATICS: a course to O Level: editorial
adviser, G.J.R. Potter. London, Nelson.
Book 4 teachers' book, (the pupils' book with addition of
special notes and answers), [by] H. J. Peake, J. Pitts,
F. E. Chettle. 15/-. Oct 1967. [8],256p. illus., tables,
diagrs. 22½cm.

(B67-22890)

Book 4: [Pupil's book, by] H. J. Peake, J. Pitts, F. E.
Chettle. 10/6. Oct 1967. [8],215p. illus., tables, diagrs.
22½cm.

FIG. 13. *British National Bibliography*. (By kind permission of the
Council of the British National Bibliography Ltd.)

Conference Board of the Mathematical Sciences.
The role of axiomatics and problem solving in mathematics. ₁Boston₁ Ginn ₁1966₁
v, 137 p. illus. 24 cm. (Ginn modern mathematics series)

QA11.C635   510.0712   66–24286

Goodstein, Reuben Louis.
Essays in the philosophy of mathematics, by R. L. Goodstein. ₁Leicester, Eng.₁ Leicester University Press, 1965.
167 p. illus. 23 cm.
QA9.G679   65–87340

—POPULAR WORKS

Featheringill, Eve Stanton.
Your child's new math, by Eve Featheringill. Line drawings and visual aids by Philip Featheringill. New York, Morrow, 1966.
94 p. illus. 21 cm.
QA93.F4   510   66–27208

Ganelius, Tord, 1925–
Introduktion till matematiken. Stockholm, Natur o. kultur, 1966.
251, (1) p. illus. 22 cm. (₁Introduktion till vetenskaperna₁)
30.– skr (unb. 24.– skr)
QA93.G3   510   66–76193

Klinger, Fred.
Mathematics for everyone ₁by₁ F. Klinger. New York, Philosophical Library ₁1966, *1965₁
195 p. illus. 20 cm.
QA93.K5513   1966   510   66–31951

May, Lola J
New math for adults only ₁by₁ Lola May and Ruth Moss. Illus. by Roy Doty. ₁1st ed.₁ New York, Harcourt, Brace & World ₁1966₁
vii, 88 p. illus. 15 x 21 cm.
QA93.M36   510   66–12371

Quast, W   Garfield.
New math for parents and pupils ₁by₁ W. Garfield Quast. New York, Arc Books ₁1966₁
180 p. illus. 19 cm.
QA93.Q3   510   66–13834

—PROBLEMS, EXERCISES, ETC.

Abramovich, Mikhail Il'ich.
Сборник задач по математике с образцами решений; дополнение к учебному пособию. Под ред. Е. В. Вороновской. Ленинград, 1965.
205 p. illus. 22 cm.
QA43.A15   66–90998

Burkill, John Charles, 1900–   *comp.*
Mathematical scholarship problems; ₁a selection from recent papers compiled₁ by J. C. Burkill and H. M. Cundy. Cambridge ₁Eng.₁ University Press, 1962.
viii, 118 p. 22 cm.
QA43.B8   510.76   65–8426

FIG. 14. *Library of Congress Catalog—Books: subjects.* (By kind permission of the Library of Congress.)

ALGEBRA
—Problems, Exercises, etc.
—Study and Teaching
ALGEBRA, ABSTRACT
ALGEBRA, UNIVERSAL
ALGEBRA OF LOGIC see Logic, symbolic and mathematical
ALGEBRAIC CONFIGURATIONS IN HYPERSPACE
    see Hyperspace
ALGEBRAIC FIELDS see Fields, Algebraic
ALGEBRAIC FUNCTIONS see Functions, Algebraic
ALGEBRAIC TOPOLOGY
ALGEBRAS, LIE see Lie algebras

Of less impressive proportions, but none the less very useful, is the *Aslib Book List: a monthly list of recommended scientific and technical books with annotations* (Aslib, London). The subtitle is an accurate description, for brief notes on each work amplify the grading which has been made by the reviewing subject specialist, according to the following plan:

A: Books suitable for general readers; treating their subject in an introductory, elementary, or general manner.

B: Books of intermediate technical standard or students' textbooks.

C: Books of an advanced or highly technical character.

D: Directories, dictionaries, handbooks, lists and catalogues, encyclopedias, yearbooks, and similar publications.

The entries are arranged by the Universal Decimal Classification.

National bibliographies are available for most of the major countries. They are particularly useful when their entries are grouped by subject, as in the German (FGR) *Deutsche Bibliographie. Biblio* covers all books published in French, and its arrangement resembles that of the *Cumulative Book Index*.

Publishers' Announcements

Any publisher will be pleased to keep a potential customer informed of his new mathematical books by mailing descriptive sheets and subject lists. All that is necessary is to notify the publisher of one's particular interests. Selection of publishers should be carefully made so as to avoid unnecessary wastage on both sides. This can be done by seeing who has published existing works on one's subject. References in the present volume include the names of most of the major mathematical publishers whose addresses may be obtained from directories or from the national bibliographies.

Exercises

1. How would you evaluate a new mathematical book which you are personally able to examine?
2. Describe two bibliographical services which may be used for checking whether or not particular books are in print.
3. What steps would you take to ensure that you were kept informed of new mathematical books published in the English language?

CHAPTER 11

Probability and Statistics

519: Probabilities and mathematical statistics

Introduction

The recent National Science Foundation report on *Employment in Professional Mathematical Work in Industry and Government*, having briefly defined statistics as "a science based on the mathematical theory of probability", goes on to state that: "In recent years there has been a growth in the application of mathematical statistics to physical and biological phenomena and to business management problems by the use of statistical theory in the design of experiments, in quality control, and in evaluating the likely result of proposed plans of action." This activity is reflected in the vast amount of relevant literature which continues to be published. Both this chapter and the one which follows provide a guide to sources of information in the field.

Current trends are reported in the *Proceedings of the Berkeley Symposia on Mathematical Statistics and Probability*, published by the University of California Press. The Berkeley Symposia have been held every 5 years since 1945. They have now become fully international, and the *Proceedings* of the fifth symposium, held in 1965–66, comprise 5 large volumes covering the theory of statistics; probability theory; physical sciences; biology and health; and weather modifications. Contributions are followed by lists of references.

A series of "Studies in the history of probability and statistics" in *Biometrika* was started by F. N. David in vol. 42.

Before proceeding to an examination of the various sources of information, attention should be drawn to the Dewey 310 class. This caters for statistics under the social sciences heading, and is divided into 311 Statistical Method; 312 Demography; 313–319 General Statistics of Specific Countries.

Books and Bibliographies

Maurice G. Kendall is a name which recurs in the literature of statistics. Kendall has published many important contributions to the subject individually, and has also combined with Alison G. Doig in the joint production of *A Bibliography of Statistical Literature*, a work designed to cover the literature of probability and statistics from the 16th Century to 1958. Volume 1 (Oliver & Boyd, Edinburgh, 1962) deals with the period 1950–8. Volume 2 (published in 1965) covers 1940–9, and vol. 3 (1968) deals with pre-1940 literature. The references—some 30,000 in the complete work—are arranged alphabetically by author.

As part of its programme of education the International Statistical Institute undertakes bibliographical work. Its *Bibliography of Basic Texts and Monographs on Statistical Methods* by William R. Buckland and Ronald A. Fox (Oliver & Boyd, Edinburgh, 2nd edn., 1963) is a result of this policy. It comprises about 190 English-language items dealing both with methods and their application. In each case, bibliographical details are followed by a list of chapter headings and extracts from reviews which have appeared (Fig. 15). The *Bibliography on Time Series and Stochastic Processes*, compiled by Herman O. A. Wold, was published for the Institute by Oliver & Boyd in 1965 (Fig. 16). Other useful bibliographies in the field are:

BUROS, OSCAR KRISEN (ed.), *Statistical Methodology Reviews 1941–1950*, Wiley, New York, 1951. Quotes reviews of 342 books, written in English and published or reviewed in the period, dealing with statistical methods "and such closely related subjects as probability and mathematics of statistics". Earlier years are covered by *Research and*

MATHEMATICAL STATISTICS

GENERAL TEXTS

AITKEN, A. C. *STATISTICAL MATHEMATICS*
1957. 8th Ed., Oliver and Boyd, vii + 153 pp.

Chapter	Contents
1	Statistics as a Science: Axioms of Probability
2	Probability and Frequency Distributions: Graphical Representation: Calculation of Moments
3	Special Probability Distributions
4	Practical Curve Fitting with Standard Curves
5	Probability and Frequency in Two Variates
6	The Method of Least Squares: Multivariate Correlation: Polynomial and Harmonic Regression
7	Probability Distributions of Statistical Coefficients

J. Inst. Actuar., 1943, **71**, 172.

" The book postulates a maturity of thought which makes it unsuitable for the beginner. Each sentence has been carefully weighed and appraised and Dr. Aitken has succeeded in providing the maximum of information in the minimum of space; like the mathematical text-book, it needs perseverance or an accompanying series of lectures for its full appreciation. . . .

" The book is entirely mathematical in character but occasionally Dr. Aitken works through a numerical example. . . . To the serious student who is prepared to do more than scratch at the surface of a difficult subject, the book is a scholarly, and, in its way, complete, introduction to that " advanced " statistical technique which is so fascinating to its adepts and so irritating to its detractors."

J. R. Statist. Soc., 1940, **103**, 400.

" . . . the need for a treatise on Mathematical Statistics as opposed to the existing Method textbooks. . . . With his *Statistical Mathematics* Dr. Aitken has made a first step towards producing

61

FIG. 15. Buckland, *Bibliography of Basic Texts and Monographs on Statistical Methods.* (By kind permission of Oliver & Boyd Ltd.)

1951—1959

8	A	20	p	20	Jap	**Abe, O.** Regression analysis and errors in variables in economic time-series. *Kyodai Keizai Ronsō*, 78, 1, 55–69.	1956
42	A	18	o	—	Ru	**Abramov, L. M.** . The entropy of flow. *Dokl. Akad. Nauk SSSR*, 128:5, 873–875.	1959
42	A	18	o	—	Ru	**Abramov, L. M.** The entropy of productional automorphism. *Dokl. Akad. Nauk SSSR*, 128: 4, 647–650.	1959
41	A	10	o	—	Ru (En)	**Abramov, L. M.** The entropy of solenoid group automorphism. *Teor. Verojatnost. i Primenen.*, 4:3, 249–254.	1959
8	A	23	x	23	En	**Abramovitz, M.** Resource and output trends in the U.S. since 1870. National Bureau of Economic Research, Occ. Papers No. 52, 23 pp.	1956
42	A	16	o	22	En	**Abrams, I. J.** Contribution to the stochastic theory of inventory. Dissertation, Univ. Calif.	1957
8	C	15	x	23	En	**Abramson, A. G., Russel, H. M.** Business forecasting in practice: Principles and cases. Wiley, New York, 275 pp.	1956
21	D	11	p	1	Ge	**Ackermann, W.-G.** Einführung in die Wahrscheinlichkeitsrechnung. Hirzel, Leipzig, 182 pp. *Math. Rev.*, 16, 1127.	1955
42 51	A B	10	–	1	Hu	**Aczél, J.** Funktionalgleichungen in der angewandten Mathematik. *Magyar Tud. Akad. Mat. Fiz. Oszt. Közl.*, 1, 131–142.	1951
51	A	10 19	o	—	En (Ru)	**Aczél, J.** On composed Poisson distributions, III. *Acta Math. Acad. Sci. Hungar.*, 3, 219–224. *Math. Rev.*, 14, 770.	1952
52	A	10 19	p	1 9 10 11	Ge (Ru)	**Aczél, J.** Lösung der Vektor-Funktionalgleichung der homogenen und inhomogenen *n*-dimensionalen einparametrigen "Translation" der erzeugenden Funktion von Kettenreaktionen und des stationären und nicht-stationären Bewegungsintegrals. *Acta Math. Acta. Sci. Hungar.*, 6, 131–141. *Math. Rev.*, 17, 272.	1955
9	A	10	o	—	En (Ru)	**Aczél, J.** A solution of some problems of K. Borsuk and L. Jánossy. *Acta Phys. Acad. Sci. Hungar.*, 4, 351–362. *Math. Rev.*, 16, 1128.	1955
32	A	10 19	p	1	Fr	**Aczél, J.** Remarques algébriques sur la solution donnée par M. Fréchet a l'équation de Kolmogoroff, I. *Publ. Math. Debrecen*, 4, 33–42. *Math. Rev.*, 16, 989.	1955

239

FIG. 16. Wold, *Bibliography on Time Series and Stochastic Processes.*
(By kind permission of Oliver & Boyd Ltd.)

Statistical Methodology Books and Reviews, 1933–38 and
The Second Yearbook of Research and Statistical Methodology, Books and Reviews, published by Gryphon, High-land Park, New Jersey, in 1938 and 1941 respectively.

DEMING, LOLA S., Selected bibliography of statistical literature, 1930–1957, *Journal of Research of the National Bureau of Standards—B: Mathematics and Mathematical Physics*. This is a series of bibliographies of which the first, on correlation and regression theory, appeared on pages 55–68 of the Jan.–Mar. 1960 issue. The entries are based on the NBS Statistical Engineering Laboratory's card file of abstracts compiled from the *Zentralblatt für Mathematik* (for the years 1930–39) and *Mathematical Reviews* (from 1940 onward).

SAVAGE, I. RICHARD, *Bibliography of Nonparametric Statistics*, Harvard University Press, Cambridge, Mass., 1962. This revised edition contains about 3000 entries published up to April 1961.

Additional works are listed in H. O. Lancaster's *Bibliography of Statistical Bibliographies* (Oliver & Boyd, 1968).

A statistical bibliography of a different kind, published by Cambridge University Press, is *A Bibliography of the Statistical and Other Writings of Karl Pearson*, compiled by G. M. Morant and B. L. Welch. The same publisher has also issued *Karl Pearson: an appreciation of some aspects of his life and work* by E. S. Pearson; and in 1966 the University of California Press published *The Selected Papers of E. S. Pearson*, issued by the *Biometrika* Trustees to celebrate his 30 years as editor.

As far as textbooks and monographs are concerned, no more can be done here than to indicate a few important new or recently revised works. The International Statistical Institute's *Bibliography of Basic Texts* contains details of many of the well-tried books. Textbooks dealing with the various applications of statistics are mentioned in their appropriate sections in both this chapter and the next.

A useful and reasonably-priced work on elementary statistics suitable for early university work is *Elementary Statistical Exercises* by F. N. David and E. S. Pearson (Cambridge University Press, 1961). Other titles are *Statistics: an introduction* by D. A. S.

Fraser (Wiley, 1959); *An Introduction to Mathematical Statistics* by H. D. Brunk (Blaisdell, 2nd edn., 1965); *Introduction to Mathematical Statistics* by P. Hoel, Wiley (3rd edn., 1962); *Introduction to Statistics* by J. M. Bevan (Newnes, London, 1968); and *A First Course in Mathematical Statistics* by C. E. Weatherburn (Cambridge University Press, London, 1961). Two books by Samuel S. Wilks should also be noted: *Elementary Statistical Analysis* (Princeton University Press, 1948), and *Mathematical Statistics* (Wiley, New York, 2nd edn., 1962). In 1967 Wiley published a memorial volume entitled *S. S. Wilks: collected papers; contributions to mathematical statistics* which was sponsored by the Institute of Mathematical Statistics and edited by T. W. Anderson. Especially suitable for non-mathematicians is *Statistical Methods* by George W. Snedecor and William G. Cochran (Iowa State University Press), now in its sixth edition, 1967.

Of particular importance is *The Advanced Theory of Statistics* by M. G. Kendall and Alan Stuart (Griffin, London). Its 3 volumes cover: 1, *Distribution Theory* (2nd edn., 1961); 2, *Inference and Relationship* (2nd edn., 1967); and 3, *Design and Analysis, and Time-Series* (1966).

Good introductions to probability theory which may be listed together are:

> BOREL, ÉMILE, *Elements of the Theory of Probability*, Prentice-Hall, Englewood Cliffs, NJ, 1965. (Translated by John E. Freund this is in a way an introduction to Borel's *Traité du calcul des probabilités et de ses applications*.)

> FELLER, WILLIAM, *An Introduction to Probability Theory and its Applications*, 2 vols., Wiley, New York. (Vol. 1, which covered only discrete probability spaces, has remained a standard text since it was first published in 1950. Volume 2 (1966), involving more advanced work, maintains the high quality.)

> GNEDENKO, B. V., *The Theory of Probability*, Chelsea, New York, 4th edn., 1967.

> KOLMOGOROV, A. N., *Foundations of Probability Theory*, Chelsea, New York, 1950.

PARZEN, EMANUEL, *Modern Probability Theory and its Applications*. Wiley, New York, 1960. (Parzen's *Stochastic Processes* was published by Holden-Day, San Francisco, in 1962.)

PRABHU, N. U., *Stochastic Processes*, Macmillan, New York, 1965. (An intermediate level text.)

TUCKER, HOWARD G., *An Introduction to Probability and Mathematical Statistics*. Academic Press, New York, 1962.

Probability Theory by M. Loeve (Van Nostrand, Princeton, New Jersey, 3rd edn., 1963), covers more advanced work, as does Tucker's *Graduate Course in Probability* published by Academic Press in 1967. D. R. Cox and H. D. Miller's *The Theory of Stochastic Processes* (Methuen, 1965), deals exclusively with mathematical theory; it has already become a standard reference work and graduate text.˙

Dictionaries

The second edition of *A Dictionary of Statistical Terms* by Maurice G. Kendall and William R. Buckland (Fig. 17) was published for the International Statistical Institute by Oliver & Boyd, Edinburgh, in 1960. Following the definitions in the first edition, 1957, were four glossaries of equivalent terms arranged in alphabetical order for each of the four other working languages of the Institute: French, German, Italian, and Spanish. The new work has in addition a *combined* glossary in which the English terms determine the alphabetical order. This new revised, combined glossary is also available separately. Much less complete, but still a handy volume for the student, is the *Dictionary Outline of Basic Statistics* by John E. Freund and Frank J. Williams (McGraw-Hill, New York, 1966). A *Statistical Dictionary*, published in 1960 by the Hungarian Central Statistical Office, contains some 1700 terms in seven languages. The volume consists of 2 parts: 1, basic tables arranged in columns, with the Russian terms determining the alphabetical (Cyrillic) sequence, and 2, alphabetical indices of the individual languages—Bulgarian, Czech

Stationary Population

See Stationary Distribution.

Stationary Process

A stochastic process $\{x_t\}$ is said to be strictly stationary if the multivariate distribution of x_{t_1+h}, x_{t_2+h}, ... x_{t_n+h} is independent of h for any finite set of parameter values t_{1+h}, ... t_{n+h}, t_1, t_2, ... t_n.

The process is said to be stationary in the wide sense if the mean and variance exist and are independent of t.

Statistic

A summary value calculated from a sample of observations, usually but not necessarily as an estimator of some population parameter ; a function of sample values.

Statistical Decision Function

See Decision Function.

Statistics

Numerical data relating to an aggregate of individuals ; the science of collecting, analysing and interpreting such data.

Stereogram

A general class of diagram which purports to show a three-dimensional figure on a plane surface. In particular, the name is given to the three-dimensional form of the histogram (q.v.), namely the diagram showing the frequencies of a bivariate distribution.

Stochastic

The adjective " stochastic " implies the presence of a random variable ; e.g. stochastic variation is variation in which at least one of the elements is a variate and a stochastic process is one wherein the system incorporates an element of randomness as opposed to a deterministic system.

The word derives from Greek στόχος, a target, and a *stochastiches* was a person who forecast a future event in the sense of aiming at the truth. In this sense it occurs in sixteenth-century English writers. Bernoulli in the *Ars Conjectandi* (1719) refers to the " *ars conjectandi sive stochastice* ". The word passed out of usage until revived in the twentieth century.

Stochastic Continuity

See Stochastic Process.

FIG. 17. Kendal and Buckland, *A Dictionary of Statistical Terms*.
(By kind permission of Oliver & Boyd Ltd.)

English, German, Hungarian, Polish, and Russian. In 1964, Oliver & Boyd published a 100-page *Glossary of Terms in Official Statistics: English–French; French–English*, compiled for the International Statistical Institute by J. W. Nixon, and in the same year the University of North Carolina Press published Samuel Kotz's *Russian–English Dictionary of Statistical Terms and Expressions and Russian Reader in Statistics*.

Abstracts and Periodicals

Mathematical statistics is well covered by scientific periodicals, most of which are published by professional societies. More recently inaugurated titles represent those fields of application in which there has been particularly rapid development in recent years. A good example is *Technometrics*.

Statisticians are fortunate in having the bulk of this literature abstracted for them in the *International Journal of Abstracts: Statistical Theory and Method* (a journal of the International Statistical Institute), Oliver & Boyd, Edinburgh, quarterly. Its expressed aim is to give complete coverage of published papers in the field of statistical theory and newly published contributions to statistical method. It is a key to the contents of such journals as the *Annals of Mathematical Statistics* (Institute of Mathematical Statistics, Stanford University); *Biometrika, Journal of the Royal Statistical Society* (Series B); *Bulletin of Mathematical Statistics;* and *Annals of the Institute of Statistical Mathematics*. Selective abstracting of other journals is also undertaken: *Biometrics, Metrika, Metron, International Statistical Institute Review, Technometrics*, and *Sankhyā*. The abstracts, which are all in English irrespective of the language of the original, are about 400 words long. Addresses of authors are given so as to facilitate communications.

Individual statistical journals do, of course, include review sections of their own. The Royal Statistical Society's *Applied Statistics* (Oliver & Boyd) and the *Journal of the American Statistical Association* are examples.

Translations

Foreign works, particularly Russian, on probability and statistics are currently receiving attention commensurate with that already described for mathematics in general.

The Society for Industrial and Applied Mathematics, for example, is producing a translation of the Russian journal *Teoriya Veroyatnosti i ee Primeneniye* under the title *Theory of Probability and its Applications*. It contains papers on the theory and application of probability, statistics, and stochastic processes, and is published through a grant-in-aid by the National Science Foundation. *Selected Translation in Mathematical Statistics and Probability* also benefits from an NSF grant. The translations are published for the Institute of Mathematical Statistics by the American Mathematical Society.

Statistical Tables

An important contribution to this subject, and one which should be in the personal library of every statistician, is J. Arthur Greenwood, and H. O. Hartley, *Guide to Tables in Mathematical Statistics* (Princeton University Press, 1962). It is a catalogue of tables of which the majority were published between 1900 and 1954. Full references are provided for each entry, and the work is well equipped with detailed author and subject indexes. A useful feature is an appendix which contains the contents lists of a number of books of tables. This check list is of such practical value, as well as being a handy bibliography, that the works represented are given here:

ARKIN, H. and COLTON, R. R., *Tables for Statisticians*.
BURINGTON, R. S. and MAY, D. C., *Probability and Statistics*.
CZECHOWSKI, T. *et al.*, *Tablice Statystyczne*.
DIXON, W. J. and MASSEY, F. J., *Introduction to Statistical Analysis*.
FISHER, R. A. and YATES, F., *Statistical Tables*.
GLOVER, J. W., *Tables of Applied Mathematics*.

GRAF, U. and HENNING, H. J., *Formeln und Tabellen.*

HALD, A., *Statistical Tables and Formulas.*

JAHNKE, P. R. E. and EMDE, F., *Tables of Higher Functions.*

KELLEY, T. L., *The Kelley Statistical Tables.*

KITAGAWA, T. and MITOME, M., *Tables for the Design of Factorial Experiments.*

LINDLEY, D. V. and MILLER, J. C. P., *Cambridge Elementary Statistical Tables.*

PEARSON, E. S. and HARTLEY, H. O., *Biometrika Tables for Statisticians.*

PEARSON, K., *Tables for S. & B. I & II.*

SIEGEL, S., *Nonparametric Statistics.*

VIANELLI, S., *Prontuari per Calcoli Statistici.*

An important collection is D. B. Owen's *Handbook of Statistical Tables* (Addison–Wesley, Reading, Mass., 1962).

Tables are frequently published for the first time in scientific journals. It is, however, worth noting here that copies of such tables can sometimes be purchased separately. An example is the series of *New Statistical Tables: separates re-issued from Biometrika*, which can be obtained from the Biometrika Office, University College, London.

Societies

The International Statistical Institute (The Hague, Holland) was founded in 1855. Its aims are the improvement of statistical methods, to which end it arranges conferences, assists with educational programmes and publishes several periodicals, including a *Bulletin* and *Review*. Details of its other titles are given elsewhere in this chapter. Further information will be found in J. W. Nixon's *A History of the International Statistical Institute, 1885–1960* (The Institute, Hague, 1960), which supplements F. Zahn's *50 Années de l'Institut International de Statistique* (1934).

National societies exist in most industrial countries of the world. As a minimum activity they publish contributions to the subject, and are able to supply current information on all aspects of statistics in their respective countries.

Important American bodies are:

American Statistical Association
810 18th Street, N.W.
Washington DC, 20006.
Institute of Mathematical Statistics
Department of Statistics
University of North Carolina
Inter American Statistical Institute
Pan American Union,
Washington 6, DC

In Britain the Royal Statistical Society, 21 Bentinck Street, London, W 1, was founded in 1834. Its *Journal* and *Applied Statistics* are important publications. The Society possesses a fine library. The Yule Library, based on books bequeathed by George Udny Yule, includes rare and early statistical publications. Among Yule's writings, his *Statistical Study of Literary Vocabulary* (1944) may be noted. In it the author discusses the application of statistical methods to the study of vocabulary in cases of disputed authorship.

Applications

The applications of statistical techniques are very numerous and have rightly occasioned a substantial amount of literature. As far as the organization of this information is concerned, its distribution in the schedules of the Dewey Decimal Classification is dependent upon the way the material is treated. Where the emphasis is on the subject to which statistics is applied, it may be classified with that subject. The sixteenth edition of Dewey, for example, allocates the notation 330.18, in the Economics class, to Econometrics. Similarly, Quality Control in Production is classified in the Production Management class at 658.562. On the other hand, such books as R. A. Fisher's *Statistical Methods for*

Research Workers, although emphasizing statistical methods in biology, are so fundamental as to be more properly classified in the 519 class. The same could also be said for R. G. D. Steel and J. H. Torrie's *Principles and Procedures of Statistics*.

Provided these considerations are borne in mind it is appropriate to describe some of the sources of information on applied statistics here. They may be conveniently grouped under three broad headings, but it should be stressed that the following is not intended as a complete survey. The object is rather to indicate the type of material which is available.

BIOLOGY, AGRICULTURE, MEDICINE

Biometrics is the application of statistical methods to the field of biological research, and is a subject which has seen a rapid growth in recent years. Undoubtedly one of the best books on statistical methods in biology is Sir Ronald Aylmer Fisher's *Statistical Methods for Research Workers*, published by Oliver & Boyd. The latest edition should always be sought. It includes a bibliography. Another classic by the same author is *The Design of Experiments* (Oliver & Boyd, Edinburgh, 8th edn., 1966); and a third book of the same high calibre is *Statistical Methods and Scientific Inference* (Oliver & Boyd, Edinburgh, 2nd edn., 1959). *Biomathematics: the principles of mathematics for students of biological science* by Cedric Austen Bardell Smith (Griffin London) is a standard reference work suitable for university courses. Now in its fourth edition, its 2 volumes deal with *Algebra, Geometry, Calculus* (1965), and *Numerical Methods, Matrices, Probability, Statistics* (1968). Also noteworthy are K. Mather's *Statistical Analysis in Biology* (Methuen, London, 4th edn., 1965) and Huldah Bancroft's *Introduction to Biostatistics* (Hoeber (Harrap), New York, 1957).

A special application is in the study of genetics. The proceedings of an international symposium, held at Ottawa in 1958 and jointly sponsored by the Biometrics Society and the International Union of Biological Sciences, were edited by Oscar Kempthorne and published in 1960 under the title *Biometrical Genetics*. Kempthorne's *An Introduction to Genetic Statistics* was published by Wiley, New York, in 1957. A more recent publication, which contains a good bibliography, is Norman T. J. Bailey, *Introduction to the Mathematical Theory of Genetic Linkage* (Oxford University Press, 1961). Bailey's *Statistical Methods in Biology* (English Universities Press, London, 1959) is a good introductory textbook. Two more works specifically on genetics are D. S. Falconer, *Introduction to Quantitative Genetics* (Oliver & Boyd, Edinburgh, 1960), and P. A. P. Moran, *The Statistical Processes of Evolutionary Theory* (Oxford University Press, 1962).

Covering another specialized field is *An Annotated Bibliography on the Uses of Statistics in Ecology: a search of 31 periodicals*, by Vincent Schultz (US Atomic Energy Commission, Washington, DC, 1961).

Agricultural applications are dealt with in D. J. Finney's *An Introduction to Statistical Science in Agriculture* (Oliver & Boyd, Edinburgh, 2nd edn., 1962). Other useful textbooks include M. R. Sampford, *An Introduction to Sampling Theory with Applications to Agriculture*, also published by Oliver & Boyd in 1962, and *Statistical Methods for Agricultural Workers* by V. G. Panse and P. V. Sukhatme, published by the Indian Council of Agricultural Research in 1955.

Tables primarily used in the above applications have been collected together by R. A. Fisher and F. Yates in *Statistical Tables for Biological, Agricultural and Medical Research* (Oliver & Boyd), of which several editions have appeared.

The importance of agricultural statistics is recognized at government level. In Britain, for instance, the Agricultural Research Council controls the Statistics Service of Cambridge University's School of Agriculture. Similar functions are performed in the United States by the Agricultural Research Service of the US Department of Agriculture.

Reviewing Austin Bradford Hill's *Principles of Medical Statistics* (*The Lancet*, London, 8th edn., 1966), in the *Journal of the Royal Statistical Society*, P. D. Oldham described it as "unquestionably the best source of statistical knowledge for those working in the field of medicine".

ECONOMICS AND BUSINESS

According to Lange (see below), econometrics "tries by mathematical and statistical methods to give concrete quantitative expression to the general schematic laws established by economic theory". Further elaboration of its nature and functions will be found in two recent articles in *Applied Statistics*. The first, by Eric Shankleman, is entitled "What is econometrics?" (**3**, 85–89, 1954), and the second "The scope and limitations of econometrics", by L. R. Klein (**6**, 1–17, 1957). Although Klein wishes to rule out pure mathematical economics as being simply economic theory in a particular form, there is one recent work on the subject which may be noted in passing. It is Reghinos Theocharis's *Early Developments in Mathematical Economics* (Macmillan, London, 1961). This covers the history up to the mid-nineteenth century work of the famous A. A. Cournot, and includes a bibliography. Also worthy of note is the second edition of R. G. D. Allen's *Mathematical Economics*, which was published by Macmillan in 1959.

Klein's paper, which considers econometrics as the theory and application of measurement in economics guided by an underlying mathematically expressed model, is followed by a useful selective bibliography in which the entries are grouped under headings: general, demand analysis, production and cost functions, export–import functions, aggregative models, sample survey information, and miscellaneous. The following are all useful works on the subject:

BEACH, E. F., *Economic Models: an exposition*, Wiley, New York, 1957. Includes many references to further readings.

KLEIN, L. R., *Introduction to Econometrics*, Prentice-Hall, 1962.

LANGE, OSKAR, *Introduction to Econometrics*, Pergamon, 2nd edn., 1963.

TINBERGEN, JAN, *Introduction to Econometrics*, London, 1953.
TINBERGEN, J. and BOS, H. C., *Mathematical Models of Economic Growth*, New York, 1962.

Econometrica, published quarterly by the Econometric Society, includes a regular book review feature and periodically includes a list of the Society's members.

A good textbook for beginners in business applications is John E. Freund and Frank J. Williams, *Modern Business Statistics* (Pitman, London, 1959). Freund has also authored *Mathematical Statistics*, published by Prentice-Hall in 1962.

INDUSTRY

Technometrics is the title of a quarterly journal of statistics for the physical, chemical, and engineering sciences. It was jointly launched by the American Statistical Association and the American Society for Quality Control, in 1959. In the following year, Academic Press commenced publication of the *Journal of Mathematical Analysis and Applications* which covers the mathematical treatment of questions arising in physics, chemistry, biology, and engineering.

Before further mention is made of statistical quality control, it must be noted that the sixteenth edition of the Dewey Decimal Classification has allocated the notation 658.562 in the Production Management class to Quality Control in Production. The inquirer should, therefore, not fail to examine this section of a library's stock.

The Industrial Statistics Committee of the Eastman Kodak Company has produced a volume of *Symbols, Definitions and Tables for Industrial Statistics and Quality Control* (Institute of Technology, Rochester, New York). Rapid progress has been made in the field of statistical quality control and the application of sampling theory to the question of life expectancy of products, and several good textbooks such as Acheson J. Duncan's *Quality Control and Industrial Statistics* are available.

Among more general works are Kenneth A. Brownlee's

Statistical Theory and Methodology in Science and Engineering (Wiley, 2nd edn., 1965), A. Hald's *Statistical Theory with Engineering Applications*, and Paradine and Rivett's *Statistical Methods for Technologists*.

A compact and inexpensive *Chemist's Introduction to Statistics; theory of error and design of experiments* by D. A. Pantony was published by the Royal Institute of Chemistry, London, in 1961.

OTHER APPLICATIONS

It would be possible to fill many more pages with information on yet more applications of statistics. As the scientific validity of statistical techniques becomes more generally accepted, their circle of influence becomes more widespread. Psychology and education, for example, can claim such works as A. L. Edwards's *Statistical Analysis for Students of Psychology and Education*, J. P. Guilford's *Fundamental Statistics in Psychology and Education*, Quinn McNemar's *Psychological Statistics*, William S. Ray's *Statistics in Psychological Research*, and the 3-volume *Handbook of Mathematical Psychology* by Robert D. Luce and others (Wiley, 1963–65).

As new applications are elaborated they are reported in the periodical press, and those who wish to keep abreast must make a habit of perusing the current issues of some of the titles listed earlier in the chapter.

Exercises

1. Explain why not all books on the application of statistics appear in class 519 of the Dewey Decimal Classification. Illustrate your answer with examples.
2. Describe any *three* of the following:
   (a) a statistical dictionary
   (b) a statistical bibliography
   (c) a statistical society
   (d) a statistical abstracting publication
   (e) a guide to statistical tables.
3. What sources would you use in compiling a bibliography of statistical quality control?

CHAPTER 12

Operational Research and Related Techniques

Dewey 519.9 Class

Operational Research

INTRODUCTION

Operational research is the term used to describe the application of scientific principles to business problems. In the United States it is more commonly referred to as *operations research*. Perhaps it is in order to overcome this diversity of terminology that the initials OR have been commonly adopted. Of the many definitions of OR that have been proposed, the following, contained in a brochure issued by the Operations Research Society of America, is as helpful as any: "Operations research is the science that is devoted to describing, understanding, and predicting the behavior of . . . man-machine systems operating in natural environments."

Management is often defined as a decision-making process. The manager's function is the efficient blending of men, materials and money. So many variable factors are involved in the succession of alternative lines of action which confront him that in making his choice he is driven to rely on his intuition and what is commonly called *business acumen*. He is compelled to follow the road he knows to be safe. Whilst safe, it may not necessarily be the most efficient and profitable. However, to grasp the implications of all the alternatives would be about as difficult as fixing in his mind the pattern produced by a constantly changing kaleidoscope. Such problems are found, for example, in the allocation of resources, in inventory management, and in the scheduling of production and shipments. It is in these and similar fields that OR has come to the aid of management by applying systematic quantitative analysis to the decision-making process. All the many factors involved in each set of alternatives are reduced by statistical techniques to a scientifically organized array of data. From these data a mathematical model may be produced and subjected to tests and analysis in order to establish a reliable basis upon which decisions may be founded.

This, of course, is only a very brief outline of the essential nature of OR. Its breadth is extending rapidly, and its implications are as wide as those of management itself. They range from military strategy to coal mining, from hospital administration to the steel industry.

Introductions to the subject are indicated later, in the section on books. A concise background periodical article which is well worth consulting is "Mathematics for decision makers" by R. K. Gaumnitz and O. H. Brownlee. It appeared on pages 48–56 of the May–June 1956 issue of *Harvard Business Review*. An account of the early history of OR is given in McCloskey and Trefethen's *Operations Research for Management* (see p. 151).

SOCIETIES

The International Federation of Operational Research Societies was founded in 1959 through the co-operation of the national societies of France, the United Kingdom, and the United States. A further seven members were accepted in 1960, and at present, there are twenty members, namely:

Argentina:	Sociedad Argentina de Investigación Operativa (Buenos Aires)
Australia:	Australian Joint Council for Operational Research (Sydney)
Belgium:	Société Belge pour l'Application des Méthodes Scientifiques de Gestion (Brussels)

Canada:	Canadian Operational Research Society (Ottawa)
Denmark:	Danish Operations Research Society (Lyngby)
France:	Association Française d'Information et de Recherche Opérationnelle (Paris)
Germany:	Deutsche Gesellschaft für Unternehmens-forschung (Bonn)
Greece:	Hellenic Operational Research Society (Athens)
Holland:	Sectie Operationele Research, of the Dutch Statistical Society (Vinklaan 1, Son)
India:	Operational Research Society of India (New Delhi)
Ireland:	Operations Research Society of Ireland (Dublin)
Italy:	Associazione Italiana di Ricerca Operativa (Rome)
Japan:	Operations Research Society of Japan (Tokyo)
Mexico:	Asociación Mexicana de Investigación de Operaciones y Administración Científica (Mexico)
Norway:	Avdeling for Operasjonsanalyse (Oslo)
Spain:	Operations Research Society of Spain (Madrid)
Sweden:	Swedish Operations Research Society (Stockholm)
Switzerland:	Schweizerische Vereinigung für Operations Research (Zürich)
United Kingdom:	Operational Research Society Ltd. (London)
United States:	Operations Research Society of America (Chicago, Illinois)

Every 3 years an International Conference on Operational Research is held; the first was held in Oxford, England, in 1957. *Proceedings* are published and form an important contribution to the literature of the subject.

From the outset the need for an effective medium for dis-seminating OR information was recognized. The very first

Annual Report of IFORS states that: "Preliminary consideration has been given to methods of compiling and distributing international abstracts." The first issue of *International Abstracts in Operations Research* (Fig. 18) appeared in 1961. In order to provide world-wide coverage, reviewers in the member societies of the Federation analyse the literature of their respective countries, and additional countries are also represented. The abstracts are written in English, and there are comprehensive subject and author indexes. The subject indexes are based on "index titles" prepared by the abstractors.

The Operations Research Society of America (428 East Preston Street, Baltimore, Maryland 21202) was founded in 1952, and in its first 10 years attracted a membership of over 3500. It now has nearly twice that number. Normally, two national meetings are held annually in different cities of the United States and Canada, as well as sectional meetings. The Society is responsible each year for awarding the $1000 Lanchester Prize for the best English-language paper in OR. In addition to the periodicals included in the list below a *Directory* of members is published annually, and the Society also sponsors the publication of a series of books called *Publications in Operations Research*.

The Operational Research Society Ltd. (64 Cannon Street, London, EC 4, England) developed from a club founded in 1947. It now has approaching 2000 members whose common interest is "the development and extension of operational research as a branch of science". The activities of the Society include the publication of *Operational Research Quarterly* (Pergamon Press), a comprehensive library, an information service dealing with all aspects of OR technical meetings, and assistance in the setting up of courses in OR at universities and colleges of technology.

BIBLIOGRAPHIES

Operational Research, being a recently developed science, has been able to profit from the experience of the older sciences in tackling the problem of communication. As has already been

Abstracts and Reviews

MODELS OF COMMON PROCESSES

modeling of operations, functions or activities common
to various enterprises

ACCIDENTS—*see:* 6932

BEHAVIORAL—*see also:* 6812, 6933

6807

Applications of Complex Behavioral Models to Regional and Organizational Analysis. DAVID L. RAPHAEL. *J. Ind'l. Engrg.* (U. S.) **18** (1967) 1 (Jan.), pp. 123–130.

This article describes two behavioral models and gives examples of how these models can be used. The first is a microregional model of Clinton County, Pennsylvania, which had been used to simulate economic and technological changes and analyze the effects of these changes on the regional economy. It is also being used to analyze the effects of air pollution, water supply and demand, and water quality on regional economy. The second model is an input-output model of the Pennsylvania State University. This model can be used for controlling the operations of the university, studying the effects of changes on the operations, and for management decision-making by simulating alternative courses of action.

(Author. Reprinted from *J. Ind'l. Engrg.*)

6808

A Stochastic Approach to Goal Programming. BRUNO CONTINI. *Opns. Res.* (U. S.) **16** (1968) 3 (May–June), pp. 576–586.

This paper deals with the problem of attaining a set of targets (goals) by means of a set of instruments (subgoals) when the relation between the

334

FIG. 18. *International Abstracts in Operations Research.* (By kind permission of the International Federation of Operational Research Societies.)

noted, provision was made soon after the foundation of the International Federation for the publication of an abstracting journal of world-wide scope. There still remained, however, a need for retrospective indexing, and the following two publications fill in a large part of the gap:

BATCHELOR, JAMES H., *Operations Research: an annotated bibliography*, St. Louis Academy Press, 1959–64. (Covers materials to 1961 in 4 volumes.)

CASE INSTITUTE, OPERATIONS RESEARCH GROUP, *A Comprehensive Bibliography on Operations Research*, Wiley, New York, 1958–63. (Covers materials to 1958 in 2 volumes.)

Apart from the *International Abstracts*, information on current literature is given in many of the periodicals which cover the subject and in some of the abstracting and indexing publications described in Chapter 4. A review of progress in the field is also published: *Progress in Operations Research* (Wiley, New York) vol. 1 (Russell L. Ackoff, ed.), 1961; vol. 2 (David B. Hertz and Roger T. Eddison, eds.), 1964.

BOOKS

Students and practitioners of OR are fortunate in having a hard-core of fundamental works on the subject. All are naturally of recent origin and are readily available. The following list is necessarily selective and does not imply that other literature is less important.

CHURCHMAN, C. WEST, *et al.*, *Introduction to Operations Research*, Wiley, New York, 1957.

EDDISON, R. T., *et al.*, *Operational Research in Management*, English Universities Press, London, 1962.

FLAGLE, CHARLES D., *et al.*, *Operations Research and Systems Engineering*, Johns Hopkins University Press, Baltimore, 1960.

GODDARD, L. S., *Mathematical Techniques of Operational Research*, Pergamon, 1963.

HILLIER, FREDERICK S. and LIEBERMAN, GERALD J., *Introduction to Operations Research*, Holden Day, 1967.

MCCLOSKEY, J. F. and TREFETHEN, F. N. (eds.), *Operations Research for Management*, Johns Hopkins University Press, Baltimore. 1: 1954, 2: 1956.

MORSE, PHILIP M., *Queues, Inventories and Maintenance*, Wiley, New York, 1958.

MORSE, PHILIP M. and KIMBALL, G. E., *Methods of Operations Research*, Wiley, New York, 1951.

RAIFFA, HOWARD and SCHLAIFER, ROBERT, *Applied Statistical Decision Theory*, Harvard Business School, Div. of Research, 1961. An introduction to the mathematical analysis of decision making.

SAATY, THOMAS L., *Mathematical Methods of Operations Research*, McGraw-Hill, New York, 1959. A graduate-level work which includes scientific method, mathematical models, optimization, programming, game theory, probability, statistics, and queuing theory. Each chapter has a useful bibliography.

STOLLER, DAVID S., *Operations Research: Process and Strategy*. University of California Press, Berkeley, 1964.

PERIODICALS

Most of the periodicals which deal specifically with OR are published by national societies. They report the results of experiment and practice, review new literature, and give particulars of meetings and other relevant activities. Through their advertisements some of them also provide useful information on job opportunities.

The oldest established scientific journal on the subject is *Operational Research Quarterly*, published for the Operational Research Society (London) by Pergamon Press. The United States counterpart is *Operations Research, the Journal of the Operations*

Research Society of America, which is published six times a year and is devoted principally to contributions to the field. The Society also publishes a *Bulletin* twice yearly which includes complete programmes of its national meetings.

Other titles containing a significant amount of OR material are:

> *Management Science*, Institute of Management Science, Ann Arbor, Michigan.
>
> *Naval Research Logistics Quarterly*, Superintendent of Documents, Washington, DC.
>
> *Operations Research/Management Science International Literature Digest Service*, Interscience.

The names of additional periodicals may be readily ascertained by perusing the references provided in *International Abstracts in Operations Research*.

Queueing Theory

"Queueing theory is a branch of applied mathematics utilizing concepts from the field of stochastic processes. It has been developed in an attempt to predict fluctuating demands from observational data and to enable an enterprise to provide adequate service for its customers with tolerable waiting." This definition is quoted from Thomas L. Saaty's valuable *Elements of Queueing Theory, with Applications*, which was published by McGraw-Hill, New York, in 1961. The work is arranged in 4 parts as follows: 1, Structure, technique and basic theory; 2, Poisson queues; 3, Non-Poisson queues; 4, Queueing ramifications, applications, and renewal theory. A most useful feature is an extensive bibliography of 910 items.

Another good bibliography of the subject is Alison Doig's "A bibliography on the theory of queues", *Biometrika*, **44,** 490–514, Dec. 1957. Each of the papers listed is followed by a classification symbol. Capital letters are used to denote the following headings which are reproduced here as they give an insight into some of the applications of the mathematical theory of queues:

C: Problems dealing with storage (content)
F: Problems relating to flow through a network
G: Applications not covered by other categories
I: Inventory problems
M: Problems arising in servicing automatic machines
P: Point processes and counter problems
Q: The general theory of queues
R: Road traffic and related problems
S: Stochastic processes directly related to the study of queues
T: Problems in telephone traffic.

Most of the literature so far published on queueing theory is in non-book form, and it is greatly to the credit of those who have written books on the subject that they have included bibliographical guides. In addition to Saaty there are three other recent works which may be mentioned:

COX, D. R. and SMITH, WALTER L., *Queues*, Methuen, London, 1961. (*Methuen's Monographs on Applied Probability and Statistics.*) Appendix 1 comprises bibliographical notes whose purpose is "to indicate key papers on which the treatment in this monograph is based and from which references to other work may be obtained".

KHINTCHINE, A. Y., *Mathematical Methods in the Theory of Queueing*, Griffin, London, 2nd edn., 1968.

TAKACS, LAJOS, *Introduction to the Theory of Queues*, Oxford University Press, New York, 1962. Each chapter is followed by a bibliography; and the principal mathematical theorems used—Markov chains, Markov processes, recurrent processes—are treated in an appendix.

Theory of Games

The theory of games is another example of a recently developed subject where the bulk of the literature is in the form of papers and periodical articles. Early work was done by John Von Neumann who, together with Oskar Morgenstern, wrote *Theory of Games and Economic Behaviour*, Princeton University Press, 3rd edn., 1953. In 1967 Princeton University Press published *Essays in Mathematical Economics* in honour of Morgenstern, edited by Martin Shubik. The volume includes twenty-seven papers on game theory and other topics, and includes a bibliography of Morgenstern's works.

Handy introductions are S. Vajda's *An Introduction to Linear Programming and the Theory of Games* (Methuen, London, 2nd, edn., 1966), and *The Elementary Ideas of Game Theory* by Maurice Peston and Alan Coddington (HM Treasury, HMSO, 1967).

Works which contain guides to further study are:

> DRESHER, MELVIN, *Games of Strategy: theory and applications*, Prentice-Hall, Englewood Cliffs, NJ, 1961.
>
> LUCE, R. DUCAN and RAIFFA, HOWARD, *Games and Decisions*, Wiley, New York, 1957. A college textbook which includes a very extensive reading list.
>
> McKINSEY, J. C. C., *Introduction to the Theory of Games*, McGraw-Hill, New York, 1952.

Contributions from the world's leading researchers are contained in *Advances in Game Theory*, edited by M. Dresher, L. S. Shapley, and A. W. Tucker, and published in 1964 by Princeton University Press (*Annals of Mathematics Studies*, no. 52).

Mathematical Programming

Books on mathematical programming differ in treatment according to whether they are aimed at the mathematician or the businessman. On the one hand the theoretical aspects of the subject are emphasized; on the other hand the results which can be achieved by practical application of the theory.

An introduction is provided by Robert W. Metzger's *Elementary Mathematical Programming* (Wiley, New York, 1958). It contains a guide to further reading in which the entries are broadly graded. S. Vajda's *Mathematical Programming* (Addison-Wesley, Reading,

Mass., and London, 1961), is a graduate-level textbook of linear and nonlinear programming. Its bibliography contains 128 references.

A most important source book is *Linear Programming and Associated Techniques: a comprehensive bibliography on linear, nonlinear, and dynamic programming* by Vera Riley and Saul I. Gass (Johns Hopkins Press, Baltimore, revised edn., 1958). It comprises references to over 1000 items, including articles, books, monographs, documents, theses, and conference proceedings. Part I contains an annotated list of basic references, Part II covers general theory, Part III applications, and Part IV nonlinear and dynamic programming. Useful, too, is *Dynamic Programming: a bibliography of theory and application*, compiled by Richard Bellman and others. First issued by the Rand Corporation, Santa Monica, California, in 1964, this has achieved wider distribution through the Clearinghouse for Federal Scientific and Technical Information.

"A linear programming problem differs from the general variety in that a *mathematical model* or description of the problem can be stated, using relationships which are called 'straight-line', or linear." This is how Gass puts it in the introduction to his textbook *Linear Programming: methods and applications* (McGraw-Hill, 1958, now in its 2nd edn., 1964).

The first general linear programming problem was formulated by George B. Dantzig, who also developed the simplex method for its solution. The bulk of his writings on all aspects of the subject are contained in his *Linear Programming and Extension*, a valuable compendium published by Princeton University Press in 1963. Furthermore, this volume includes a good bibliography covering the literature to 1963.

The following is a selection of useful titles:

FICKEN, F. A., *The Simplex Method of Linear Programming*, Holt, Rinehart & Winston, New York, 1961.

GARVIN, WALTER W., *Introduction to Linear Programming*, McGraw-Hill, New York, 1960. It is stressed in the preface that this book is an *introduction* to the subject and has been limited in size to make it suitable as a textbook for a one-semester course.

GEARY, R. C. and MCCARTHY, M.D., *Elements of Linear Programming with Economic Applications*, Griffin, London, 1964.

GLICKSMAN, A. M., *An Introduction to Linear Programming and the Theory of Games*, Wiley, 1963.

GREENWALD, DAKOTA ULRICH, *Linear Programming*, Ronald Press, New York, 1957. This is intended as an explanatory text for those not possessing advanced mathematics. It is a concise book of only 75 pages.

Exercises

1. From a bibliographical point of view, how can a new science seek to avoid the problems of communication experienced by workers in the older sciences? Illustrate your answer with examples from operational research.
2. What reasons would you give to a person entering the field of OR in advising him to join the national society?
3. What steps would you take to ensure that you did not miss the appearance of any important new writings on applications of the theory of games?

Sources of Russian Mathematical Information

Russian-Language Literature

There are two main ways in which collections of Russian literature are built up: by exchange and by purchase. Many libraries and other organizations which can offer publications for exchange enter into agreements with similarly-placed Russian bodies and are thereby able to augment their stocks acquired by purchase.

PERIODICALS

The purchase of Russian periodicals by subscription is carried out through agents appointed by the central Russian distributing organization, Mezhdunarodnaya Kniga. These agents issue, free of charge, an annual list of available material entitled *Newspapers and Magazines of the USSR*.

An analysis of Russian mathematical journals appears in *Russian Journals of Mathematics: a survey and checklist* by H. A. Steeves (New York Public Library, 1961). It contains three lists as follows: 51 journals which published more than 20 mathematical papers in a specified period; 68 with 7 to 20 papers; and 131 with 3 to 6 papers. There is also a Russian–English glossary.

BOOKS

As for books, the printings are strictly limited, and the supply of any particular title in non-Soviet-bloc countries cannot be guaranteed. Often the entire stock of a book is exhausted in

157

Russia itself, leaving no copies available for foreign booksellers.

Newly published books are listed in the monthly Russian publication *Knizhnaya Letopis'* and *Novye Knigi SSSR* which have sections devoted to mathematics. Another Russian-language work of value is A. M. Lukomskaya *Bibliography of Domestic Literature in Mathematics and Physics* (Akad. Nauk SSSR, Biblioteka, Moscow–Leningrad, 1961) which contains an appendix listing sources of bibliographical information. The most exhaustive single bibliography, however, is contained in the second volume of *Matematika v SSSR za Sorok Let 1917–1957* (Fizmatgiz, Moscow, 1959), in which some 22,000 articles and books are listed.

Two volumes published in the United States which may be examined are George Elmer Forsythe, *Bibliography of Russian Mathematical Books* (Chelsea Publishing Co., New York, 1956) and Joseph Pierre La Salle, and Solomon Lefschetz, *Recent Soviet Contributions to Mathematics* (Macmillan, New York, 1962), a state-of-the-art review, with references.

GUIDES TO COLLECTIONS

The best English-language publication for keeping abreast of Russian mathematical literature is undoubtedly *Mathematical Reviews*, which is described in Chapter 4. However, in order to ascertain the availability of any particular document, recourse has to be made to union catalogues, holding lists, and library accessions lists.

The *Monthly Index of Russian Accessions*, published by the Library of Congress, is a record of publications in the Russian language received by the Library of Congress and a group of co-operating libraries. It is arranged in three parts:

Part A: Monographic Works. (In each entry, the title in the original language is preceded by its English translation in brackets.)

Part B: Periodicals. (Simply indicates which issues of each

title have been received. The entries are arranged in subject groups.)

Part C: Subject Index to Monographs and Periodicals. (This is the largest section and is an extremely valuable analytical index. Items are arranged under subject headings, and the whole is preceded by a list of periodicals indexed and abbreviations used.)

Various cumulations are available, and orders for microfilm or photostat copies of the listed items which are in the collections of the Library of Congress may be placed with the Library's Photo-duplication Service.

Another valuable Library of Congress guide is its *Serial Publications of the Soviet Union, 1937–1957: a bibliographic checklist*.

In Britain the National Lending Library of Science and Tech-nology (NLL) is rapidly expanding its collection of Russian literature. Book accessions are announced in its *List of Books Received from the USSR and Translated Books*, which is available free of charge on request. The list is arranged in broad subject groups, approximately corresponding to the Universal Decimal Classification. The NLL's translation activities are described in the following section.

Literature in Translation

GENERAL GUIDES

Since only a relatively small number of mathematicians in Western countries are able to read Russian, it has become imperative for translation on a large scale to be undertaken if important Russian contributions are to be fully exploited. Until a few years ago Western scientists remained largely unaware of the scientific progress being made in the Soviet Union. Today, how-ever, there is little excuse for not keeping informed. Translation activity is so widespread and virile that it has become necessary to produce guides to its ramifications.

One of the first of these bibliographical signposts was entitled *Providing US Scientists with Soviet Scientific Information.* It proved useful enough to warrant a second edition in 1962. Compiled by B. I. Gorokohoff and published by the Massachusetts Institute of Technology, it covers such things as translations of periodicals, books and other documents, abstracting services, and the availability of this material in United States libraries.

An international survey is "Translations of Russian scientific and technical literature" by Alice Frank, *Revue de la Documentation,* **28,** 47–51, May 1961. The areas dealt with are Austria, Belgium, Canada, France, Germany, Netherlands, Scandinavia, Spain, United Kingdom, and the United States.

Also worth noting is *Notes on Searching Russian Scientific Literature in Translation* by Harry La Plante, published by the University of Detroit Library in 1961.

GOVERNMENT SUPPORT

In the United States the Office of Scientific Information Services of the National Science Foundation is particularly concerned in this field. It was established in 1958. Two of its designated programmes are Support of Scientific Publications and Foreign Science Information. The Office supports, through grants, the cover-to-cover translation of some fifty periodicals and a large number of Russian books. Recently, for example, it awarded $16,687 for the translation of *Mathematics: its contents, methods and meaning* by A. D. Aleksandrov *et al.* With NSF support the Battelle Memorial Institute is preparing a guide to East European scientific and technical literature available to US scientists, and a directory of scientific institutions in the USSR, which will include details of personnel and publications. All US government-sponsored translations are listed in *US Government Research and Development Reports* and indexed in *USGR & DR Index.* Those not sponsored by US Government agencies are announced in the *Translations Register—Index,* published by the Special Libraries Association Translations Center (John Crerar

Library, Chicago), and in the *ETC Quarterly Index,* published by the European Translations Centre (Delft, Netherlands).

Anglo-American co-operation is firmly established; the National Lending Library for Science and Technology has a loan collection of translations which includes microfilms of all Russian translations held by the SLA Translations Center. The Library issues, each month, its own *NLL Translations Bulletin* (HMSO, London) which reports on a variety of translation activities.

Translation Services

Translators and Translations: Services and Sources, first published by the Special Libraries Association, New York, in 1959 and since revised, offers a unique guide. A register is also maintained by the American Translators Association. In Britain, such bodies as the Institute of Linguists and Aslib can advise on mathematical translation services, and Patricia Millard has compiled a *Directory of Technical and Scientific Translators and Services* (Crosby, Lockwood, London, 1968).

TRANSLATION OF PERIODICAL LITERATURE

Over 2500 Russian scientific and technical periodicals are published currently. The information they contain is being made increasingly available to English-speaking scientists and technologists through translation and review services of many kinds. These range from complete cover-to-cover translations of important periodicals to individually commissioned translations of specific items.

Translations are initially expensive, but through co-operative enterprises costs have been shared, thus making the translated material more readily accessible.

Well over a hundred Russian periodicals are now being completely translated into English on a continuing basis. Many of the publishers receive government subsidies to enable them to maintain subscription rates at a reasonable level.

"Cover-to-cover translations", according to Aslib, "have two main functions. On the one hand they help to overcome ignorance of foreign scientific literature generally, to spread awareness of its importance, and stimulate a demand for it. For languages which few people can read this function is particularly important. At the same time they provide in published form translations of individual articles which some people wish to read and should thus reduce the number of translations which might otherwise be made specially by or for the individuals who want them." (*The Foreign Language Barrier*, Aslib, London, 1962.)

The following is a selection of titles and suppliers of those journals in translation which are of particular interest to mathematicians:

Automation and Remote Control (*Avtomatika i Telemekhanika*)
Instrument Society of America
530 William Penn Place
Pittsburgh, Pennsylvania 15219, USA

Computer Elements and Systems: Collection of Papers (*Vychislitelnye Sistemy*)
Clearinghouse for Federal Scientific and Technical Information
Port Royal and Braddock Roads
Springfield, Virginia, 22151, USA

Differential Equations (*Differentsialnye Uravneniya*)
The Faraday Press, Inc.
84 Fifth Avenue
New York, NY, 10011, USA

Journal of Applied Mathematics and Mechanics (*Prikladnaya Matematika i Mekhanika*)
Pergamon Press Ltd   Pergamon Press Inc.
4 Fitzroy Square   44-01 21st Street
London, W 1, England   Long Island City
                                   NY 11101, USA

Mathematical Notes (*Matematicheskie Zametki*)
Plenum Publishing Corporation
227 West 17th Street
New York, NY 10011, USA

Moscow Mathematical Society. Transactions (Moskovskoe Matematicheskoe Obshchestvo. Trudy)
American Mathematical Society
PO Box 6248, Providence
Rhode Island 02904, USA

Problems of Cybernetics (Problemy Kibernetiki)
Pergamon Press Ltd.,
(as above)

Russian Mathematical Surveys (Uspekhi Matematicheskikh Nauk)
Macmillan & Company Ltd
10–15 St. Martin's Street
London, WC 21, England

Soviet Mathematics—Doklady (Akademiya Nauk SSSR Doklady—Otdel. Matematiki; Proceedings of the Academy of Sciences of the USSR
Mathematics Section)
American Mathematical Society,
(as above)

Soviet Physics—JETP (Zhurnal Eksperimental'noi i Teoreticheskoi Fiziki; Journal of Experimental and Theoretical Physics)
American Institute of Physics
335 East 45th Street
New York 17, NY, USA

Theory of Probability and its Applications (Teoriya Veroyatnosti i ee Primeneniye)
Society for Industrial and Applied Mathematics
Box 7541, Philadelphia 1, Pennsylvania, USA

USSR Computational Mathematics and Mathematical Physics Journal (Selections from *Zhurnal Vychislitel'noi Matematiki i Matematicheskoi Fiziki*)
Pergamon Press Ltd.,
(as above)

TRANSLATION OF MONOGRAPHIC LITERATURE

There is a plentiful supply of good mathematical books published in the USSR, though it is not always easy to obtain copies outside the Soviet bloc countries. For the most part they are quite inexpensive. Every effort is being made to produce translations of those which are considered to have the greatest value, and the American Mathematical Society is very active in this area. An Index to AMS Selected Translations is available from the Society. Book translations are made either as part of a planned publishing programme or as individually commissioned work. Two examples of the former are the Israel Program for Scientific Translations, which includes mathematics, and the scheme of the Hindustan Publishing Corporation of Delhi, India, under which a number of English translations of Russian mathematical books have been marketed.

Russian–English Mathematics Dictionaries

More English-speaking scientists than ever are learning Russian as the educational facilities increase and suitable textbooks are produced. The mathematician is at somewhat of an advantage. The vocabulary of each branch of his subject is to some extent international. It is relatively limited and highly specialized. With a good knowledge of the subject, a reading ability of mathematical texts in the important languages is not as difficult as might at first be imagined.

In dealing with Russian, the Cyrillic alphabet has first to be mastered, but a little practice in transliteration (converting Russian characters into English characters) soon renders this less frightening. There are several different systems of transliteration which vary slightly in certain respects.

A knowledge of the order of the letters in the Cyrillic alphabet is necessary for locating Russian works in a dictionary.

Many Russian–English dictionaries have been produced in the past few years to meet the rising demand. A New York University study which includes the dictionary needs in the fields of mathematics and allied subjects is *Russian–English Scientific and Technical Dictionaries—a Survey.*

A

a, and, but, while; **не...**, **а...**, not..., but...; **а именно**, namely; **а не то**, or else; **а так же**, just as; **а так как**, and since, now as

а-, *prefix*, non-

абака, *f.*, abacus

абгомотопический, *adj.*, abhomotopy

абелевость, *f.*, commutativity

абелевский, *adj.*, abelian

абелевый, *adj.*, abelian

аберрационный, *adj.*, aberrational

аберрация, *f.*, aberration, deviation, error

абзац, *m.*, indentation, paragraph, item

абонент, *m.*, subscriber

абрис, *m.*, contour, outline, sketch

абсолютный, *adj.*, absolute

абсолютно, *adv.*, absolutely; **абсолютно наименьший вычет**, least positive residue

абсорбент, *m.*, absorbent

абсорбер, *m.*, absorber

абсорбировать, *v.*, absorb

абсорбирующий, *adj.*, absorbing, absorptive

абсорбция, *f.*, absorption

абстрагировать, *v.*, abstract

абстрагируясь, *adv. part.*, abstracting, generalizing, if we abstract

абстрактность, *f.*, abstractness, abstraction

абстрактный, *adj.*, abstract

абстракция, *f.*, abstraction

абсурд, *m.*, absurdity

абсурдность, *f.*, absurdity; **абсурдность допущенного очевидна**, the absurdity of the assumption is obvious

абсурдный, *adj.*, absurd, inept, preposterous

абсцисса, *f.*, abscissa, *x*-coordinate

авария, *f.*, accident, wreck, damage, mishap

авиа-, *prefix*, air-, aero-; e.g. **авиабаза**, airbase

авиационный, *adj.*, aviation

авиация, *f.*, aviation

автоблокировка, *f.*, automatic block system

автодистрибутивность, *f.*, autodistributivity

автодуальный, *adj.*, autodual; **автодуальное отображение**, autoduality

автоколебание, *n.*, auto-oscillation

автоколебательный, *adj.*, self-vibrating, self-oscillating

автокоррелированный, *adj.*, autocorrelated

автокорреляционный, *adj.*, autocorrelated, self-correlated

автокорреляция, *f.*, auto-correlation

автомат, *m.*, automatic machine; **автоматы**, *pl.*, automata

автоматизация, *f.*, automation

автоматизированный, *adj.*, automatized, automated

автоматизировать, *v.*, automatize

автоматизм, *m.*, automatism

автоматика, *f.*, automation

автоматически, *adv.*, automatically

автоматический, *adj.*, automatic

автомашина, *f.*, truck, motor vehicle, lorry

автомобиль, *m.*, motor car, autocar, automobile

автоморфизм, *m.*, automorphism

автоморфность, *f.*, automorphism, automorphy

автоморфный, *adj.*, automorphic

автономность, *f.*, autonomy, self-regulation

автономный, *adj.*, autonomous, self-governing, self-regulating

автопараллельный, *adj.*, self-parallel, autoparallel

автопилот, *m.*, automatic pilot, mechanical pilot, robot pilot

автополярность, *f.*, self-polarity

Fig. 19. Lohwater, A. J., *Russian–English Dictionary of the Mathematical Sciences*. (Reprinted with permission of the publisher, the American Mathematical Society, Copyright © 1961, p. 1.)

матема́тик, mathematician

матема́тика, mathematics

ма́тери, *see* мать

материа́л, material, stuff, fabric

матри́са, матри́ца, matrix

   квазидиагона́льная матри́ца, reducible diagonal matrix

   матри́ца коэффицие́нтов корреля́ции, correlation matrix

   матри́ца перехо́да, transition matrix

матри́чный, matrix-, pertaining to a matrix

мать (*gen.* ма́тери, *pl.* ма́тери, *gen. pl.* матере́й), mother

Матье́, Mathieu

маха́ть, ма_ у́ть (машу́, ма́шешь), wave, wag, flap

ма́чта, mast, tower

ма́шет, *see* маха́ть

маши́на, machine, engine, mechanism, car

   вычисли́тельная / счётная / числова́я маши́на, calculating/computing machine

   маши́на непреры́вного де́йствия, analogue machine/device

маши́нка, typewriter

ма́ятник, pendulum

мгнове́ние, instant, moment

мгнове́нный, instantaneous, momentary, prompt

Мёбиус, Möbius

медиа́н, медиа́на (*n.*), median

медиатри́са, mid-perpendicular

ме́дленный, slow

медли́тельный, sluggish

ме́длить (ме́длю, ме́длишь, ме́длят), linger

меж = ме́жду

междоу́зельный, intersticial

ме́жду (+*instr.*), between, among(st)

   ме́жду про́чим, by the way, among other things, besides

   ме́жду собо́й, among themselves

   ме́жду тем, meanwhile

   ме́жду тем как, while, whereas

   ме́жду тем э́то так, nevertheless it is so

междунаро́дный, international

мезо́н, mason

ме́лкий, small, fine, shallow, minor, petty

ме́лочь, trifle(s)

мелька́ние, flashes, glimpses; (*pl.*) beams

ме́льком, in passing, cursorily

мельча́йший, smallest, finest

мембра́на, membrane, diaphragm

ме́на, exchange, barter

ме́нее, less

   ме́нее всего́, least of all

   не бо́лее не ме́нее, как, neither more nor less than, no less . . . than

   тем не ме́нее, nevertheless

мени́ск, meniscus

Менье́, Meusnier

ме́ньше, smaller, less

   ме́ньше всего́, least of all

   не бо́льше не ме́ньше как, neither more nor less than

FIG. 20. Burlak, *Russian–English Mathematical Vocabulary.* (By kind permission of Oliver & Boyd Ltd.)

A *Russian–English Dictionary of the Mathematical Sciences*, compiled and edited by A. J. Lohwater with the collaboration of S. H. Gould (Fig. 19), was published by the American Mathematical Society in 1961. It includes a concise grammar of the Russian language. Its English–Russian companion is the *Anglo-Russkiĭ Slovar' Matematicheskikh Terminov*, edited by P. S. Aleksandrov (Izdatel'stvo Inostrannoĭ Literatury, 1962).

Also available are the *Mathematical Dictionary: Russian and English*, published by the VEB Deutscher Verlag der Wissenschaften, Berlin, in 1959 (see p. 38), and Louis Melville Milne-Thomson's *Russian–English Mathematical Dictionary*, published by the University of Wisconsin Press in 1962. The *Russian–English Mathematical Vocabulary* of J. Burlak and K. Brooke (Oliver & Boyd, Edinburgh, 1963), which includes a grammatical introduction of its own, is illustrated in Fig. 20. Its editors recommend the use of P. H. Nidditch's *Russian Reader in Pure and Applied Mathematics* to those mathematicians who wish to acquire a reading knowledge of the language; it provides interlinear word-for-word translations. The intending translator will also wish to know of the existence of S. H. Gould's 24-page *Manual for Translators of Mathematical Russian*, published by the American Mathematical Society in 1966.

As regards statistics, there is Samuel Kotz's *Russian–English Dictionary of Statistical Terms and Expressions and Russian Reader in Statistics* (University of North Carolina Press, 1964).

Abbreviations are commonly used in Russian works and can often cause difficulty. This applies both to the actual subject matter and to the names of institutions. The Scientific Research Institute of Mathematics, for instance, is quoted as NIIM. This stands for Nauchno-Issledovatel'skii Institut Matematiki. Such contingencies are catered for by the *Glossary of Russian Abbreviations and Acronyms* (Library of Congress, Washington, DC, 1967) which lists 23,000 items in Cyrillic alphabetical order and for each provides both the full Russian version and an English translation.

Mathematics and the Government

Introduction

The cost of research in science and technology mounts steadily every year. It is borne by every sector of the community, either directly as taxation or indirectly in the price of goods—the end products of research. The three principal sponsors of research are industry, universities, and governments, though this is an extreme simplification of the picture. Often, the constituent members of a particular industry co-operate to form research associations through which the cost of research can be shared. Co-operation can be international as well as national, and the individual participating organizations vary considerably in size. Generally speaking, the idea of free exchange of information is becoming much more widely accepted as beneficial to all parties concerned, though of course the practice of patenting new inventions continues to flourish as vigorously as ever.

A unique example of altruism is to be seen in the Battelle Memorial Institute in the United States. The Institute, which carries out a very substantial volume of research, was expressly founded by the will of Gordon Battelle as a practical means for producing social benefits through scientific research and the making of scientific discoveries and inventions. Every month it issues the *Battelle Technical Review* which reports its activities and contains abstracts of an international range of technical documents reflecting the primary areas of its research. Among these are control systems, computers, automation, cybernetics, systems engineering, and operational research.

In Great Britain there are some fifty research associations which carry out research projects for member organizations, but which also make available to the public at large the results of the greater part of their research activities. Their organization is described below. For the present, it will be noted that they have strong financial backing from the British Government. Government support does in fact pervade the whole domain of scientific research. It can be seen in the financial backing of research publications, in the placement of research contracts in industry, in the maintenance of its own research laboratories, in grants to universities, and in many other forms. Some of these activities, and in particular those which affect mathematics, are worth a closer look.

United States

The activities of the majority of US Government departments and agencies are described in the *United States Government Organization Manual*, which is published annually by the General Services Administration. This is a most useful source for obtaining the addresses of offices whose names only are usually quoted in articles and references. A guide to US Government publications themselves is *Government Publications and Their Use* by Laurence F. Schmeckebier and Roy B. Eastin. It was first published in 1936, but a revised edition was published by The Brookings Institution, Washington, DC in 1961.

The government maintains several research groups of its own which, by designation, are mathematically orientated. They include the following:

> Air Force Applied Mathematics Research Branch at the Wright Aeronautical Development Center.
> Office of Naval Research Mathematical Sciences Division.
> Naval Ordnance Laboratory Mathematics Department.
> David Taylor Model Basin Applied Mathematics Laboratory, US Navy.
> Army Research Office Mathematical Sciences Division.

In 1961 the US Naval Research Laboratory also announced the creation of an Applied Mathematics staff as part of the Office of the Director of Research, primarily to carry out research programmes on numerical analysis, mathematical physics, and optimization techniques.

In addition, numerous federal research contracts are placed with industrial companies, universities, and other organizations equipped for research or the publication of research results. A number of universities and libraries have been nominated as depositories of reports and have accepted the responsibility of making available the results of government-sponsored research and development. Each centre receives the reports of the Department of Defense, the National Aeronautics and Space Administration, and the Atomic Energy Commission, and meets demands from the public through reference, lending, and photocopying facilities. These documents are recorded in *US Government Research and Development Reports*, a semi-monthly abstract journal published by the Clearinghouse for Federal Scientific and Technical Information, Springfield, Va. 22151, USA (Fig. 21). Another useful source of current information is the *Monthly Catalog of US Government Publications* issued by the Superintendent of Documents, US Government Printing Office, Washington, DC.

The activities of the National Bureau of Standards are of particular interest to the mathematician. They are divided among various divisions and sections including Applied Mathematics, which deals with numerical analysis, computation, statistical engineering, mathematical physics, and operational research. Its *Journal of Research* is published quarterly in two parts. Section B is entitled *Mathematics and Mathematical Physics*. It reports studies and compilations designed mainly for the mathematician and the theoretical physicist, and each issue contains abstracts of the Bureau's publications.

The National Science Foundation is noted elsewhere in this book with reference to the financial support it gives to mathematical publications, especially translations of foreign-language

12/1. MATHEMATICS AND STA-TISTICS

AD-620 102   Div. 15/1
CFSTI Prices: HC $1.00 MF $0.50
*MATHEMATICS   RESEARCH   CENTER
UNIV OF WISCONSIN MADISON*
ARCS AND CHORDS, II.
Technical summary rept.,
by A. S. Besicovitch. 0 Jul 65, 16p. Rept. no.
MRC-TSR-564
Contract DA11 022ORD2059

Unclassified report

Descriptors: (*Measure theory, Set theory),
Geometry, Functional analysis

The simple arcs of Hausdorff dimension $h > 1$ are
studied from the point of view of the limits of the
ratio: (lambda to the hth power) $J(v,v')$/absolute
value $(v-v')$ to the hth power, as v' approaches v,
where $J(v, v')$ is the arc between v and v'. The
bounds of various densities at points of arcs are
given. It is shown that in all cases but one the
bounds are exact. (Author)

AD-620 103   See Fld. 20/12

AD-620 104   See Fld. 20/12

AD-620 105   See Fld. 20/4

AD-620 106   Div. 15/1
CFSTI Prices: HC $1.00 MF $0.50
*MATHEMATICS   RESEARCH   CENTER
UNIV OF WISCONSIN MADISON*
ON MONOSPLINES OF LEAST DEVIATION
AND BEST QUADRATURE FORMULAE II.
Technical summary rept.,
by I. J. Schoenberg. 0 May 65, 17p. Rept. no.
MRC-TSR-569
Contract DA11 022ORD2059

Unclassified report

Descriptors: (*Least squares method, Polynomials), Numerical analysis, Functions, Integration

A new method of constructing best quadrature
formulae is established in the form of a generaliza-
tion of Rodrigues' formula for Legendre polynomials. (Author)

FIG. 21. *US Government Research and Development Reports*. (By
kind permission of the Clearinghouse for Federal Scientific and
Technical Information.)

works. It was established by Act of Congress in 1950, and its work is described in *National Science Foundation; general review of its first 15 years*, prepared by Dorothy M. Bates and others (89th Congress, House Report 1219, 1966).

A national library service is provided by the Library of Congress whose printed author and subject catalogues are highly valued bibliographical tools. The Science and Technology Division issues specialized publications, one of which is a *List of Russian Serials Translated into English and other Western Languages.*

The US Department of Health, Education, and Welfare is another relevant government department. An example of its publications is *Facilities and Equipment for Science and Mathematics: requirements and recommendations of state departments of education* (1960).

Details of fellowships and other support for basic research in mathematics are listed in *A Selected List of Major Fellowship Opportunities and Publications for Educational Research*, which is available from the Fellowship Office of the National Academy of Sciences—National Research Council.

A book worth noting is Kidd, Charles V., *American Universities and Federal Research* (Belknap Press of Harvard University Press, 1959). More specifically Saunders Maclane discusses "The future role of the Federal Government in mathematics" and J. W. Tukey asks "What can mathematicians do for the Federal Government?" in articles contributed to the Fiftieth Anniversary Issue of the *American Mathematical Monthly*, 1967.

United Kingdom

Government support has grown considerably since the first half of the last century when Charles Babbage received money to construct his early computer on which the Registrar-General was later to produce official life tables. Under the Science and Technology Act 1965, the Science Research Council was set up to carry out and support research, to make grants and disseminate knowledge. Its annual report shows that between 1963–4 and

1967–8 research grants in mathematics rose from £14,000 to over £150,000.

The Ministry of Technology operates several research *establishments* of its own, and in addition supports the work of a number of research *associations* serving the needs of particular groups of industries. Their activities are described in their annual reports. The Department of Education and Science issues annual data on work in progress in *Scientific Research in British Universities and Colleges*. The entries are arranged in alphabetical order of institutions and within each are given the names of staff members and their topics of research. There is a complete name index and a detailed subject index.

Industrial Research in Britain (Harrap, London, 6th edn., 1968) is a very comprehensive work dealing with all aspects of the subject from sponsored research organizations to periodical abstract journals covering industrial research. There is also an article on "Government and industrial research" by E. Lee, Deputy Controller of the Ministry of Technology.

The National Lending Library for Science and Technology, in Boston Spa, Lincolnshire, became fully operational in 1962. It meets requests for loans and photocopies from a vast stock of literature which includes the largest collection of Russian serials in Western Europe. The extensive loan service previously operated by the Science Museum Library in London has been absorbed into the National Lending Library's activities. A counterpart library offering reference services is planned.

British government publications are made known in the *Daily List*, the *Monthly Catalogue*, and the *Annual Catalogue*, all issued by HMSO. A 5-year index is also published. *Sectional Lists*, which are available free on request, list departmental publications in print. Publications of interest to mathematicians are produced by a number of different government departments. Some recent examples are:

MEDICAL RESEARCH COUNCIL, *Mathematics and Computer Science in Biology and Medicine*, 1965.

SCHOOLS COUNCIL, *Mathematics for the Majority: a programme in mathematics for the young school leaver* by Philip Lloyd, 1967; and *Mathematics in Primary Schools* by E. E. Biggs, 2nd edn., 1966.

TREASURY, *The Elementary Ideas of Game Theory* by Maurice Peston and Alan Coddington, 1967, and *Introducing Computers* by F. J. M. Laver, 1965.

An important entry under the National Physical Laboratory heading concerns its *Mathematical Tables Series*, the first of which is entitled *The Use and Construction of Mathematical Tables*. When the series of *Sectional Lists* is complete it will comprise a catalogue of all current parliamentary publications. The lists are periodically revised.

Title, frequency and price of government periodicals are given in the printed catalogues.

In the United States, British government publications are obtainable from British Information Services, 45 Rockefeller Plaza, New York 20, New York. A complete list of overseas agents appears in the HMSO catalogues mentioned above. In Britain there are government bookshops in seven major cities, and their names and addresses together with booksellers who act as agents for government publications are similarly given in the catalogues.

Government Information and the Research Worker, edited by Ronald Staveley and published by the Library Association, London (2nd edn., 1965), contains a chapter dealing with HMSO and its publishing policy. Other chapters are contributed by experts on each of the major government departments.

Actuarial Science

Orientation

In the Dewey Decimal Classification, actuarial science is provided for in the Insurance class 368. The ramifications of the subject are, however, very wide. Its roots go deep into the theories of probability and statistics. Mathematics is, indeed, an essential part of the professional actuary's background. A brief summary of sources of relevant information will therefore be given.

The following definition is taken from a useful booklet entitled *The Actuarial Profession*, published by the Institute of Actuaries in London. "Actuarial science is concerned with applying the theory of probability, and statistical processes generally, to practical affairs and especially to the financial problems connected with the management and administration of life insurance, pension schemes and social insurance."

It follows that since much of the actuary's work involves statistics, the bibliographical sources described in Chapters 11 and 12 will be of considerable value.

Again, attention should be drawn to Dewey's Statistics class 310, the broad arrangement of which is as follows:

310: Statistics
e.g. United Nations, *Statistical Yearbook*, New York.
311: Statistical method
e.g. Snedecor, George W., *Statistical Methods*, Iowa State University Press, Ames, Iowa, 6th edn., 1967.
312: Demography
e.g. Cox, Peter R., *Demography*, Cambridge University Press, 3rd edn., 1959.

KEYFITZ, NATHAN, *Introduction to the Mathematics of Population*, Addison-Wesley, Reading, Mass., 1968.

314–319: General statistics of specific countries

e.g. 314.2 Central Statistical Office, *Annual Abstract of Statistics*, HMSO, London.

317.3 US Bureau of the Census, *Historical Statistics of the United States, Colonial Times to 1957*, Washington, DC, 1960.

Monographic Literature

In the absence of relevant bibliographies the monographic literature of actuarial science is best represented in the catalogues of libraries devoted to the subject. They inevitably reveal a preponderance of statistical works. This is clearly shown in the library of the Institute of Actuaries, London, which has a stock of some 10,000 volumes. Its *Additions to the Library* is issued as a separate booklet and distributed with the *Institute of Actuaries Year Book*. The latter itself includes a reading list for students. It is graded in so far as the references are related to the various examinations held by the Institute. The Institute publishes its own series of student textbooks which are available to members at preferential rates. Recent volumes, quoted as examples only, include *Finite Differences for Actuarial Students* by H. Freeman, and *Probability, an Intermediate Text-Book* by M. T. L. Bizley. Another valuable work, *Actuarial Statistics*, is in 2 volumes: 1, *Statistics and Graduation* by H. Tetley (1950), and 2, *Construction of Mortality and Other Tables* by J. L. Anderson and J. B. Dow (1958). Herbert Frederick Fisher and John Young's *Actuarial Practice of Life Assurance: a textbook for actuarial students* was published by Cambridge University Press for the Institute of Actuaries and the Faculty of Actuaries in 1965. Also in 1965 appeared a thoroughly revised second edition of *The Mathematics of Life Insurance* by Walter O. Menge and Carl H. Fisher (Macmillan, New York).

Government statistical publications are of great importance, and the guides described in Appendix II will be found useful in locating them. Two series which may be noted are the British *Government Actuary's Reports* and the *Actuarial Studies* of the US Department of Health, Education and Welfare.

Foreign actuarial terms are included in the *International Insurance Dictionary* published by the European Conference of Insurance Supervisory Services in Berne, 1959. As indicated in the preface, the international actuarial notation is a reliable link between different languages. Part IV of the Dictionary is therefore devoted to actuarial symbols with explanations of their meaning in English, German, Dutch, French, Italian, Spanish, Portuguese, Danish, Swedish, Norwegian, and Finnish.

Actuarial Notation

A full statement of the International Actuarial Notation was published in the *Journal of the Institute of Actuaries* (vol. 75, p. 121) and in the *Transactions of the Faculty of Actuaries* (vol. 19, p. 89).

Periodicals and Abstracts

Due to the diversity of subjects in which the actuary must achieve competence, his breadth of reading is quite substantial. In addition to purely actuarial journals he has considerable interest in those relating to mathematics, statistics, and computation. More recently his list has been increased to cover such fields as operational research. The following English-language titles would therefore constitute a fairly representative list:

American Mathematical Monthly
Annals of Mathematical Statistics
Annals of Mathematics
Applied Statistics
Computer Journal
Journal of the American Statistical Association

Mathematics of Computation
Operational Research Quarterly
Operations Research.

His basic English-language actuarial publications include:

Journal of the Institute of Actuaries
Transactions of the Actuarial Society of Australasia
Transactions of the Faculty of Actuaries, and
Transactions of the Society of Actuaries.

World progress in the subject, and reviews of foreign-language publications, are well covered by the above reading. For more detailed information on activities in other countries, however, the actuary may have recourse to the periodical publications of the respective national societies.

There is no central abstracting journal covering actuarial science, but by regularly consulting two or three more general services—especially in the field of statistics—the actuary can keep informed quite adequately. Some actuarial journals go much further than simply providing book reviews. The *Journal of the Institute of Actuaries* (Alden Press, Oxford), for example, includes three regular sections to assist its readers in keeping abreast:

1. Notes on other actuarial journals (covering the overseas press).
2. Notes on the *Transactions of the Faculty of Actuaries*.
3. Articles, papers and publications of actuarial interest (comprising annotated references to appropriate items selected from a wide range of peripheral publications).

Actuarial Societies

INTERNATIONAL

The profession's international organization is called the Comité Permanent des Congrès Internationaux d'Actuaires (9 rue des Chevaliers, Brussels, Belgium). The first International Congress of Actuaries was held in Brussels as far back as 1895.

Congress papers are published in a volume of *Transactions* (17th edn., 1964, in 2 volumes).

ASTIN is a section of the Comité which was established in 1957 in New York. Its purpose is the promotion of mathematical research in non-life insurance, and its periodical publication is called *Astin Bulletin*.

NATIONAL

National bodies contribute to the work of the Comité as well as acting as co-ordinating centres for their respective countries.

The Institute of Actuaries (Staple Inn Hall, High Holborn, London, WC 1), working in close co-operation with the Faculty of Actuaries in Scotland, is not only the focal point of the profession in England, but also in the Commonwealth. It was founded in 1848, and incorporated by Royal Charter in 1884. Its publications have already been described, and its educational and career activities are dealt with in Chapter 1. Another British society is the Association of Consulting Actuaries (Great Britain) whose address is 23 Blomfield Street, London Wall, London, EC 2.

In the United States, the national body is the Society of Actuaries (208 South La Salle Street, Chicago, Illinois 60604). Founded in 1949, it absorbed both the Actuarial Society of America and the American Institute of Actuaries. It has active meetings and publishing programmes, and sponsors Associateship and Fellowship examinations.

The following is a list of other national actuarial societies, arranged alphabetically by country:

Australia: Institute of Actuaries of Australia and New Zealand
c/o MLC Bldg., Victoria Cross
N. Sydney, NSW

Belgium: Association Royale des Actuaires Belges
48 rue Fossé-aux-Loups
Brussels

Canada:	Canadian Institute of Actuaries c/o Equitable Life Insurance Co. of Canada Waterloo, Ontario
France:	Institut des Actuaires Français 15 rue Bachaumont Paris 2ᵉ
Germany:	Deutsche Gesellschaft für Versicherungsmathe-matik Von Werth Strasse 4–14 Cologne
India:	Actuarial Society of India c/o Life Insurance Corporation of India "Yogakshema", Madame Cama Road Bombay
Italy:	Istituto Italiano degli Attuari Via dell' Arancio 66 Rome
Scotland:	The Faculty of Actuaries in Scotland 23 St. Andrew Square Edinburgh 2
South Africa:	Actuarial Society of South Africa PO Box 4464 Cape Town
Switzerland:	Association des Actuaires Suisses Fluhmattstrasse 1 Lucerne

Index

181

Forum Italicum
Filibrary Series
No. 24

Cover:
"Italian American laymen and women helped to perpetuate the religious tradition from the Old to the New World, frequently serving as catalysts for feast celebration transplantation. This photograph shows immigrant Carmen Bianco, organizer of the Feast of St. Liberata that was celebrated for decades by Patchogue, Long Island Italian Americans. Bianco's commitment to the saint of his village encompassed that of being custodian of the statue of the saint even constructing a chapel on his property where he sheltered the statue."

MODELS AND IMAGES
OF CATHOLICISM IN
ITALIAN AMERICANA:

ACADEMY AND SOCIETY

DISCARDED

Co-Edited by:
Joseph A. Varacalli
Salvatore Primeggia
Salvatore J. LaGumina
Donald J. D'Elia

Forward by:
Honorable Thomas P. Di Napoli

Forum Italicum Publishing
Stony Brook, NY

Library of Congress Cataloging-in-Publication Data

Models and Images of Catholicism in Italian Americana:
Academy and Science, edited by Joseph A. Varacalli,
Salvatore Primeggia, Salvatore J. La Gumina, Donald J. D'Elia

Proceedings of a conference held at Stony Brook University,
Stony Brook, NY, October 2002

p. cm.
1. Italian American studies 2. Catholic American studies
3. Sociology and distribution of knowledge
I. Title II. Series

ISBN 1-893127-24-9

FORUM ITALICUM
Center for Italian Studies
State University of New York at Stony Brook
Stony Brook, NY 11794-3358
USA
http:// www.italianstudies.org

Table of Contents

Part III: Feminist Models and Images

Part IV: Modern and Post-Modern Models and Images

Part V: Concluding Note

PREFACE

The collection of papers presented in this volume explores a subject rich in material. Though the relationship between the Roman Catholic Church and the Italian people is a long and involved one, the many twists and turns of that relationship lead to areas of academic inquiry ripe for further investigation. When you extend that study to a consideration of the connection of Italians, Italian immigrants and Americans of Italian descent to Roman Catholicism as a religious belief system as well as a church institution, you sail into murky waters with currents and rip tides that make for a tricky navigation of deeply held values, sensibilities and emotions.

Fortunately, the contributors to this volume understand this and the editors' skills as pilots through this scholarly voyage are apparent and commendable. Editors Joseph A. Varacalli, Salvatore Primeggia, Salvatore LaGumina and Donald J. D'Elia, in selecting and arranging these papers, have successfully organized this fascinating and provocative material in a way that will engage the reader and likely tap into your own experiences and views. Some of what you read will challenge you, some will surprise you and some will help explain more fully that which you already know.

As with any study of a particular community's religious traditions, one runs the risk of generalizing about experiences that, though shared, are ultimately intensely personal and varied among the members of the group. Nevertheless, I found this volume to be exceedingly useful in interpreting and enriching my own sense of my religious tradition and of how my family and church passed tradition and faith on to me. I'm sure the readers will be similarly touched by this book — closely identifying with parts of it or perhaps disagreeing with some of the observations.

These essays evoked strong recollections of my mother's parents, Angelina DiCecilia and Vito Abbondandelo, who were born in Sturno, a small city in Italy's Avellino province. When they migrated to the United States, they brought with them their dreams and their hopes for a better life. They also brought a deeply held attachment to their Catholic faith that was a mainstay of their lives. My grandparents' dedication to the church and to their special saints was unshakeable and truly sustained them through the life challenges they faced. The Blessed Mother, Saint Anthony and Saint Michael the Archangel were especially honored saint figures to them and as grandchildren we were intrigued and fascinated by the array of religious statuary displayed in our grandparents' home in Roslyn Heights, Long Island. However one would categorize my grandparents' faith, there is no doubt in my mind but that what I witnessed in them was a pure and genuine devotion to God and church that was instilled in their children and grandchildren and inspires me to this day, so many years after they've passed on.

For helping me to better understand the context of my family's religiosity, I thank the writers of these essays. For bringing this volume to publication, *Forum Italicum* of Stony Brook is to be applauded. The combined academic leadership and excellence of Adelphi University, the Center for Catholic Studies and the Center for Italian American Studies at Nassau Community College and the Center for Italian Studies at SUNY Stony Brook, as well as the support of the Order Sons of Italy in America, are all deserving of appreciation for their significant contributions to this work. On a most personal note, I join with the editors in singling out Dr. Mario Mignone of SUNY Stony Brook for his unique and inexhaustible dedication to promoting Italian and Italian American studies and for his guidance in the preparation of *Models and Images of Catholicism in Italian Americana: Academy and Society.*

HONORABLE THOMAS P. DiNAPOLI
Great Neck, Long Island
April 2004

INTRODUCTION

All intellectual activity and scholarly research is animated by some religious or cultural perspective, whether that perspective is consciously held and fully articulated (as is indicated by the use of the terms "models" or "paradigms") or, relatively speaking, more taken-for-granted and less than fully developed (as in implied in the concept of "images"). The purpose of this volume is to apply this insight to an important topic in Italian and Italian American studies. More specifically, the question that this anthology starts to address is "what are the various models and images of Catholicism qua institution, set of ideas, and community of individuals that animate scholarly discussion in that portion of the academy devoted to studying things Italian as well as drives non-scholarly discourse in the society in general?"

Whether accepted enthusiastically, rejected totally, or more likely, embraced selectively in some other manner, Catholicism has been central in the lives of both Italians and Italian Americans. Many different groups, philosophies, and academic disciplines on both sides of the Atlantic have had distinctive interpretations of the significance, for better or worse or mixed, of Catholicism for both the Italian nation and the Italian American community. As readers will quickly discern, some of the essays in the volume are purely academic while others combine the academic with the devotional. Some assume the truthfulness of the cognitive claims of the Catholic faith, others deny it through either reduction or transformation into the secular, and yet others bracket the issue altogether. Some are hostile to the religious institution, others sympathetic, and yet others descriptively neutral. All in all, the attempt has been made to be as inclusive as possible regarding the various appropriations of Catholicism utilized by scholar and lay person alike in Italian Americana.

This volume attempts to start the process of filling a major void, especially in the field of Italian American studies. Starting with an insightful Foreword by the Honorable Thomas P. DiNapoli, this volume includes essays dealing with the Irish-American (Gesualdi and LaGumina), American Protestant (Salamone), and Jewish and Jewish American critique (Marchione) of Italian American Catholic religiosity. There is also an analysis explaining the attitudes of the nationalist/fascist, liberal/ capitalist, socialist/Marxist, and romantic idealist traditions toward the role of the Catholic Church in the affairs of Italy (Rao), attitudes brought to American shores by a small and — at least according to the noted Italian American scholar, Rudolph Vecoli — not insignificant sliver of the turn into the twentieth century migration. Literary (Gardaphé), feminist (Ardito and Bona), psychoanalytical (Carroll), social constructionist (Haynor), and post-modern (Zagano) models and images of Catholicism in Italian Americana add to the intellectual pluralism and breath of this volume. The specific appropriation of the Catholic religion by the southern Italian peasant and typical first generation Italian immigrant is the subject of yet another essay (Primeggia) while there is another piece devoted to the attempt to analyze the relationship between the Catholic and Italian worldviews (Varacalli). There is also an essay which tries to lay out the Catholic Church's self-understanding of herself as devoted t o assisting the Italian Catholic and Italian Catholic American, both body and soul, as exemplified in the apostolate of Mother Cabrini (D'Elia).

The majority (nine) of these essays were first presented at a unique three day, three institution conference held under the title of this volume on the dates of October 4th, 5th, and 6th, 2002 and sponsored financially by the Office of the Provost at Adelphi University, the Center for Catholic Studies and the Center for Italian American Studies at Nassau Community College-SUNY, the Center for Italian Studies at Stony Brook University, and the Order, Sons of Italy in America. The respective sub-conferences were co-ordinated by, respectively, Dr. Salvatore Primeggia of Adelphi University, Dr. Joseph A. Varacalli and Dr. Salvatore LaGu-mi-na of Nassau Community College, and Dr. Mario M. Mignone of Stony Brook University.

Another essay (Ardito) was presented at the Annual Italian Heritage Day Conference of Nassau Community College held on October 7th, 2002 under the coordinatorship of Dr. Salvatore LaGumina, an annual conference originally initiated by Dr. Sean A. Fanelli, President of Nassau Community College, in October of 1986. Other essays were originally delivered at, respectively, a conference titled "The Life and Legacy of Mother Cabrini, Saint of Immigrants" sponsored by the Center for Italian Studies of Stony Brook University on October 13, 2001 (D'Elia) and the and the Annual Conference of the American Italian Historical Association held in Lowell, Massachusetts on November 10, 2000 (Varacalli). Three other essays (Marchione, Haynor, and Zagano) were solicited specifically for inclusion in this volume.

Special recognition for their indispensable assistance and support should go to Joseph Sciame of the Order, Sons of Italy and to Jo Fusco of the Center for Italian Studies of Stony Brook University. Finally, neither the book nor the three day conference would have been a reality without the extraordinary devotion and competence of Ms. Donna Severino, aide de camp to Dr. Mignone.

The co-editors express their hope that this anthology initiates a new line of inquiry into Italian and Italian American studies. The co-editors, furthermore, are cautiously optimistic that this work will be accepted as enthusiastically as was our previous anthology, *The Saints in the Lives of Italian Americans: An Interdisciplinary Investigation*, published by Forum Italicum in 1999 and now reprinted in 2003.

Dr. Varacalli dedicates this work to his two sons, Thomas F. X. and John Paul. Dr. Primeggia dedicates this work to his parents, Caterina and Francesco Primeggia and his wife, Pamela. Dr. LaGumina dedicates this work to his wife, Julie, and his children, Frank, John, Mary, and Christine. Dr. D'Elia dedicates this work in memory of his mother, Frances Marie Santello D'Elia.

JOSEPH A. VARACALLI
SALVATORE PRIMEGGIA
SALVATORE J. LAGUMINA
DONALD J. D'ELIA

Part I
Ethnic Models and Images

Chapter 1

✦

LA VIA VECCHIA AND ITALIAN FOLK RELIGIOSITY: THE PEASANTS AND IMMIGRANTS SPEAK

The religious belief system of Italian peasants and immigrants in the early twentieth century was quite complex. In both Italy and the United States, Italians sustained a powerful religious orientation, which included official Roman Catholic dogma and values as well as other beliefs that were unacceptable to the Catholic Church. This paper will look at the elements of this religious belief system, its deep roots in *la via vecchia* (the old way) of southern Italy, and how it was maintained or modified by Italian immigrants in the United States.

To fully understand this subject, we must turn our attention to the multi-faceted components of the southern Italian peasants' religiosity. To this end, we will examine the impact of the cults of the saints and the Madonna, the role of feasts, and the influence of superstition on the folk religion of the Italian peasants and of first-generation Italian Americans.

Church and Religion in Pre-immigration Italy

Prior to their mass migration to the United States, which began in the 1880s and continued unabated until the immigration and quota acts of 1921 and 1924, Italian *contadini* (peasants) constituted the vast majority of the population in southern Italy. Life for these peasants was extremely difficult due to the ongoing poverty and *miseria* (misery) of the South. They were constantly beset by natural calamities (e.g., earthquakes, volcanic eruptions, and crop-killing infestations), and they were expected to defer to the authority and wealth of the *prominenti* (those of prominent status).

In an atmosphere of hardship and despair, religious orientation provided solace and hope and rationalization. Second to family, LoPreato (1970: 87) pointed to the critical significance of religion as a southern Italian peasant institution. Indeed, for the peasants, the Catholic Church, "played a vital role in the rites of passages..." (Vecoli, 1977: 27). However, while they considered themselves to be very religious, the southern Italian

15

peasants were poor Catholics in the eyes of the Catholic Church (Varacalli, 1986: 47).

From a southern Italian peasant perspective, the Catholic Church was seen less as a system of spiritual, moral, and social values and more as an institution that offered protection against severe everyday realities in a world plagued by evil spirits and demons (Covello, 1972: 112). Many of the formal, and sometimes abstract, teachings of the Catholic Church had little or no appeal to these people of the soil. Rather, they found stories about the Holy Family, the saints, and the Madonna to be far more relevant than any theological supposition or papal decree (Gambino, 1991: 48). Thus emerged a folk religion that sometimes complemented — and other times diametrically opposed — official Catholic faith.

In pre-immigration Italy, women tended to be most fully immersed in religious observances (masses, novenas, pilgrimages, processions, and confessions) while men rarely participated. For the most part, male involvement in the formal practice of Catholicism was restricted to the more momentous rites of baptism, marriage, and funerals, plus feast days of patron saints and holidays such as Christmas, Palm Sunday, and Easter.

Still, these men saw themselves as *Cristiani* (Christians), believing they did not have to attend church regularly to be God-fearing and God-loving souls. This male distancing from the institutional Catholic Church was reinforced by a pervasive anti-clericalism. "Towards their village priests, whom they regarded as parasites living off their labors, the peasants often displayed attitudes of familiar contempt" (Vecoli, 1969: 229). It was — and still is — commonplace to see men accompany female family members on their way to Sunday mass, leave them off at the front door of the church, and rejoin them when religious services conclude.

In truth, the religious beliefs of the southern peasant went beyond those of the official Catholic Church. In fact, "his private conception of religion was nevertheless heavily strewn with all sorts of beliefs in the forces of good and evil and included faith in various sorts of magical practices" (LoPreato: 89). In addition to selective acceptance of Church teachings, peasant religiosity embraced quasi-religious pagan and non-religious superstitions. The strongest ties for the southern Italian peasant to the Catholic Church were primarily through the cults of the saints and the Madonna.

Il contadino (the peasant) believed in God and possessed strong devotions to the Virgin Mary and the saints. In discussing the Sicilian peasant, Salamone-Marino (1897/1981: 171) observed a blind adoration of God and the saints resulting in the creation of a deity structure that strays from Catholic convention. In this unique hierarchy, Christ the Son is revered above God and the Holy Spirit, and Mary (*La Madonna*) holds the most revered position.

The Cult of the Saints

Interestingly, Italians are more widely represented in the catalog of saints (38 % of the total) than any other national group (Weinstein and Bell, 1982: 141) and even today, Italians love their saints and what they believe those saints can do. Meanwhile, the Catholic Church has long opposed the special devotion of the saint cults, because:

> Saints, according to the Catholic tradition, have no magic power; they are intermediaries between God and humans, and only God can perform miracles. Despite this official viewpoint, people worship the saints and believe they possess the powers to perform miracles. (DiTota, 1981: 321)

In the late 1800s, southern Italian peasants had no trouble relating to the saints (those who lived for a time on earth and created miracles during their lifetime or after death and canonization). After all, the saints had experienced mortal existence. They knew pain, humiliation, and defeat. Having travailed as humans themselves, the saints could understand the suffering of those devoted to them. The saints, it was believed, were able to empathize with the peasants and their struggles to overcome anguish and misfortune.

This perceived kinship marks a telling divergence between the Church's stance and the peasant views on sainthood. The Catholic Church defines saintliness based upon heroic virtue, doctrinal purity, and evidence of miraculous intercession after death. For the southern Italian peasant, sainthood was all about miracles such as curing illness, preventing natural disasters from destroying homes and villages, or making dead flowers bloom and dried blood flow (Weinstein and Bell: 142-143). These were what Carroll (1992: 33) calls wonder-working saints. The peasant knew the difference between the *santo edificante* (the exemplary saint) favored by Church authorities and the *santo miracolonte* (the saint with miraculous powers) and was always attracted to the latter (Carroll, 1992: 35).

The term cult has been defined by Swatos (1981: 20) as a following that focuses on a real or imaginary person whose followers believe their lives are vastly improved through behaviors that honor this person. Southern Italian peasants saw their saints as powerful emissaries who could be called upon to intercede with God on behalf of oneself or loved ones. As Vecoli tells us:

> the cult of the saints thus served as the focus for their formal devotional practices. The saints of southern Italy were legion: San Rocco, Santa Lucia, San Michele, San Gennaro, la Madonna del Carmine, and many others, some whose names will not be found in any hagiology. Each saint had special powers to cure a particular disease, to render a certain favor, or to assure success in a trade or occupation. One prayed to San Biagio in case of a throat ache, to Santa Rita for women's ailments, or to San Francesco di Paolo if one were a fisherman. (1977: 27-28)

Belief in and around miraculous specialists abounded. Santa Lucia, the protector of eyesight, was often pictured holding a platter of disembodied human eyes (Gambino, 1974: 196). In Sicily, prayers to Santa Lucia were embellished with magic rituals-for example, making the sign of the cross over the eyes three times with a slice of garlic (Covello: 126). Those who offered up prayers to Santa Bologna asked for healthy teeth. San Antonio was the patron saint of animals and finder of lost objects. San Cologero cured hernias. Santa Apollonia calmed toothaches. Santa Monica brought lapsed Catholics back to the fold and Sant' Antonio *abate* cured fever. In fact, it was largely held that the saints watched over all aspects of human life. As Gambino states:

> No human activity or possible harm went uncovered. San Vito protected against the bites of animals, especially rabid dogs and Santa Barbara shielded her supplicants against the devastations of lightening. Even gamblers had their patron saint: San Pantaleone. In addition, each region, town, family and even individual had special saints, a custom which dated back to Roman household gods, or lares. Each saint had his

day in the calendar on which he was to be specially celebrated, worshipped, or placated. (Gambino, 1974: 197)

The healing or protective power with which a saint was associated was
often based on a story from the saint's life. For example, San Biagio was
thought to cure maladies of the throat because he had saved a boy who was
choking on a fishbone. Santa Agata was credited with curing diseases of the
breast because her own breasts were cut off in martyrdom; just as San
Sebastiano could heal wounds because his body was riddled with arrows
(Carroll, 1992: 43).

In the peasant tradition of saint devotion, a unique position was reserved for patron saints. While the relationship between the peasant and
generic saints was usually affectionate in nature (the saint might even be
given a pet name), the relationship to the patron saint was often characterized by submission (DiTota: 327). Patron saints watched over provinces,
cities, certain city neighborhoods, and villages; and in this context, they
were honored as community protectors from natural or unnatural disasters.
The peasants believed patron saints were intervening constantly to ward off
forces that could destroy one's locale or livelihood. San Gennaro, the patron saint of Naples, protected this city from many calamities; just as Santa
Rosalia watches over Palermo.

Underscoring the need for saints to be accessible, and crucial to the
devotion of patron saints and the Madonna, was the religious procession.
Carrying a statue or painting of the saint or Madonna through the village or
neighborhood streets on feasts days was a public demonstration by the
believers to their communities and the saints that watched over them. Furthermore, moving religious ceremony out of the church and into the streets
reaffirmed the relationship between the saint and the protected zone in
which the peasants resided. As Carroll puts it:

> If the patron is to prevent hostile forces from invading the space that
> defines a community, then he or she must be associated with that space.
> In the smaller villages, this often means literally carrying the patron's
> image (usually a statue or painting) past every home in the community.
> (Carroll, 1992:401)

The procession embodied the sacred and the profane. Customarily, members of a society dedicated to a specific saint were the ones who transported the statue from the realm of the church to the people's domain. In the streets, men and women — especially those petitioning for certain favors or those giving thanks — made offerings of money or jewelry to the saint. As the statue passed in procession, everyone felt assured that they and their homes would be blessed and would receive the protection of the saint (Primeggia, 1999: 74-75).

In addition to geography, patron saints protected occupations and those who worked in them. San Giuseppe (St. Joseph) was the patron saint of carpentry. In Palermo, San Cosimo was the patron of bricklayers. San Damiano watched over barbers, San Crispino was the patron saint of shoemakers, and San Paolino was the patron of gardeners. Even thieves had a patron saint, San Gerlando, to watch over them (Primeggia: 71-72), and Gower (1928) tells us in the commune of Modica, Santa Caterina was the patron of prostitutes.

In a world of harsh realities, whether dealing with an earthly *padrone* (master, owner or boss) or an otherworldly saint or Madonna, the southern Italian peasant expected *niente per niente* (nothing for nothing). If the peasant sought assistance for himself or his loved ones, a bargain would be made with the saint or Madonna. Such bargains were statements of what the peasant was willing to do or sacrifice for the saint or the Madonna to intercede. Pre-intercession offering might include attending mass daily, committing to regular novenas or frequent rosary recitations, participating in religious processions, or acts of self-denial, self-deprivation and self-humiliation. Some of these acts went as far as bringing self-induced physical pain; and in some cases, as Carroll (1992: 135-136) has pointed out, the acts had masochistic tendencies. Examples include walking barefoot in lengthy processions and crawling on one's knees from the back of a church to its altar or dragging one's tongue on the floor. (If lacerations and bleeding occurred, the sacrifice was seen as greater.) Bargaining-related depths of sacrifice and degrees of difficulty were calculated to demonstrate to the saint or Madonna the correspondent urgency of need and the resolve of the petitioner.

So seriously were these agreements taken, that if the saint did not live up to his end of the bargain, "...and grant the petitioner his or her request, he would severely be punished, by having his statue or image cast down and even destroyed" (Vecoli, 1977: 28). Williams provides this illustrative account of the following about Sicily:

> ...if the patron of fishing-Saint Francesco di Paolo did not produce a shoal of tuna fish at the proper time, his statue was taken from its niche down to the beach. There the fishermen gathered and threatened, "*a ca vi beddamu* (here, we'll dunk you)!" (Williams, 1938/1969: 137)

To the matter-of-fact southern Italian peasant, punishment of saints and Madonnas who did not "perform" served as a warning that, patrons or not, they could lose their devotional following. Statues of non-performing saints could be banished to closets, attics, cellars, and even the garbage. The southern Italian peasant had no qualms about putting the saints in their place (both figuratively and literally), denigrating or presenting ultimatums to those who neglected to hold up their end of a bargain. Sereni (1968: 196) identifies the practice of humiliating and insulting saints as a unique aspect of the religious practices of the Mezzogiorno (midday, a reference to southern Italy). Here, it was acceptable to teach the saint a lesson, so that he or she would not fail the peasant again, when a future bargain might be struck.

Another feature of the cult of the saints, the *ex-voto* offering, is clearly captured by Williams (139), who describes how, "each statue of St. Lucy furnished her with a platter on which supplicants might place their clay or wax models of eyes. Sometimes recipients of a saint's favors presented a picture of the miracle for which they were grateful." Most commonly, *ex-voto* offering items were models of body parts pressed from metal or candle wax, or painted on a tablet (Rossi, 1969: 167-168). Other *ex-voto* media included dough, cheese, silver, or clay shaped into body parts. In some instances, an afflicted person's clothing was laid at the foot of the statue, because it was believed that when the garments were later worn, the saint's power would flow into an ailing body and restore health (Williams: 139). However in general, unlike pre-intercession bargains, ex-voto offerings were post-performance displays of public thanks to a saint or Madonna for having cured a particular illness or injury as represented by the modeled limb or organ.

Summarizing this discussion of the cult of the saints, "...the southern Italian appropriation of the saint was used, in the peasant mind set qua folk religion, in a practical, world-affirming way as a vehicle to defend oneself from an otherwise hostile, poverty-ridden and politically oppressive social order..." (Varacalli, 1999: 6). Surrounded by unfriendly forces that could at any moment bring pain, suffering, and calamity, the peasants made these erstwhile beings into tangible allies who were to be loved, respected, and honored in order to ensure their saintly support.

The Madonna Cult

Carroll (1986: 11) has observed that the Mary cult seems to be a crucial feature of the Mediterranean countries, especially Italy and Spain; and indeed, *La Madonna* (the Blessed Mother) has long been a central figure in southern Italian religiosity. In fact, within the popular religious belief system of the *Mezzogiorno*, the Madonna overshadows her son Jesus.

Adoration of the Madonna fit neatly into the culture of southern Italy. In a society that assigned primary importance to the family and a dominant role to the mother, the Madonna easily rose to an exalted position in the southern Italian peasant religion. The holy family unit of Jesus, Mary, and Joseph was also readily accepted, with Mary being the most important of the three. Most revered and worshiped by the Italian peasant, "...she was primarily a miracle worker..." (Covello: 121) and their ultimate protector, capable of working miracles for individuals or on behalf of humankind.

Highly loved and esteemed, the Madonna was seen as a powerful intermediary and intercessor to Jesus. Peasant mothers especially related to the *Madonna con Bambino* (Madonna with child), because:

> ...Mary, who had experienced ultimate spiritual glory and earthly trag-
> edy, was seen as the one who could best understand a mortal mother's
> hopes, fears, and concerns for her family and surroundings. In the eyes
> of many, Mary was the strongest advocate for petitioners, for she alone
> could plead their cases directly to her son, the God-made man Christ.
> (Primeggia: 76)

Thus was the Madonna endowed with many attributes of an earthly mother. In fact, "most miraculous images of a madonna in Italy are images of a madonna with child" (Carroll, 1996: 25), and the strongest of madonnine image is that of the *Madonna del Latte* (depicting the infant Jesus nursing at his mother's breast). For the Italian peasants, this classic image underscored the unique relationship that existed between the divine son and the earthly mother, making the Madonna the foremost choice to intercede on their behalf to Jesus (Carroll, 1996: 157).

Although formal catechism teaches there is only one mother of Christ, the Italians worship a multitude of Madonnas. "While these different Madonnas may indeed be vaguely associated with the official Mary, each Madonna nevertheless has a separate identity, and each is the object of distinctive cultic devotions" (Carroll, 1992: 59). Among the many feasts that honor this range of Madonnas are: *Madonna dell' Arco, Madonna dell' Assunta, Madonna delle Galline, Madonna di Bagni, Madonna Avvocata, Madonna delle Pace, Madonna del Carmine, Madonna Addolorata, the Madonna of Piedigrotta, and the Madonna del Monte Virgine* (Primeggia: 75-76). Across southern Italy, all these Madonnas and more (each with her appropriate title) were prayed to for protection. As Christianity replaced goddess worship, "Hundreds of local Madonnas were endowed with the qualities of the ancient deities they supplanted" (Covello: 121).

In Italy, a clear distinction between the cult of the saints and that of the Madonna is the phenomenon of apparition. An apparition is defined as an unexpected face-to-face interaction between a human and a supernatural being. Although apparitions of Christ and certain saints have been documented, the overwhelming majority of these occurrences involve the Virgin Mary. The multitude of Marian sanctuaries in Italy (located at purported apparition sites) attests to the importance of the Madonna cult in this country (Primeggia: 76-77). In fact, Italy accounts for the largest percentage of Marian apparitions in the world (Carroll, 1986: 140).

There are a number of contradictions in how the southern Italian peasant worshipped the Madonna. While infinitely worthy of processions, feasts, and ritualistic veneration, she could be kind and forgiving, or angry and unwilling. In this context, much like the saints, she was honored as well as bargained with. Moreover, if she failed to deliver, she was not above punishment. Ultimately, however, the Madonna held special appeal to the Italian peasants because she was not only a very powerful female presence in their lives and religious beliefs, she was a mother as well (Carroll, 1996:

28). Put more precisely, for Italian Catholic peasants, the Madonna was the divine mother of us all.

The Italian Feste

The Italian *festa* (feast) was, and still is, a time-honored tradition of honoring a saint or Madonna. In fact, As LoPreato (90) states, "Nothing about religion was more important to Italians than the festa. Without it, religion was cold, formal, and lacking significance." En masse demonstrations of public devotion, feasts were a consummate melding of the sacred and profane components of religion. Formal recognition of a saint or Madonna was solemnized in the church, giving way to secular festivities that took place on the street. Both speculative and spectacular, feasts held deep religious, communal, and personal significance in the hearts and minds of people throughout the *Mezzogiorno*.

In peasant villages especially, feasts of patron saints were as eagerly awaited, in religious and social contexts, as Christmas and Easter. Here Vecoli (1977: 28) captures how the religious life of the *paese* (country, district or village) reached its zenith on the feast day:

> For the contadini, the festa (feast day) provided one of the few releases from the year round cycle of work and want. Putting aside austerity for the day, they indulged themselves in food, drink, and emotions. Dressed in their festive garb, they packed the church for the High Mass when the priest delivered the panegyric, declaiming the life and miracles of the saint. Gifts of money, candles, or grain were brought to the Church in fulfillment of vows made during the year... In the afternoon, the statue of the saint was carried in procession through the streets, accompanied by the religious confraternities in colorful robes, a brass band, and the throng of devotees. During the procession, emotions reached a high pitch among the chanting women with wailing, weeping, and trance like behavior.

Pageantry was always a key element. Often, as the mass ended and the statue was carried outside, the transition from church to street was signaled by the setting off of fireworks or the freeing of a flock of caged doves or pigeons. In some instances the pageantry was tied to physical exertion as an expression of faith. In certain traditions, the carriers of the statues executed complicated march cadences or exaggerated swaying, offering up their efforts to the saint or Madonna as penance, thanksgiving, or consecration (Primeggia: 78-79).

One feast in particular that dramatically links participants' religious devotion to physical discomfort and endurance is the *Gigli* (Lilies) Feast honoring *San Paolino* (St. Paulinus). Annually, since 434 A.D., crews of one hundred men have shouldered four-ton wood and paper maché structures of eighty-five to ninety feet in height (Field, 1990: 82). The *Gigli* Feast commemorates the return from captivity of *San Paolino*, bishop of Nola, who offered himself in exchange for a widow's son, kidnapped in a vandal raid of the town (Primeggia and Varacalli, 1996: 424). When the holy man was eventually released, the *Nolani* (people of Nola) welcomed his returning ship by throwing lilies into the water.

Today, the lilies are represented by the huge and heavy spires, which are carried and danced through the streets in daylong celebration. The painful, swollen lumps that inevitably rise on the shoulders of the men who lift and carry are seen as religious "red badges of courage" and signs of their love for *San Paolino*. "The men consider the swellings important stigmata and bear them proudly and ostentatiously as testaments to their physical strength and masculinity" (Field: 87). Also regarded as "bargaining chips," the lumps demonstrate that which devotees are willing to endure for their saint. Sacrifice is a way to request intercession or to give thanks for answered petitions.

The southern Italian feast was a unique occasion, combining religious fervor with social revelry. On feast days, the peasants took their religious responsibilities (mass, prayers, incantations, processions, and offerings) seriously. Yet the feast was also a venue of worldly treats. Feasts "...were very special days when family, friends, and *paesani* joined in a common celebration. Town streets were a special source of pride for the people and the feasts were opportunities to compete with neighboring towns" (LaRuffa, 1988: 115). Street vendors peddled specialty foods — such as salted beans, fried cakes, tomato pies, seeds, cheese, and sweet *torrone* (almond candy) — and *ex-voto* offering items. Downtrodden peasants were granted a brief respite from an otherwise oppressive routine of life. For one extraordinary day, they could focus on their dedication to a specific saint or Madonna with a passion that is neatly expressed in the Italian phrase, "*quando è festa è festa*" (when it's the feast day, it's a feast)!

Folk Religion and Superstition

Southern Italian peasants were extremely pragmatic and as such, would exercise a variety of measures to help themselves and their families survive. In religious terms, this meant that their formalized Catholicism was tightly intertwined and unabashedly augmented by superstitious beliefs and practices, which had been passed through the generations (Williams: 141). As Sartorio (1918/1974: 135) tells us, "Superstition is the step-daughter of a good mother, religion." Indeed, the fusion was so innately seamless, that in analyzing southern Italian peasant's religiosity, it can be difficult to differentiate where orthodox religious belief diverges from superstitious notion.

For example, professed Catholic peasants believed that spirits of deceased relatives, friends, or townspeople could intervene in the lives of the living, yielding positive or negatives results (Covello: 106). They also fully accepted the existence of witches and wizards, who possessed knowledge of the black arts and could concoct potions that either restored one's health or caused ill health. With regard to the casting of spells:

> ...an erring husband, a sick child, or a poor crop were all caused by malevolent spirits. To counter such curses (or perhaps cast a spell of your own), the peasants had an assortment of charms, amulets, potions, and incantations. Each daily act such as the baking of bread or sowing of grain, had its associated magical formula to ward off evil. (Vecoli, 1977: 29)

Everyday facts of life in the peasants' folk religion, amulets, potions, and magical rites enjoyed equal importance with the Church's sacraments in dealing with the spirit world (Vecoli, 1969: 229).

Another common, and one of the strongest, superstitious beliefs was in the *mal' occhio* (the evil eye). Perhaps a peasant rationale for a life full of pitfalls in which no one seemed to be able to get ahead, making others responsible for misfortune was preferable to questioning one's own shortcomings. At the center of evil-eye credence was the power of envy. It was held that certain people who resented their neighbors' achievements (a business success or the birth of a beautiful child, for example) could cast a spell through the evil eye, causing business reversals, sickness, and even death. In truth, just about every evil that befell an individual or his family members tended to be blamed on the *mal' occhio* or the *jettatore* (the one whose look cast the evil eye).

However, rather than compete with or discredit church doctrine, the superstitious beliefs and practices of southern Italian peasants complemented their Catholicism. They saw no contradiction in reciting the rosary at mass in the morning and consulting the local *strega* (witch) that afternoon, or "putting crosses in ink on different parts of the bodies of children to cure them from all kinds of diseases" (Sartorio: 102). Instead, the peasants' creative pairing of magical notion and spiritual devotion was born of their ongoing quest for reliable, well-balanced survival tactics.

Church and Religion and the Italian Immigrant

When the Italians migrated to the United States, the vast majority settled in urban neighborhoods which, when predominantly populated by Italians, became known as Little Italies. This first crop of Italian Americans, more so than any subsequent generation, held onto their peasant traditions and way of life by attempting to recreate the old village way of life in the American urban setting (Glazer and Moynihan, 1963: 194). They became what Gans (1962) describes as urban villagers. Even those small numbers who put down rural or suburban roots tended to cluster in neighborhoods labeled by others as "the Italian area," "Dago Hills," or "Woptown." While some inroads were made in their lives by the American culture and society (including a certain number of conversions to Protestantism), for the most part the immigrants staunchly maintained their old ways in the New World.

Religion was one sphere in which the immigrants endeavored to hold fast. In dealing with Catholicism in the United States, the immigrants continued to espouse many of their Old World ideas and values, including folk religious beliefs drawn from the southern Italian experience. They continued to worship and rely on the saints in a folk religiosity that fused Catholicism with certain elements of superstition, magic, and the occult (Varacalli, 1999: 6). Furthermore, the immigrants' community and family orientation fostered strong resistance to the influence of the American Church. The Italian male immigrants, for example, continued in their anti-clerical and anti-Church stance. Just as in Italy, the men stayed at home on Sundays while the women in the family went to religious services (Peroni, 1979: 34).

The Italian immigrants were turned off by the too legalistic and moralistic Catholic Church they found in America. The religion that they brought with them centered on relationships with the persons of the Trinity, the Holy Family, and the saints (Gambino, 1991: 41). By the same token, their folk religious practices and exotic interpretations of Catholicism disturbed the leaders of the Catholic Church in America. In fact a mutual offense developed almost immediately upon the immigrants arrival. The American Church was reluctant to embrace the Italian newcomers, and the immigrants did not take readily to the Catholic Church (Tricarico:10). Diametrical differences between their folk religion and the Catholic faith practiced in the United States became startlingly evident when,

> Those Italians who ventured into the Irish or German churches found them as alien as Protestant chapels. The coldly rational atmosphere, the discipline, the attentive congregation were foreign to the Italians. (Vecoli, 1969: 230)

At the same time, what Catholic Church leadership termed the "Italian Problem" was manifested, in part, in the immigrants' anti-Catholic beliefs and practices. Catholic priests saw the refusal to follow purely doctrinal Church tenets as a basic failure of immigrant religiosity (Primeggia: 79). Their transplanted form of faith "...did not so much embrace Catholic Christianity as encompass it" (McBride, 1981: 339). The immigrants were viewed as "outsiders" because too much of their faith was based on superstition. Italian religiosity was seen as too pagan, and caught up with idolatry-not grounded in official teachings. Or as Vecoli (1969:25) portrays the situation, "...the Italian immigrants were characterized by ignorance of Christian doctrine, image worship, and superstitious emotionalism. In short, they were not true Christians."

One reason the Italian immigrants were looked down on by the American Catholic Church was that its leadership was comprised of Irish bishops who were predominantly Americanists (Quinn, 1999: 98). These bishops, who were on a mission to prove that all Catholics could assimilate into the mainstream culture, were duly dismayed by the Italians who tenaciously held to their native tongue and perpetuated Old World values. The

Church's top echelon was further distressed by the Italian males' lack of attendance at mass, except on special religious holidays and feast days. Priests found it incomprehensible that the major connection Italian men had to their faith was the feast-an occasion dismissed by the Church as a non-worthwhile religious observance. This disapproval is borne out by a 1914 article in the Jesuit journal, *America* that Tricarico (11) quotes, "no matter how numerous be the Italian processions, no matter how heavy the candles, no matter how many lights they carry..." the Church's view was that Italian religiosity lacked piety.

During the mass migration period (1880-1924), the most positive thing Church leaders accomplished for the immigrants and themselves was the creation of national parishes. The ethnic orientation of these parishes sustained the immigrants by maintaining what was culturally important to them. Italian-speaking priests were brought in to say mass and oversee the flock. These priests were familiar with the unique religiosity of the Italian laity and sanctioned the feast as centerpiece of their religious participation. While effectively keeping rapid assimilation in check, the institution of the national parish did, in fact, lead to eventual assimilation of the second and subsequent generations (Russo, 1977: 197).

Saint Devotion in America

While their priests did not migrate with the immigrants initially, their saints did. As part of their attempt to recreate things familiar in a great unknown, the Italians continued their unique style of worshipping patron saints and Madonnas in America. Statues of revered saints and Madonnas, exact copies of those left behind in far away villages, were imported from Italy to be placed in the local church or society headquarters. Many images of different saints and Madonnas could be found on altars or in separate chapels. To a knowledgeable observer, "one could determine the regional composition of an Italian parish by the saints and Madonnas who were venerated there" (Vecoli, 1977: 30).

The Italian immigrants believed that the saints would provide protection and respond to pleas in America just as they had in Italy. The women, especially, lighted the candles before the saints and Madonnas, reciting special prayers. Smaller statues or images, illuminated by votive candles, were even placed in the immigrants' homes as personal shrines that made private prayer and devotion to the patron possible at any time.

When the cult of the saints reemerged across the Atlantic in America, it was "the strongest bond, outside the family which tied the immigrants to each other and to the distant *paese*" (Vecoli, 1977: 30). Honoring their saints, Madonnas and God with customary displays of respect, love and deference helped the immigrants find small havens of comfort in a vastly alien land.

The Madonna Cult in America

The strong cult of the Blessed Virgin Mary in Italy, where she is seen as the Mother of all, was not to be eclipsed in America. Just as in the old country, most Italian immigrants found it easier to pray to Mary, who seemed so human to them (Russo: 204). Indeed, the Madonna was viewed as representative of all Italian and Italian American mothers, who were revered family matriarchs. In strange, new surroundings, devotion to the Madonna was an emotional, spiritual, moral, and communal congruity that provided a much-needed sense of the familiar. Thus, the immigrants chose to believe that the Madonna, as well as the saints, had made the trans-- Atlantic trip along with their devotees and would continue to watch over them in America.

Different immigrant neighborhoods and parishes worshipped specific Madonnas, but *La Madonna del Carmine* (Our Lady of Mount Carmel) held a special place in the hearts and prayers of southern Italian immigrants. Many erected new churches in her honor, or at the very least, organized feasts at which to celebrate and venerate her name. In New York City alone, Mount Carmel churches arose in Manhattan's East Harlem, Brooklyn's Williamsburg, and the Bronx's Arthur Avenue section (Belmont). These parishes were all born of the desire to replicate Italian devotion to the Madonna in America and the hope of passing on their love of The Lady to subsequent generations.

The Italian immigrant women especially, felt a human connection to the Madonna. In the popular Mariology of East Harlem, for example, the women, "believed that Mary had suffered the pain of childbirth, that she had menstruated, and that she worried constantly about her child" (Orsi, 1985: 227). Maintaining views that were outside the official Church, but consistent with their European tradition, women would promise the Blessed Mother to perform acts of penance in payment for supplications answered or blessings received (Orsi: 9-10).

Devotion to the Madonna endured resolutely among the first generation Italian Americans. Because of her humanized qualities and perceived eternal power, the Madonna continued to speak in the loudest voice and remained the major force in the religious life of the immigrants. Through the lens of their transplanted folk religion, she was still the strongest intercessor, intermediary and conduit to her son, Jesus Christ. Hence, they placed themselves and their families in her watchful care and demonstrated their respect in the time-honored ways, brought from Italy into the crowded slums of the New World.

Italian American Feasts

Seeking to maintain their village customs, the Italians who came to America directed much of their religious devotion, fervor, and energy to feasts honoring a particular patron saint or Madonna. The elaborate planning and extraordinary work these occasions involved underscored the immigrants' need to concretize their religious practices and reinforce their sense of *Italianità* ("Italianicity"), albeit in *paese* mind-set terms. During the immigrant era, feasts-a distinctive element of Italian Catholicism-were organized and held by either a church or one of the many festa societies established in Italian American neighborhoods. So ardent were the first generation's efforts, that many feasts are still held today in yet-solid Italian American communities throughout the United States.

Moreover, several of these feasts retain aspects first introduced by immigrant forebears (e.g., masses honoring the patron, barefoot processions, street bands playing throughout religious ceremony and profane celebration). Among such customs that endure in the Italian North End of Boston is the feast of Our Lady of Soccorso. Here, the pageantry of suspending a praying child (outfitted with white gown, halo and wings) on a wire and lowering her to the street is a ritual steeped in tradition. Williams (140) discusses a similar spectacle with Sicilian origins having taken place in the town of Polizzi Generosa, where children dressed as angels descended on ropes from balconies as the patron saint statue passed on feast day. To this day, to be asked to enact the Bostonian "flight of the angels" is considered an honor for the chosen child and her entire family. Likewise, in terms of social solidarity, the recurring event is a way for the local Italian American community to salute those who transplanted this cultural component in the United States five generations ago.

For immigrants suffering homesickness for *la bella Italia* (beautiful Italy), the feasts not only offered opportunities to pay homage to a saint or Madonna, but to recapture a bit of their Italian way of life in America. According to Vecoli (1969: 232):

> the festa was the most authentic expression of southern Italian culture transplanted to the New World. No effort or expense was spared in the effort to recreate the feast in every detail.... During the festa, the streets of the Italian quarter took on the aspect of a village fair: streets and houses were decorated with banners, flags, and lanterns, while streets were lined with shrines and booths. Delicacies to titillate the Southern Italian palate were dispensed from sidewalk stands as were religious items and amulets against the evil eye. Meanwhile, brass bands played arias from the well loved operas of Verdi. Everything was continued to create the illusion of being once more in the Old Country.

While feast food featured basic staples, each *festa* offered distinctive edible specialties, typical of the paese or region from which local residents had emigrated. Italian feast treats included fried and sugared batter (*zeppoli*), fried dough pockets stuffed with meat and cheese (*calzoni*), boiled beans seasoned with oil and red pepper, baked corn, grilled sausage, peppers and sweet breads, as well as countless pasta variations. Certain feast foods, such as the *chiambelli* (ring-shaped bread eaten during the feast of St. Rocco) held symbolic meaning related to the saint being honored. The *chiambelli* tradition arose from the story of a dog bringing bread to the leprous saint while he was living as a recluse (Malpezzi and Clements, 1992: 105). In fact, quite commonly, specially shaped breads, baked for feast occasions only, were designed to commemorate a significant saintly act or anecdote.

In the Italian immigrant enclaves, the Neapolitans still honored *San Gennaro*, the patron of Naples and *la Madonna del Carmine*. Sicilians continued to pray to *Santa Rosalia*, the patron of Palermo and their beloved protectress *Santa Agata*. Numerous feasts were held for popular saints and

Madonnas, as well as for less familiar and even unknown saints to whom the people of a particular *paese* attributed special meaning. In some instances, the immigrants honored village or regional protectors and patrons not recognized officially by the Roman Catholic Church. Regardless of any sanctioned authenticity, the feasts kept the immigrants mindful of their historical roots, their current connections and their future priorities. Perpetuating these rituals instilled a confidence among these new Americans that they could go on asking their Italian saints and Madonnas for the assurance, assistance and miracles necessary to make a better life on alien shores.

It mattered little that the Church leaders, Catholics of other nationalities, and Protestants denounced Italian American feasts (and what they stood for) as yet another form of superstition. The unashamed immigrants' focus was on replicating the feasts celebrated in their Italian villages — right down to pinning money, notes of petition, or expressions of thanks to the passing statue. Just as in Italy, the immigrants erected high arches, which spanned the streets at regular intervals while feasts took place below. The arches often reflected a religious motif or might be decorated with green, white, and red streamers (the colors of the Italian flag). Over time, in America, the arches were illuminated with electric lights, replacing yet reproducing the glow of torches that burned on the feast arches in the Italy of their past. In fact, the format of the immigrants' Italian American feast, "from the stringing of the lights to the exploding fireworks," was built upon that which was "an ancient art in Italy" (Swiderski, 1986: 62).

The *ex-voto* tradition was also sustained in the New World, as Orsi confirms in the following description of a first-generation Italian American feast in East Harlem:

> ...the booths were filled with wax replicas of internal organs and with models of human limbs and heads. Someone who had been healed-or hoped to be healed-by the Madonna of headaches or arthritis would carry wax models of the afflicted limbs or head, painted to make them look realistic, in the big procession. The devout could also buy little wax statues of infants. Charms to ward off the evil eye, such as little horns to wear around the neck and little red hunchbacks were sold alongside the holy cards, statues of Jesus, Mary, and the saints, and the wax body parts. (Orsi, 3)

Thus, dramatic demonstrations of pre-immigration beliefs in combined power of sacred and profane forces continued to play out on the streets of New York City and elsewhere on Italian feast days.

In East Harlem, and at other neighborhood feasts, devout Italian immigrant women displayed their learned intensity and depth of devotion to the Madonna. With a new American backdrop, female "Penitents crawled up the steps" of the church "on their hands and knees, some of them dragging their tongues along the stone" (Orsi: 4). Other Italian women walked barefoot for miles to church in painful processions that proclaimed their love for the Madonna. Each prolonged pedestrian pilgrimage from household to house of worship was considered "...an act of penance, a demonstration of *rispetto* (respect) for the Virgin, and because they considered the place holy" (Orsi: 4).

Meanwhile, in the Williamsburg section of Brooklyn, immigrants from Nola, established a neighborhood *San Paolino* Society almost immediately upon arrival. By the turn of the century, the neighborhood residents were annually re-enacting the lifting and dancing of the *giglio* structure in their new home (Primeggia and Varacalli, 1996: 423). Constructing their *giglio* spire to the exact specifications of its Italian counterpart was an untarnished attempt to replicate this 1600-year-old feast observance. Over the years, additional *Giglio* feasts began to take place in the Italian immigrant enclaves of East Harlem and Long Island City, where residents recognized and embraced the carried-over practice of calling on a patron saint to bless or "sacralize" their neighborhoods (Primeggia and Varacalli, 1993: 52). The Brooklyn *Giglio* feast[1] has endured for over 100 years-a testimony to the tenacity of devotion to *San Paolino* and Our Lady of Mount Carmel that this parish has maintained. To this day, the sons, grandsons, and great grandsons of the *Nolani* immigrants carry the four-ton weight of the *giglio* spire proudly on their shoulders and their inherited religious belief system lovingly in their hearts.

In strange surroundings, and made to feel unwelcome by the *Americani* (anyone in America who was not Italian) and their Church, feasts were one important mechanism through which Italian immigrants successfully sustained their distinctive brand of saint devotion in the New World. As the second and subsequent generations of Italian Americans adapted to the American way of life making many concessions along the way, the immigrants held on steadfastly to most of their religious practices and social traditions. Making only minimal adjustments as necessary to survive in their new environment, they were able to keep alive, as Varacalli (1986) puts it, the "sacredness" *of la via vecchia.*

The Continuation of Folk Superstitions

In a less-than-hospitable and often confounding new environment, the ever pragmatic Italian immigrants clung staunchly to superstitious beliefs and practices. Just as these superstitions had served their needs in Italy (Iorizzo and Mondello, 1971: 179), they trusted the same application would help them survive in the New World. "Precautions then had to be taken against the evil eye; amulets were worn, rituals performed, and incantations chanted to fend off the power of witches" (Vecoli, 1974: 33). In city tenements, the immigrants still gave credence to black magic and the supernatural powers that could provide protection against unseen forces that threatened them in all walks of life. Even those who settled outside the urban areas followed the pattern. According to LaGumina, (1987:10) in the predominantly Italian American suburb of Glen Cove, New York, immigrants mirrored the religious belief structure of their Old World and urban counterparts with aspects of their ancestral peasant culture, which had its roots in pagan and magical practices.

In the United States, the Italian immigrants continued to believe that certain people could bring harm, pain, and sorrow into their lives with a mere glance of the evil eye. That's why, from the moment they left Italy and long after disembarking on America's shores, the immigrants wore a *corno* (pointed coral charm) on their person to protect against the *mal' occhio.* A hidden, but well-aimed *corno* could pierce the evil eye and render it impotent. Caught without a *corno,* one could improvise by folding the middle fingers to the palm and surreptitiously pointing the index and pinkie fingers at a suspicious character. It was believed this *mano cornuta* (hand charm) could absorb incoming harm through one of the extended fingers and redirect it back through the other toward its ill-intentioned initiator. Superstition had it that a left-handed *mano cornuta* was a stronger weapon, because the left side of the body was controlled by the Devil. In fact, the Italian translation of "to the left" is a *sinistra* (the sinister side). So, folk logic dictated that the left-handed gesture fought the power of evil with an equally powerful counter force.

At any time, an unsuspecting individual could be "looked over" by a person who possessed "the eyes," as the *mal' occhio* was described in America (Williams: 142). Even partially assimilated and quasi-skeptical second-generation Italian Americans thought twice about unexplained ailments or misfortune, consulting the elders who knew curative "eyes" prayers and rituals. In an archetypal intertwining of religion and superstition, evil eye remedies could only be learned on Christmas Eve, and "curers" could only teach two others without losing their own power. (In Catholicism, the number three relates to the Holy Trinity.) To assess if someone was suffering from an evil eye-related affliction, the following process was repeated three times: Drip three drops of oil into a pan of water. If the oil coagulates, the person is a victim of the evil eye. If the oil disperses, the malady is due to other causes. Evil-eye antidotes involved the recitation of special prayers, and making signs of the cross on the sufferer's forehead, chest and stomach as the curer called out to God. The tap water and olive oil applied in various ways can be directly tied to the holy water and sacred oils used in Catholic Church rites.

While many superstitious and magical components of the Italian peasant belief system continued to flourish in the new American setting, it is important to remember that immigrant folk religiosity retained a pragmatic counterbalance in formal Catholicism. Newly arrived Italian Americans deliberately selected — both from Church teachings and pagan ways — those aspects that offered an optimal shield of defense for themselves and those they loved. Wearing a cross and a *corno* on the same chain, for example, effectively covered more bases.

As was the case of the peasants back home, the Italian immigrants had many hardships to face and difficulties to overcome in America. As they met new challenges that included disparagement by elements within the Catholic Church and those outside it, the southern Italian immigrants found solace, faith and validity in the folk religiosity they brought from Italy to America. In a crucial sense, shared religious beliefs and practices served as a bridge that empowered the peasants in Italy and immigrants in America to speak in one voice and find strength in numbers.

SALVATORE PRIMEGGIA

¹ For a more detailed description and analysis of the Brooklyn Giglio Feast, see Primeggia, Salvatore and Joseph A. Varacalli. 1996. "The Sacred and Profane Among Italian-American Catholics: The Giglio Feast," *International Journal of Politics, Culture and Society*, 9, no. 3: 423-450 and the video, "Heaven Touches Brooklyn in July," (produced and directed Tony DeNonno) DeNonno Productions, Inc., 7119 Shore Road, Brooklyn, NY 11209.

References

Caroll, Michael P. 1986. *The Cult of the Virgin Mary: Psychological Origins* (Princeton, NJ: Princeton UP).

_____. 1992. *Madonnas that Maim: Popular Catholicism in Italy Since the Fifteenth Century*. (Baltimore: Johns Hopkins UP).

_____. 1996. *Veiled Threats: The Logic of Popular Catholicism in Italy* (Baltimore and London: Johns Hopkins UP).

Covello, Leonard. 1972. *The Social Background of the Italian American School Child: A Study of the Southern Italian Family Mores and Their Effect on the School Situation in Italy and America* (Totowa: Rowman and Littlefield).

DiTota, Mia. 1981. "Saint Cults and Political Alignments in Southern Italy," *Dialectical Anthropology*, 5: 317-329.

Field, Carol. 1990. *Celebrating Italy* (New York: William Morrow).

Gambino, Richard. 1974. *Blood of My Blood: The Dilemma of the Italian-Americans* (Garden City, NY: Doubleday).

Gambino, Richard. 1991. "Italian American Religious Experience: An Evaluation," *Italian Americana*, 10, no.1: 38-60.

Gans, Herbert. 1962. *The Urban Villagers* (New York: Free Press).

Glazer, Nathan and Daniel P. Moynihan. 1963. *Beyond the Melting Pot* (Cambridge: M.I.T Press).

Gower, Charlotte Day. 1928. *The Supernatural Patron in Sicilian Life* (Ph.D. dissertation, University of Chicago).

Iorizzo Luciano J. and Salvatore Mondello. 1980. *The Italian Americans* (New York: Twayne Publishers).

LaGumina, Salvatore. 1987. "Immigrants and the Church in Suburbia: The Long Island Italian American Experience," Records of the American Catholic Historical Society of Philadelphia, 98, no. 1-4: 3-20.

LaRuffa, Anthony. 1988. *Monte Carmelo: An Italian American Community in the Bronx* (New York: Gordon and Breach Science Publishers).

LoPreato, Joseph. 1970. *Italian Americans* (New York: Random House).

Malpezzi, Frances M. and William M. Clements. 1992. *Italian American Folklore* (Little Rock, Arkansas: August House Publishers).

McBride, Paul M. 1981. *The Solitary Christians: Italian Americans and Their Church* (Great Britain: Gordon and Breach Science Publishers).

Chapter 2

◆

A COMPARISON OF THE ATTITUDES AND PRACTICES OF THE IRISH AMERICAN AND ITALIAN AMERICAN CATHOLICS

This paper reviews some of the literature on Irish American and Italian American Catholics. It investigates the attitudes and practices of these two American Catholic groups, and argues that socioeconomic conditions along with enduring cultural traits affect the behavior of Italian American Catholics and that under certain socioeconomic conditions the behavior of Italian American Catholics is similar to the Irish American Catholics.

Some of the sociological literature indicates that there are differences in cultural traits such as mistrust of clergy (Greeley, 1969: 68), a de-emphasis in church attendance (Abramson, 1973), church attendance perceived as an activity for women (Gambino, 1974), veneration to the saints and celebration of religious feasts (LoPreato, 1970, Nelli, 1967 and Vecoli, 1964) and a Catholicism that is not considered to be very strong (Sartorio, 1918) between Irish American and Italian American Catholics. Specifically, past research points to the cultural traits mentioned above as emerging from Italian history and maintained by today's Italian Americans, at least to some degree. Some of the literature indicates that the maintenance of these traits by Italian Americans is the reason why this group attends religious services less often than Irish American Catholics, among whom weekly church attendance is considered an important religious practice (Gambino, 1974). However, there are works indicating that since World War II Italian American Catholics, in their attitudes and practices tend to resemble the Irish American Catholics (Russo, 1970, Glazer and Moynihan, 1971 and Crispino, 1980).

The Church and Religious Practices in Southern Italy

To be able to understand the religious way of life of Italian Americans it is important to become cognizant of the people who lived in southern Italy (since most Italian Americans are descendants of Italians from southern Italy) and the religious life style they brought with them to this country. Let's examine the nature of religious life in the Italian South.

Joseph LoPreato (1970: 87) indicated that the people in the *Mezzogiorno* (Southern Italy) have never been strong Catholics and that except for very special occasions few people attended religious services. He noted that when young and middle-aged males attend Mass on these occasions, they were likely to assemble outside the church while the religious service was being held inside (LoPreato, 1970:87).

Richard Gambino stated that the s outhern Italian peasant (before WWII) viewed religious life such as prayer and regular church attendance as *cose feminili* (women's things). He also mentioned that the male peasant limited his religious participation to the *feste* (feasts), baptisms, marriages and funerals (Gambino, 1974: 232-33). Only elderly males participated in weekly church attendance (LoPreato, 1970: 87). All these aside, there existed a mistrust of the Church and clergy (Gambino, 1974: 229-34).

Autobiographical reflections corroborated Gambino's description of Italian religious practices in southern Italy. Constantine Pannunzio described the memories of his boyhood in the Italian South before the turn of the twentieth century. He stated that religious participation was considered a female's role, not important to males, and a big part of the southern Italian humor. He noted that his relatives constantly spoke of the Church's corruption (Pannunzio, 1921: 18). Negative feelings by the peasants toward the Church could be found throughout southern Italy.

The spiritual influence of the clergy was not strong among southern Italians, as it was not based on trust (LoPreato, 1970: 88-89). Banfield, in his discussion of "amoral familism" in southern Italy, indicated that little, if any, trust persisted beyond the nuclear family. This mistrust involved the clergy and the church. He stated that southern Italians showed a casual attitude toward religion, they did not take seriously the doctrinal beliefs of the Catholic Church and very little formal religious instruction existed in the villages of Southern Italy (Banfield, 1958: 83).

In his writings of the religious practices of the Italians in the early 1900s, Enrico C. Sartorio stated that most Italians were indifferent Roman Catholics who seemed to go through life without religious feelings. Clearly, the Italians were not very strong Catholics. He indicated that the Italians for traditional reasons would have a religious ceremonial wedding and would baptize their children in the Catholic Church but would not participate in other religious activities (Sartorio, 1918: 82-83).

The *contadino's* (peasant's) private conception of religion at the turn of the 20th century and even during and after the 1970s in some places of southern Italy, was a very personal type of faith which involved all sorts of beliefs of good and evil forces and magical practices. Richard Alba noted that the Southern Italians believed in the "evil eye" (*il mal' occhio*) and used a horn (*cornuto*) from a bull which is hung over a doorway and /or a specially made necklace containing a symbol of a bull's horn to ward off evil spirits (Alba, 1985: 90-91). The heart of the *contadini's* beliefs, according to LoPreato, was the religious *festa*. He pointed out that the *festa* was a joyful and a religious celebration for the people (LoPreato, 1970: 89).

Richard Gambino mentioned that there were many priests, in southern Italy, who used the peasant's version of Catholicism to their own advantage in order to exploit them, and these priests portrayed the Italians as being helpless and needing the authority of the clergy and landowners to guide their lives (Gambino, 1974: 230). Thus, the poor in southern Italy often showed a strong distrust towards formal religion. Rudolph Vecoli noted that the southern Italians had little respect for the Church and viewed the clergy as exploiters living off the peasants' hard work (Vecoli, 1969: 229).

Harold J. Abramson indicated that the historic lack of involvement by the southern Italians in the Church was due to an absence of competition between distinct cultural and religious systems. Unlike in Ireland throughout the 1800s to the early 1900s where the Irish were under English control, religious and cultural differences were absent in southern Italy and so the Church was held responsible for social problems that did not emanate from some foreign cultural system (which did not exist in southern Italy) (Abramson, 1973: 139). Traits that involve mistrust of people in general, a mistrust of clergy, church attendance not emphasized, churchgoing as an activity for women, a veneration to the saints and a celebration of religious feasts, and a Catholicism that is not considered to be very strong emerged among the Italian peasants. This was life in the *Mezzogiorno* (southern Italy) before World War II and it was this life style that the Italians in the South presumably brought with them to America.

The Irish Relationship with the Catholic Church

Catholicism in Ireland was different from the Catholicism in southern Italy. Harold Abramson stated that much of Irish Catholicism from the 1800s to the 1920s was an outcome of England's political domination of Ireland. Abramson indicated that there was always-to-be- blamed-England for Ireland's poverty and he noted that the persecution of the Irish Catholics by the English Protestants led to the integration of the Irish nationality with the Catholic religion and to the development of nationalistic feelings by the Irish against the English (Abramson, 1973: 132). Therefore, the religio-ethnic conflict in Ireland between the English and the Irish was an important factor to the development of Irish Catholicism in Ireland (which was brought over by the Irish immigrants to America). What developed in Ireland was a strong trust of clergy, frequent Mass attendance by both men and women, and with the Irish considering themselves to be strong Catholics. Religion was political (Catholic vs Protestant), thus the Irish followed all the religious requirements as part of a political symbol against the English.

The Italian and Irish Americans and the Catholic Church

The Irish Americans have been described, according to Shannon (1963) and Abramson (1973), as being the important group in the development of the Catholic Church in the United States. Abramson noted that explanations for the Irish organizational success in the institutionalization of the Catholic Church in America, referring to the Irish's facility with the English language and their relatively early arrival to the United States in the mid-nineteenth century (Abramson, 1973: 106). On the other hand, in relation to the Catholic Church, the Italians in America were often treated with prejudice and contempt by fellow Catholics (especially the Irish). There were several social and economic factors which fueled the development and/or maintenance of this prejudice and contempt of the Italians by many Irish. Many Irish by 1900 to 1930 had been in the United States for one, two, or more generations a nd ha d int ernalized na tivist v alues inc luding x enophobic attitudes toward the Italians. Second, many Irish at this time period still worked in construction jobs for which Italian immigrants competed (Tomasi, 1970: 163-93). Third, some of the Irish had reached middle class status and saw no advantage in being identified with impoverished, illiterate

Italians. Finally, many Irish American Catholics were appalled by the Italians' de-emphasis on church attendance, mistrust of clergy, a Catholicism which was not considered to be very strong, veneration of their saints and celebration of the *feste* (Femminella and Quadagno, 1976: 68).

Silvano M. Tomasi (1975: 3) contrasted Italian and Irish religious practices in America before World War II. He observed that for the Italian immigrant religion was not a dictated set of beliefs and practices.

Irish belief, according to LoPreato, was difficult for Italians to comprehend (LoPreato, 1970: 89-90). The Irish were stricter in their religious practices, such as attending weekly church services (Alba, 1985: 91). LoPreato stated that the Irish conception of religion tended to devalue or de-emphasize the Italian's interest in the *feste*. The most important part of religion to the Italians was the celebration of religious festivals (LoPreato: 1970:90).

Robert Orsi (1985) states that the Irish clergy in New York from the Archbishops on down were never enthusiastic about the *feste*. He points out that church authority, though, made no effort to clamp down on any of the festivities for fear of driving the Italians out of the Church.

Rudolph Vecoli indicated that the feast celebrations (1900 to 1940) among Italian immigrants in Chicago were fond remembrances of the life left in southern Italy. He stated that although the Catholic clergy tried to stop the *festa*, the Italian immigrants insisted on their celebrations (Vecoli, 1964: 416-417).

Drawing from the WPA Federal Writers' Project Files of Connecticut, Louis Gesualdi (1997a) examines the data on the Italian immigrants' religious participation in Connecticut from 1880 to 1940. He points out that the Italians had encountered extreme prejudice from the Irish clergy along with the many Irish parishioners' unwillingness to associate with Italians.

Henry J. Browne indicated that between 1885 and 1915 there existed in America few Italian priests, inadequate churches, ethnic conflicts, and Protestant proselytizing. He also showed that the Italians were often treated with prejudice especially by fellow Catholics; and that because of the hatred which existed between the Church and the Kingdom of Italy during the 1870s, Italian men had anticlerical views (Browne, 1946: 46-72).

Aurelio Palmieri stated that during the turn of the 20th century, Irish priests, the predominant group in the American clergy, were unable to deal with Italian immigrants.

Palmieri indicated that little respect or understanding existed between Irish priests and Italian immigrants and that these immigrants came to America with a mistrust of clergy attitude and did not consider themselves to be strong Catholics. This behavior of the Italians and social and economic problems of immigration, according to Palmieri, caused a great discomfort among the Irish American clergy (Palmieri, 1921:14-15).

Because of the political situation after 1870 in Italy resulting from the seizure of the Papal States, there existed a great bitterness between the Italian government and the Church. Richard Gambino stated that the Italian Americans were being punished by the Irish American Catholic clergy for the pro-state attitude held by Italian in the church-state conflict in Italy and for the anticlerical and antichurch attitudes held by these same Italians. Gambino indicated:

> Parochial schools in the United States refused to teach the Italian language until 1929, when Mussolini finally ended the church-state hostility by negotiating the Lateran Treaty with the Vatican. (Gambino, 1974: 238)

Even after the negotiation of the Lateran Treaty with the Vatican, the Irish American clergy, according to Gambino, tried to educate the Italian Americans in their brand of Catholicism. This angered many Italian American Catholics.

The Italian American Catholics had been and remain under represented in the Church hierarchy. According to The Official Catholic Directory of the 1980s, only 17 of the country's 309 bishops had Italian surnames; one of the 39 archbishops was Italian (Hall, 1983: 42). In the 1980s it should be noted that while 15 percent of the Catholics were Italian Americans, 12 percent of the priests had Italian ancestry (Alba, 1985: 92). On the other hand, the Irish American Catholics had been and remained over represented in the Church hierarchy. Although the Irish Americans during the 1970s represented less than 20 percent of all American Catholics, 30 percent of the clergy and over half the hierarchy were Irish (Fallows, 1979: 131).

The Gianelli-Cardillo Report: Italian American Political and Cultural Failure in the United States (2002) concludes that an Italian surname is apparently an impediment to reaching the upper ranks of Roman Catholic Church hierarchy. According to the report, while Italian Americans make up approximately 26% of all Catholic Americans, Italian surnames represent only 8% of Church leaders (that is, 2 out of 8 cardinals, 3 out of 39 archbishops and 26 out of 309 bishops with Italian surnames).

Andrew M. Greeley (1969: 46) in his survey of American Catholic ethnics of the 1960s reported that the Italians were the least pious of all Catholic groups (with the Irish being one of the most pious). In agreement with Greeley's findings, Harold Abramson in his research, which dealt with a comparison of third and later generations of Italian and Irish American Catholics of the 1960s (as well as other Catholics), indicated that the low attendance of weekly religious services by the Italian American Catholics in comparison to Irish American Catholics was an outcome of enduring cultural values (Abramson, 1973: 111-115).

Rudolph J. Vecoli, stated that the Catholic Church did not provide the community building for the Italians as it did for the Irish and that the Italian immigrants found an American Catholicism hostile to their traditions (including the veneration of their saints and celebrating the *feste*) and to them. He also indicated that the Italian immigrants at the turn of the 20th century brought with them a mistrust of the Catholic Church as result of a neglectful and abusive clergy in southern Italy. Vecoli, in his discussion of the attitudes and behavior among the different Catholic ethnic groups in the 1970s, suggested that the Italian cultural trait (mistrust of the clergy) persists beneath an apparently homogenous surface, and noted that the Italian American Catholics were less regular in church attendance than the Irish American Catholics (Vecoli, 1978: 132-34).

Salvatore Primeggia (1999) points out that many Italian immigrants transplanted folk-religious beliefs drawn from the southern Italian experience to the United States. He argues that many Italian American Catholics today continue to keep alive many of the devotional practices involving saint and Madonna worship put forth by their immigrant forbears.

A study from the Order Sons of Italy in America (2002) reveals that Italian American communities across the United States are keeping alive the tradition of ceremonies honoring a favorite saint brought to America by early Italian immigrants at the turn of the last century. The study lists over 300 Italian religious festivals in 23 states and the District of Columbia that are held between March and November annually.

Several of the social scientists mentioned above state that there is an association between past cultural traits and today's Italian Americans' behavior toward religion. They suggest that the lack of religious piety among Italian Americans was an outcome of past cultural traits. These traits, as it has been stated previously, involved a mistrust of the Church's authority, church attendance not emphasized, the belief that religion is an activity (churchgoing) for women, a Catholicism that is not considered to be very strong, veneration for their saints and the celebration of *feste*. However, there also exist studies showing that since World War II Italian American Catholics tend to become similar to the Irish American Catholic in attitudes and practices.

Gerhard Lenski in his 1961 study stated that among white Catholics, the third generation was more active in weekly church attendance than the second and first generations, and suggested that an increase in religious participation by Catholics is associated with an increased pattern of Americanization (Lenski, 1961: 41-45). Nicholas J. Russo's findings of Italian American Catholics in the 1960s provided support for Lenski's study. Russo's work indicated that the longer Italian American Catholics have remained in America (especially third-generation) the more they tended to resemble Irish American Catholics (Russo, 1970: 201-09). Similarly, James Crispino (1980: 142), in his research of the Italian Americans in Bridgeport, Connecticut during the 1970s also supported Lenski's research of increased religiosity by second and third generation ethnic groups.

Continuing on this theme of Italian Americans and Irish Americans resembling each other, Gesualdi (1997b) indicates that there was no difference in trust levels of the clergy between Irish Americans and Italian Americans. His study showed that approximately 60% of Irish American Catholics and 60% Italian American Catholics have a lack of confidence concerning the clergy.

Glazer and Moynihan (1971: 204) observed that the Irish and Italian Americans of the 1950s and 1960s who were in constant conflict with each other in the city, worked together in the development of the Catholic Church in the suburbs and that their separate ethnic identities had practically disappeared and were being united in an American Catholic identity. Joseph LoPreato stated that religious participation is often a symbol of "social respectability" in middle class America, and the Italian Americans' participation in the Catholic Church of the suburbs signified their new middle class status (LoPreato, 1970: 92).

Nicholas J. Russo noted that since World War II, the descendants of the Italian immigrants have left the churches in working class neighborhoods and have moved to the suburbs. These Italian Americans have been influenced by many Irish American priests and teaching nuns and have become like the Irish (Russo, 1970: 198). He shows that the Italian cultural trait of devotion to the saints, especially when it comes to participation in a feast day, lighting candles before saints' statues, and having Mass said in gratitude for favors received from patron saints, does not persist across generations of Italian American Catholics.

Rocco Caporale found Italian American professionals of the late 1970s active in religious participation (church attendance). He noted that the general attitude of the Italian American professional appeared to be an acceptance of things religious, as a component of the American way of life (Caporale, 1983: 281). Joseph A. Varacalli (1992) states that Italian Americans have now become thoroughly americanized-and thus secularized. Italian folk customs (reverence to saints, including feasts celebration in honor of saints) have lost much of their appeal, especially for upper-middle class Italian Americans.

After reviewing the literature by Lenski, Glazer and Moynihan, Russo, LoPreato, Crispino, Caporale and Varacalli, one can argue that today's Italian American Catholics are becoming more like the Irish American Catholics in religious attitudes and practices (especially in church attendance).

Conclusions

Overall, after examining some of the literature on Irish and Italian American Catholics, one can argue that there exist two groups of Italian Americans. These two groups are different as an outcome of different socioeconomic conditions and their impact on the maintenance and non-maintenance of southern Italian cultural traits.

One group of Italian American Catholics has the following characteristics: 1) either a blue collar worker in general or a white collar worker with a high school education or below; 2) churchgoing is perceived as an activity for women; 3) they tend not to consider themselves to be very strong Catholics; 4) they have a veneration for their saints and celebrate the *feste*; and 5) they have a mistrust of the clergy. The other group can be categorized in the following ways: 1) white collar and college educated; 2) weekly church attendance is high among males and females; 3) a high percentage consider themselves to be very strong Catholics; 4) they do not have a veneration for their saints or celebrate the *feste*; and 5) they still have a mistrust of the clergy. Two groups, in a sense, have emerged from this ethnic group (that is major differences exist between middle class Italian American and working/lower-middle class Italian Americans). The working/lower middle-class Italian Americans seem to maintain southern Italian cultural traits while middle class Italian Americans begin to resemble middle class Irish Americans (for that matter, probably middle class Americans in general).

It can be argued that the higher level of educational and occupational achievements, the more tenuous the connection Italian Americans have with southern Italian cultural traits. This is not surprising considering southern Italian culture developed as a response to a poverty situation (Silverman, 1968) not a middle class one. Educational and occupational mobility can be seen as a major cause of assimilation. As second, third and fourth generation Italian Americans move into the white collar professional middle class, major cultural traits seem to disappear.

The continuation or non-continuation of Italian Americans' participation in higher education will have an effect on the Italian Americans' maintenance of cultural traits. For instance if the American economy suffers and there is a slowing down of Italian Americans attending college, a substantial number of Italian Americans will remain and be working and lower middle class, and more than likely remain separate from mainstream America (that is living Italian American working class neighborhoods in the suburbs and/or cities) and thus to some degree will maintain many of the cultural traits discussed throughout this study. On the other hand, if an increasing number of Italian Americans become white collar workers and college educated, they will be very similar in behavior to Irish American (and other Americans for that matter) who are under the same socioeconomic conditions and then most of the Southern Italian cultural traits would probably disappear.

LOUIS GESUALDI

References

Abramson, Harold J. 1973. *Ethnic Diversity in Catholic America* (New York: John Wiley and Sons).

Alba, Richard. 1985. *Italian Americans Into the Twilight of Ethnicity* (Englewood Cliffs, New Jersey: Prentice-Hall).

Banfield, Edward, C. 1958. *The Moral Basis of a Backward Society* (New York: The Free Press).

Browne, Henry, J. 1946. "The Italian Problem and the Catholic Church of the United States, 880-1900," *United States Catholic Historical Records and Studies*, 35: 46-72.

Caporale, Rocco. 1983. "The Value System of Southern Italian and Italian American Professionals: A C omparative P rofile" in R emegio U . P ane's (editor) *Italian-Americans in the Professions* (Staten Island: Center for Migration).

Crispino, James. 1980. *The Assimilation of Ethnic Groups: The Italian Case* (Staten Island: Center for Migration Studies).

Fallows, Marjorie, R. 1979. *Irish Americans* (Englewood Cliffs, NJ: Prentice-Hall, Inc.).

Gambino, Richard. 1974. *Blood of My Blood: the Dilemma of the Italian American* (New York: Doubleday and Company, Inc.).

Gans, Herbert. 1962. *The Urban Villagers: Group and Class in the Life of Italian Americans* (New York: Free Press).

Gesualdi, Louis. 1997a. *The Italian Immigrants of Connecticut, 1880-1940* (Connecticut: The Connecticut Academy of Arts and Sciences).

Gesualdi, Louis. 1997b. *The Religious Acculturation of the Italian American Catholics: Cultural and Socioeconomic Factors* (New York: The John D. Calandra Italian American Institute, City University of New York).

The Gianelli-Cardillo Report: Italian American Political and Cultural Failure in the United States. 2002. (New York: Italic Institute of America).

Glazer, Nathan and Moynihan, Daniel, P. 1971. *Beyond the Melting Pot* (Cambridge: The M.I.T. Press).

Greeley, Andrew, M. 1969. *Why Can't They Be Like Us?* (New York: Institute of Human Relations Press).

LaGumina, Salvatore. 1999. "Saints, Suburbs and Parish Life in Long Island's Italian American Communities" in Joseph A. Varacalli, Salvatore Primeggia, Salvatore J. LaGumina and Donald J. D'Elia's (editors) *The Saints in the Lives of Italian*

Americans: An Interdisciplinary Investigation (Stony Brook, New York: Forum Italicum).

Lenski, Gehard. 1961. *The Religious Factor* (Garden City: Doubleday and Co.).

LoPreato, Joseph. 1970. *Italian Americans* (New York: Random House).

Nelli, Humbert, S. 1967. "Italian in Urban America: A Study of Ethnic Adjustment" *The International Migration Review*, 1: 38-55.

Order Sons of Italy in America. 2002. *Italian Festival Directory*.

Orsi, Robert. 1985. *The Madonna on 115th Street* (New Haven, CT: Yale University Press).

Palmieri, Aurelio. 1921. *Il grave problema religioso italiano negli Stati Uniti* (Florence: Libreria Editrice Fiorentina).

Pannuzio, Constance M. 1921. *The Soul of an Immigrant* (New York: Macmillan).

Primeggia, Salvatore. 1999. "The Social contexts of Religious Devotion: How Saint Worship Expresses Popular Religiousity" in Joseph A. Varacalli, Salvatore Primeggia, Salvatore J. LaGumina, and Donald D'Elia's (editors) *The Saints in the Lives of Italian Americans: An Interdisciplinary Investigation* (Stony Brook, NY: Forum Italicum).

Russo, Nicholas J. 1977. "Three Generations of Italians in New York City: Their Religious Acculturation" in Silvano M. Tomasi and Madeline H. Engel (editors) *The Italian Experience in the United States* (New York: Center for Migration Studies).

Sartorio, Enrico. 1918. *Social and Religious Life of Italians in America* (Boston, Massachusetts: Christopher Publishing House).

Shannon, William, V. 1963. *The American Irish* (New York: Macmillan).

Tomasi, Silvano, M. 1975. *Piety and Power* (Staten Island: Center for Migration Studies).

Varacalli, Joseph. 1992. "Italian American-Catholic: How Compatible?," *Social Justice Review* 83: 82-5.

Vecoli, Rudolph J. 1978. "The Coming Age of the Italian Americans: 1945-1974," *Ethnicity* 5: 119-147.

_____. 1969. "Prelates and Peasants, Italian Immigrants and the Catholic Church," *Journal of Social History* 2: 217-268.

_____. 1964. "Contadini in Chicago: A Critique of the Uprooted," *The Journal of American History*, 51: 404- 417.

Chapter 3

✦

ITALIAN AMERICAN AND IRISH AMERICAN CATHOLIC INTERACTION

This paper is designed as an effort to re-examine the phenomenon of Irish American and Italian American Catholic interaction by focusing on that part of the historical record that sheds light on positive aspects of the encounter. It should be stressed that this presentation is not proffered as definitive but rather as a very preliminary inquiry. It is hoped furthermore, that this might encourage more research on the issue of antagonism and partnership between the two major Catholic ethnic sub-groups. The paper is divided into three main sections, to wit: Introduction and acknowledgment of the prevailing interpretation, an honest attempt to highlight the bright side of the encounter, and a conclusion.

Acknowledgment of tension in historical record

There seems little debate among scholars who have studied this terrain that the engagement of Irish and Italian-descended people in the United States was marked by tension, especially in the early period. Most researchers have judged that Irish American leadership, which was predominant in the American Catholic Church at the time of mass Italian immigration to this country, regarded newcomer Italians as weak Catholics who evinced a form of religiosity lacking in an understanding of fundamental theological truths and replete with superstition. Infrequent attendance at Sunday Mass, miserly financial support, disrespect for clergy, abysmal ignorance of basic Catholic dogma, and exaggerated saint devotion that approached worship, were some of the criticisms leveled against the Italians (Di Giovanni, 1994: 20-3). In the wake of Italian political developments attributed to that nation's unification movement that alienated many Italians from the Papacy, it appeared that transplanted anti-clericalism was even more pronounced. To Irish American Catholic leaders this background could only exacerbate the problem of American acceptance of Catholicism. Having struggled for years against nativists' charges that Catholics were not truly loyal to America, the overwhelming view of the hierarchy was that newcomer Catholics would do well to shed their distinctive nationality traits and rapidly become Americans, that is to acculturate and assimilate a Hibernian type of Catholicism.

The debate that was shaping up within American Catholicism over how to assimilate assorted newcomers became a dispute between two opposing views. The Cahenslyists were the proponents of the maintenance of culture, language and traditions of incoming groups, arguing they should be respected and accommodated even including appointment of separate nationality members of hierarchy to represent individual groups. Americanists, on the other hand, together with the majority of the American Catholic hierarchy, determined that expeditious assimilation was the most desirable policy both for church unity and to validate their loyalty to the land of their adoption. Americanists, furthermore, were in a position to, and utilized their considerable authority to, promote their view.

An Attempt to Highlight the Bright Side

Notwithstanding the summary of the prevailing view just recited, perhaps it is time to re-examine *positive* aspects of experiences to ascertain if there has been an overstating of the negative case, and to determine whether there were important nourishing and bolstering encounters.

From 1885 to 1902 Michael Augustine Corrigan, was Archbishop of New York during a period that saw such a huge influx of Italian immigrants that they were becoming the largest non-English speaking Catholic group. From his contact with priests and laymen who interacted with Italian immigrants, Corrigan concluded that the newcomers possessed a woefully inadequate conception of their Catholic faith and that they could easily be converted to Protestant religions or apostasy. Stephen M. DiGiovanni's biography of Archbishop Corrigan is rather revealing when he cites instances wherein Corrigan held negative views about Italian immigrants, however, he also credits him with sincerity in attempting to help them. Thus while Corrigan responded to calls for Italian clergy to minister to Italian immigrants, he also desired more than warm bodies. He urged the appointment of dedicated priests who were interested in more than a cozy sinecure. "In New York, however, the problem was not simply a lack of Italian priests... In great part the problem was the lack of 'good, zealous and pious priests ready to save souls,' who had not come to America for 'temporal gains'" (DiGiovanni, 82-3). In his effort to provide Italian immigrants with priests familiar with their language, Corrigan demonstrated unusual resourcefulness even exploring the desirability of assigning American born seminarians to years of study in Rome before ordination as a possibility of deepening their understanding of the Italian mentality (DiGiovanni, 85).

Locally the prevailing view was that unless Italian Americans divested themselves of their ethnic characteristics, they would not become assimilated. Corrigan's more favorable view, however, was that they would inevitably Americanize and that in the meantime Italian-speaking clergy be enlisted to serve them. Accordingly this led to the creation of "national" parishes frequently staffed by Italian religious orders. Archbishop Corrigan also "solicited religious orders to send Italian-speaking clergy and sisters, and gave from his own funds to support them during the difficult early years. For example, in 1884 he enlisted priests from the Pious Society of the Missions (Pallotines) for Our Lady of Mount Carmel Parish in East Harlem" (DiGiovanni, 118). The religious mission to Italian immigrants on the part of Archbishop Corrigan was a huge challenge, one that would tax the patience of the most forbearing. Given the failure of the Italian immigrant populace to financially support their own parishes, it is not surprising therefore that Corrigan vacillated between the notion of individual Italian parishes and traditional ones (DiGiovanni, 139-143). DiGiovanni concludes that

> Corrigan was more than willing and generous to provide the traditional Catholic structures and institutions in abundance to assist the Italians. But when faced with the vast numbers of Catholics who cared little for the Church, and who desired no assistance from any traditional Catholic institutions or agencies, Corrigan was left helpless and frustrated. (298)

When Corrigan heard about the Scalabrinians' founding, he sent congratulations, a donation, and requests for two missionaries (Brown, 1992: 6). Scalabrini's biographers readily acknowledge the support Corrigan gave to the Italian missionary leader.

> Archbishop Corrigan is an invaluable collaborator of our bishop. The American bishop is an alumnus of Rome where he has nurtured himself

on the love of the Catholic Faith and of Italy. Today this American bishop loves the Italians in his own land, helps them, and welcomes the Italian priests who go to America. He opened schools for our people and is offering great help to the missionaries of Bishop Scalabrini. (Felici, 1955: 186)

National Parish: Our Lady of Pompei in downtown Manhattan is a case in point of a national parish. It began as chapel for Italians under the hard-working and dedicated Father Pietro Bandini who wanted to expand the house of worship into a full-fledged parish church but faced an intimidating debt. Although at first skeptical that this could be a successful parish, Archbishop Corrigan acquiesced to the requests of Fr. Bandini, even making gifts from his personal funds to satisfy the creditors that let Fr. Bandini settle on the church building that became Our Lady of Pompeii. Before Archbishop Corrigan's death in 1902 several more Italian national parishes had been established.

Edward Dunne, Bishop of Peoria, Illinois, was one of the few members of the Irish hierarchy to demonstrate a genuine defense of Italian immigrant practices. His empathy toward the newcomers was evident while a priest in the Chicago diocese even before he became a bishop. In 1899, he was instrumental in founding Guardian Angel Church for Italians on Chicago's the West Side where he served as pastor for five years (Nelli, 1985: 190). As Peoria bishop in 1914 he engaged in a debate reported in the papers of the respected Jesuit magazine *America*, on the "Italian Problem" that found many clergymen highly critical of the weakness of their religious practice. Dunne claimed to speak from his experience as rector of the largest Italian parish in the country where he witnessed the infectious piety of laymen and women on special feast days that were not days of obligation. He also commented on their adherence to the sacrament of penance. "As to approaching the Sacraments, I have been kept busy hearing confessions of Italian men until after two o'clock Holy Thursday morning" (Quinn, 1999: 101).

Archbishop J.E. Quigley. Buffalo Scalabrini priest Fr. Giacomo Gambera identified Bishop Quigley as unusually sympathetic to Italian immigrants. Gambera cited Quigley's encouragement of Italian parishioners to assemble in the lower church of the cathedral even "before the Scalabrinians began their apostolate there, and who eventually approved wholeheartedly the founding of our church of St. Anthony which was opened by Father Gibelli, who was a companion of mine when we first left for the missions" (Gambera, 1994: 129). Bishop Quigley followed his ascension to archbishop of Chicago by establishing several more Italian parishes.

Concurrence with a positive view of Quigley comes from Bishop John Scalabrini, apostle to immigrants who wrote of Bishop Quigley, "The Bishop is a modest, cordial man, who is truly apostolic and admirable. If they were all like him, full of respect and affectionate attitude for the Italians, our colonies would soon advance and would soon acquire a very important status" (Caliaro and Francesconi, 1974: 208). Bishop Scalabrini also expressed his gratefulness to a group of Irish women of Boston "who cared for young Italian girls who were warmly received by the bishop, who was highly impressed by their spiritual goodness and their religious fervor (Felici, 192).

Still another instance of members of the Irish hierarchy who were well disposed to Italians is the case of Philadelphia's Fr. Philip R. McDevitt, later the Bishop of Harrisburg. In response to typical nativist denigration of Italians McDevitt explained "the plain truth is that the immigrant often has the qualities which the native American lacks and when the scales are balanced it is wholly probable that the foreigner gives to America as much as America gives to him."[1]

Annie Leary. A unique instance of Irish help for Italians is that rendered by Annie Leary, wealthy Irish woman and patron of Catholic causes. Her munificence to Mother Francesca Cabrini in establishing an "industrial school" in Manhattan was warmly acknowledged. "Now we have an industrial school in New York in the midst of the Italians. There are almost 20 girls who learn every type of work and can keep for themselves what they have made. The expenses are paid by Miss Leary" (Sullivan, 1992:176). Mother Cabrini and her Missionaries were also invited by Brooklyn's Bishop Charles E. McDonnell to establish an institution in that diocese. "McDonnell was moved to compassion when he saw in his new diocese an Italian colony deprived of Catholic religious and educational assistance" (Sullivan, 195). Also to be noted is very substantial welcome to Mother Cabrini and her sisters by Los Angeles Bishop Conaty (Sullivan, 234-235).

The remarkable Annie Leary was of immense help to the newly arrived Italian immigrants and the Scalabrini fathers who were trying to establish the Italian parish of Our Lady of Pompeii, in lower Manhattan. Originally a chapel connected with the Saint Raphael Society (Società San Raffaele) to aid Catholic Italian immigrants, the religious place of worship was experiencing such great financial difficulty that it was on the verge of closing. Help came from Ms. Leary, who in addition to her own generosity, also tried to assemble an auxiliary organization of women to aid the newcomers. Although the latter was of limited success, the chapel was able to pay its rent and continue its mission (DiGiovanni, 158). Her continued generosity from her own personal funds and fund-raising concerts, led her to assume all of Our Lady of Pompeii's debts thereby enabling the Scalabrinians to operate the church (DiGiovanni, 162).

Humphrey Desmond. Over and beyond views and policies of hierarchy it is useful to look at attitudes of some laymen who were in positions to influence relationships between the Irish and the Italians. Humphrey J. Desmond (1858-1932), lawyer, legislator, author and journalist, for example, was an Irish Catholic progressive from Milwaukee, Wisconsin, who became a highly successful lawyer and an active Democrat who served in the Wisconsin state legislature where he played a major role in repealing an anti-Catholic piece of legislation.

For a time he served as the editorial writer for *The Catholic Citizen*, the leading Catholic weekly newspaper in Wisconsin. In 1891 he bought the newspaper, then an independent organ not under the church hierarchy, thus allowing him to continue engaging in journalism in a serious vein for the next twenty years. Considered a liberal reformer who favored assimilation, he nevertheless ranged far and wide in the thousands of editorials he wrote. Among the topics that he addressed frequently was the influx of that era's "new" immigrants, particularly Italians, the most numerous of the entrants whose emergence in Milwaukee would lead to a significant enclave (Stibili, 1999: 181-2).

A committed Catholic, Desmond's concern was the religious welfare of millions of Catholic Italian immigrants who, seemingly indifferent to religious tenets and unaccompanied by sufficient Italian-speaking clergy, were vulnerable to Protestant proselytism. His repeated denunciations of the neglect and calls for rectification of "the biggest Catholic question" were addressed to the Italian hierarchy and other church officials. They also elicited reaction from the secular press, specifically *The Literary Digest* on

October 11, 1913. Although Desmond was, as Ed Stibili wrote, "an ardent Americanizer, he wanted to see the "new" immigrants, including the Italians, remain active within Catholicism and eventually adapt to the American culture. Desmond's concern for the "neglected" Italians helped stimulate a more aggressive response on the part of church authorities to the "Italian Problem."[2]

Long Island: Perhaps it is on the micro individual parish level where one can derive more meaningful insights into Irish-Italian religious interaction. Because of my extensive research in this region, I will now focus on the Long Island Italian American experience, to examine some revealing aspects of this connection. Oyster Bay, Long Island constitutes one of the surprising and approving examples. Italian immigrants began to immigrate to Oyster Bay in late nineteenth and early twentieth centuries following a systematic if informal pattern of settlement from Benevento and Paduli, towns in Naples vicinity, to their new homes on Long Island. So substantial was the influx that a St. Rocco Society was formed in 1910 in St. Dominic Catholic Church, then under the leadership of Monsignor Canivan and continues to remain active in Oyster Bay as does an Italian American civic club created in 1911. An Irish American pastor, Canivan nevertheless was so caring and so intent on accommodating his new Italian parishioners that he made extraordinary efforts to become well versed in Italian language and customs (LaGumina, 1988: 12). As a further demonstration of genuine pastoral concern for his Italian parishioners Monsignor Canivan undertook a few trips to the Benevento region of Italy to study first hand their background — an obviously rare gesture and one in sharp contrast to the Irish American pastor of St. Patrick's Church who apparently alienated Italians in nearby Glen Cove.

Beginning in 1911 St. Rocco's feast, sponsored by Oyster Bay's St. Dominic Church, attracted Glen Cove's Italian immigrants with band concerts, parades, bicycle races, grand fireworks displays and High Mass celebrated by an Italian priest (LaGumina, 92). Furthermore, the local feast made a deep impression on the rather reserved Oyster Bay natives who marveled at the immigrants' ability to conduct an affair of this magnitude and that were surpassed yearly. An editorial in the *Oyster Bay Guardian* of 1911 is instructive.

It takes the Italians to set the pace for the ordinary Oyster Bay Americans. For two days and nights Italians have given the residents of Oyster Bay some food for thought. Oyster Bay is a quaint, old-fashioned country town; its populace has been content with their lot; they do not believe in giving much toward making a show — their forte being in making a show at the bank. Not so with the Italian populace. These people are patriotic. They believe in their patron saint, St. Rocco, and are willing to place their earnings in support of their ideas, and as a consequence they have done things which should bring the blush of shame to the American populace, because of their liberality. "What is a dollar in comparison to the support of what we think if for our own benefit" has been their motto.

In 1937, Glen Cove's Italian Americans succeeded in creating their own national parish. With no support from the Irish pastor of Glen Cove's St Patrick's Church — indeed he actively opposed it, nevertheless, Italian Americans were overjoyed at their new Church of St. Rocco where their heritage would be reflected in liturgy, customs, sermons, etc. Once more Monsignor Canivan of St. Dominick proved supportive, even assigning Fr. Dante Fiorentino, his assistant pastor, to be present to greet the new pastor at St. Rocco.

Perhaps the best Long Island example of Irish support and assistance to Italian Catholics is the Parish of Our Lady of Good Counsel in Inwood. In 1900 this hamlet, tucked away in the southwestern corner of Nassau County, was home to 204 people, more Italian immigrants than any other community on Long Island. Southern Italians were attracted there by job opportunities in nearby estates and by employment in the developing recreational facilities in Far Rockaway that was emerging as a favorite resort for New York City residents. There was an absence of a Catholic church in the hamlet when the first Italians arrived. However when its Italian population increased to 762 in 1910, it became evident that a Catholic parish would be needed. Acknowledging the growth, Fr. Herbert F. Farrell, pastor of Star of the Sea in Far Rockaway, secured a store in Inwood as a Sunday school for instruction of the town's Italian children. Also available on Saturday mornings were sewing classes for the mothers — thus the humble beginnings of Our Lady of Good Counsel. Never meant to be exclusively for one ethnic group since there was a considerable number of non-Italians, the parish was popularly regarded an "Italian parish," albeit a defacto one.

It is important to note that the mother church and sponsor of Our Lady of Good Counsel, namely St. Mary Star-of-the-Sea, was a predominantly Irish American parish but was remarkably unstinting in its efforts to enable the new, largely Italian parish to get off to a good start. The contrast of ethnic names joining the common effort is most revealing. Thus St. Mary Star-of-the-Sea contributors for the tabernacle, vestments, and other furnishings in the Inwood parish included names such as Healy, Morris, Brennan, Shanley, Dowling, Castle, Desmond, Cunningham, and O'Rourke, while Our Lady of Good Counsel parishioners who provided gifts possessed names like Provenzano, D'Agostino, Capobianco, D'Elia, DiCroce, and an O'Rourke (LaGumina, 130).

In 1909 Fr. Farrell purchased six lots of land at an intersection in Inwood's Italian enclave to erect the Our Lady of Good Counsel Chapel (La Capella della Madonna del Buon Consiglio). The utilization of lumber from the original St. Mary Star-of-the-Sea, that was superceded by a new building, rendered the connection with Far Rockaway's Irish church even more striking. While under the pastorate of Irish-descended Fr. John Mahon, the congregation of chapel grew so rapidly that it overtaxed the small chapel's facility necessitating the erection of a new church building in 1914. That the establishment of Our Lady of Good Counsel was regarded as an extraordinary benefit to the community and especially to Italian Americans, was effectively recorded by a keen observer.

> The establishment of Our Lady of Good Counsel has been a benediction to the neighborhood none would deny.... To accomplish results among any class of people, a clergyman must be very near to his flock, must come to actually touch with them daily, his administration must be part of their everyday life, and this particularly so with regard to the Italian people, who are accustomed to have the offices of religion in their very doors.... The easterly section of Inwood was almost left to itself, bereft of those benign and sympathetic influences which gave a transplanted population a foot to stand on. The result was that the place had attained a distressing notoriety for all kinds of rascality and crime....Once, however, the church organization was perfected, the place took on a decided improvement....And there are few who will deny that such assimilation is the result of bringing divine influence into the hearts of the people (Bellot, 1918: 5-6).

Although this review of the more positive aspects of the Irish-Italian religious experience does not ignore the negative preponderance of the interaction, it is hoped that there will be a greater acceptance of a record of meaningful support and assistance that was also part of the historical chronicle. Notwithstanding a background of tension and misunderstanding between the nationalities, these instances cited above of genuine Christian outreach and comfort point to the real possibility of many other similar kinds of examples. This was, after all, only a preliminary study, one that was far from exhaustive, yet even in its brevity it indicates an area of research that is promising.

It is sobering to realize that some of the early commentators of Irish-Italian experience emphasized an assuring commonality that would mark their association. In 1905 for instance, H. Lord described relations between Irish and Italian Catholics as being moderately mild. That is, although there was some contention when Italian immigrants first moved into Irish neighborhoods, hostility was of short duration.

> There is less clashing between the two nationalities than might be expected. This is largely attributable, probably, to the essential good nature of both. The common religion is also a bond of union, and Italians are usually attracted to Irish-American churches and parish schools while they are too few or too poor to establish their own. The influences of a Catholic church organization are steadfastly bent against racial antagonisms, and for the promotion of Christian fellowship of its followers. Its chief directors and many of its priests of all nationalities have been trained in Italian seminaries or have visited Italy more or less frequently, and all look to Rome as the prime seat of their church. (Lord, Trenor, and Barrows, 1970: 69)

One also should consider the fascinating phenomenon of Irish Americans attracted to Italian American religious orders as an antidote to an image of unease. From the outset, Mother Cabrini's missionary order captivated Irish Americans in the work of aiding Italian immigrants. Among Cabrini's earliest recruits to her sisterhood in the United States were many young women of Irish background. "These included the first two postulants in New York — Elizabeth Garvey and Loretta Desmond — who accompanied Mother Cabrini back to Italy in July, 1889 for their novitiate" (Sullivan, 1992: 103). Cabrini's biographer cites other Irish names such as

McDermott, Sullivan, Minihan, Whalen, Rice, Donovan, that joined Mother Cabrini's religious order early on. Indeed by the time of Mother Cabrini's death in 1917, it was speculated that most of the 25% of the non-Italian Missionary Sisters were Irish.

Intermarriage between individuals of Irish and Italian background seem to offer another promising area of investigation. Although primarily based on anecdotal observation, there is some impression that substantial intermarriage exists between the two. A study of intermarriage between Irish-Americans and Italian Americans in Middletown, Connecticut in the first half of the twentieth century seems to support a close association. If this impression is borne out by research, then it constitutes another affirming element to confirm the validity of the H. Lord comment.

In conclusion my thesis is a simple one-that an examination of the interaction of Catholic Italian Americans and Irish Americans reveals that notwithstanding the tension and stress that characterized the encounter, there were a number of positive points of conjunction. It is perhaps time to acknowledge and accent the positive while not eliminating the negative; or at least make an effort to balance the exemplary and worthy aspects of the religious encounter.

SALVATORE J. LAGUMINA

---

[1] Philip R. McDevitt, Review of *Our Philadelphia* by Elizabeth Robins Pennell, *Records of the American Catholic Historical Society*, Vol. 27, pp. 181-190, 1916, quoted by Richard N. Juliani, "The Interaction of Irish and Italians From Conflict to Integration," *Italians and Irish in America*, American Italian Historical Association, 1985, pp. 27-34.

[2] Salvatore J. LaGumina, et. al., *The Italian American Experience: An Encyclopedia* p. 182. See also Steven M. Avella, Desmond, Humphrey J. (1858-1932), *The Encyclopedia of the Irish in America*, (ed. Michael Glazier, 1999, p. 210). Although not concerned explicitly with the religious dimension, it is instructive to note that some Irish American laymen such as Philadelphia humorist, poet, and journalist Thomas Augustine Daly became voices for Italian immigrants. In contradistinction to the prevailing mindset, Daly offered the view that Italian Americans were a humane and interesting people. See Gaetano Cipolla "Thomas Augustine Daly: An Early Voice of Italian Immigrants, *Italian Americana*, Vol. Six, No. 1 (Fall/Winter 1980), pp. 45-49.

References

Bellot, Alfred H. 1918. *A History of the Rockaways* (Far Rockaway: no publisher).

Brown, Mary E. 1992. *From Italian Villagers to Greenwich Village, Our Lady of Pompeii 1892-1992* (Staten Island, NY: Center for Migration Studies).

Brown, Mary E. ed., 1994. *A Migrant Missionary Story, The Autobiography of Giacomo Gambera* (Staten Island, NY: Center for Migration Studies).

Caliaro, Marco and Mario Francesconi. 1974. *John Baptist Scalabrini, Apostle to the Immigrants* (Staten Island, NY: Center for Migration Studies).

DiGiovanni, Stephen M. 1994. *Archbishop Corrigan and the Italian Immigrants*, (Staten Island, NY: Center for Migration Studies).

Felici, Icilio. 1955. *Father to the Immigrants* (Staten Island, NY: Center for Migration Studies).

Glazier, Michael, ed. 1997. *The Encyclopedia of the Irish in America* (Notre Dame: University of Notre Dame Press).

LaGumina, Salvatore. 1988. *From Steerage To Suburb: Long Island Italians* (Staten Island, NY: Center for Migration Studies).

LaGumina, Salvatore et. al. eds. 2000. *The Italian American Experience: An Encyclopedia* (New York: Garland Publishers).

Lord, Eliot, John J.D. Trenor, Samuel J. Barrows. 1970. *The Italian in America* (Freeport, NY: Ayer Company Publishers).

Nelli, Humbert. 1985. *Italians in Chicago, 1830-1930: A Study in Ethnic Mobility* (Oxford: Oxford University Press).

Quinn, John F. 1999. "Saints Over Sacraments?: Italian and Italian American Spiritual Traditions," in Joseph A. Varacalli, Salvatore Primeggia, Salvatore J. LaGumina, Donald J. D'Elia's (editors), *The Saints in the Lives of Italian Americans* (Stony Brook, NY: Forum Italicum).

Stibili, Edward C. 2000. "Desmond Humphrey," in Salvatore LaGumina, et. al. eds., *The Italian American Experience: An Encyclopedia* (New York: Garland Publishers).

Sullivan, Mary Louise. 1992. *Mother Cabrini, Italian Immigrant of the Century* (Staten Island, NY: Center for Migration Studies).

Part II
Religious Models and Images

Chapter 4

✦

PROTESTANT IMAGERY OF THE ITALIAN CATHOLIC: THE ITALIAN PROTESTANTS IN ROCHESTER, NEW YORK

There were two major sources for Rochester, New York's Protestant community. Whether long-time Protestants or converts, however, they looked down upon their fellow Italians who were Catholic. They deemed them to be priest and superstition ridden, not ones to get along in the promised land of America. They knew that changes had to be made to fit into the Rochester setting. Even to be an accepted Catholic meant adapting to the ways of Catholics who were close to the ruling WASPS. Why not go a bit further and become Protestants or if a Waldensian why not join a mainline Protestant Church? Failure to do so meant that one would not get ahead and face obstacles that could not so easily be overcome. Some of the migrants moved into the Protestant church to attain their goals, feeling that Catholicism had neglected their material and status needs. Others had been Waldensians in Italy, followers of Peter Valdez, a twelfth century preacher, whose followers were condemned as heretics in the thirteenth century. Valdez translated the Bible into the vernacular and inspired wandering preachers, greatly influencing St. Francis of Assisi. The Waldensians also condemned the corruption and wealth of the Catholic Church, influencing Luther and other reformers. These two sources merged in Rochester's Protestant churches.

The first source is amply illustrated in this excerpt from an interview with Paul Kay, in recalling his boyhood:

> Mother happened to be in a group of people and they started a church. This is the Protestant Church... so we've been Protestant all our lives. And of course that, in a way, has set us apart from or kept us away from relatives to a certain extent, which have all been Catholics. So it wasn't a strong tradition in the family, going back to the Reformation or anything, but it seemed to have come from anti-clericalism, you know... they wanted to try the Protestant faith, so they started this church. And there was a Verna family who had, well maybe four brothers... they had

67

the appearance of Greeks, you know. The classical Greek type of a face.
But in their way they were quite smart and so they set up this church.
There were three brothers who were involved here and one brother in-
volved in the same type of church in Pennsylvania. And so they left their
imprint on this city and all these families who were part of this church.
The Christian Apostolic Church, on Goodman Street, North Goodman
Street is where they're located.

On the West Side, there's a small church there. It didn't start there, it
finally moved there and that's where they're still worshipping, there's
still a group there. But we became affiliated and the word you know and
the church took on meaning. And we would go to revival meetings and
have revival meetings that was the type of a church, the Billy Graham or
the Evangelical Church, they called it.

All Italians. This was an Italian-American Protestant Church. And I believe at the time, I remember one other ... similar church, and there were
other churches. Maybe another one on the west side of the city, Italian-American also, and they just decided to make a go of it and they
were doing quite well to fit their own needs. But there was a lot of socializing between families... just having an interest in each other's families that type of thing. So later, my wife and I decided, or I decided to
go mainstream, so we joined the Baptist Church and at present we're
Presbyterians. And the Baptist Church, through our adolescent years, or
pre-adolescent, left quite a mark in that we were being taught other
things beside just the religion, you know. How to adjust to our society,
how to get along with people and different phases of life like learning
about marriage and sociological things in that respect.

It was like a mission for these churches, and people who were devoting
themselves to helping Americanize these kids who were coming up. And
so the ministers were excellent and they were quite highly trained people and they saw the needs and tried to meet the needs of all these young
people who were coming up, as far as helping them to get into the type
of society we were in. (Kay, 1992: 1)

Kay's subsequent career illustrates the value of Protestantism in his
life. Certainly, this excerpt highlights the manner in which he and his family used the evangelical church to become more familiar with American
ways. Even their trip to pick fruit provided an opportunity to become more
familiar with American practices. It is clear throughout this passage that the
groups who formed and joined the Italian-American churches in Rochester,

New York, were both proud of their ethnic identities and upwardly mobile. Kay continues.

> When I got out of college, I had problems getting a job because of the acceptance of the Italians wasn't very great then, so I knocked around. I wanted to stay in this area because I'd been out of town for so long, just to be near the family, so I tried very hard, all angles. I wanted to get a job, I studied chemistry at school. I wanted to get a job in my field, of course, and I didn't have very good success. I think I might have tried for one, maybe two, years. And finally I decided to go this route and the doors were opened pretty quickly. So I started my career and it went along pretty good after that. So I shortened the name. Capogravi. Well, I, my work career was at Bausch and Lomb Optical, and then from there in my twenties, might have been when I was twenty-seven or so, I switched over to Kodak and finished my career there. I worked all my life at Eastman Kodak. Retired when I was not quite sixty-five and since then I've been taking courses at Nazareth, continuing ed, studying Italian and Italian culture and following up on my music which I enjoy so much. (Kay, 1992: 2-3)

It is interesting that once he changed his name job offers came in. Places that had not hired him before his name change found him employable with no need for further qualifications. Kay also noted that having an Anglo surname aided in upward mobility within Kodak. At the time, Kodak was the key employer in Rochester and was known for its great pay scale and benefits. It is evocative, however, that once retired Kay found it necessary to reestablish his ethnic identity openly. He had never denied being Italian once he obtained a job for which he was qualified. He had never openly acknowledged it, either, as he moved from Italian- American Protestant churches into more "mainline" ones. He avoided both Catholicism and association with Italian-American evangelicism. Inwardly, however, he maintained his feeling of being Italian and returned to an open acknowledgment quickly when no prejudice or discrimination could any longer harm him.

The evangelical missions, in actuality, sought converts among people such as the Capogravi family who were disillusioned with Catholicism, and were religious, and upwardly mobile. In particular, they targeted strong women such as his mother who, in Kay's words, was the "bossman" in the family. Kay goes further and terms his family a matriarchy. As Seller (1978) notes, Protestant missionaries had a rather complex agenda. They sought to combat the growing influence of Catholicism in America, which the incursion of Italians represented. The slums in which most Italian lived personified social evils, which the missionaries abhorred. There was a symbolic connection in most missionaries' minds, moreover, between physical and moral cleansing. Seller (1978: 125), for example, cites the Presbyterian missionary, Mary Remington, who "would like to put... these miserable wretches into a bath tub and clean them up for once." In common with most Americans, and Italians for that matter, the missionaries sought to combat atheism and socialism. They erroneously believed that all Italian men were atheists and socialists and filled with uncontrollable violence. Finally, the evangelicals wanted to "Americanize" the immigrants.[1]

The Evangelical Missions

The first center of Italian settlement in the 1880s was called "Sleepy Hollow", the same area in which the Genoese immigrant Domenico Sturla settled. It was in a run-down district between St. Paul Street and the Genesee River. The next major settlement area, still near the river, was on Front and Mill Streets. Problems that marked later settlement were present in these early ones. For example, there was a language problem and *padroni* served as labor recruiters and go-betweens, with all the evils inherent in the system. When Italian immigrants sought to learn English in night school, they were frequently turned away because they did not know the language. The irony of the situation led to a group of prominent Rochester women founding "a mission to teach English, arithmetic, and 'Americaness' to Italian men" (Ray, 1976: 8).

As early as 1889, when the Italian population of Rochester, New York was less than 600, several prominent Rochester women had founded an Italian mission. *The Union Advertiser* notes:

> First and Second Readers Wanted — The Italian Mission has secured a room at No. 61 State St. for the purpose of establishing a school for the education of the Italians. Those in charge of the affair solicit contributions of school books. First and Second Readers are needed. Children who wish to contribute them will please leave them at the room on State Street. (March 21,1889: 3-4)

Evangelical missions sought converts among those Italians who were unhappy with the Catholic Church for whatever reason and those who were members of various groups, such as the Waldensians. They viewed Italian Catholics as more or less pagans. White, (1940: 15ff.), for example, lists a number of reasons why the Baptist Church had such difficulty in converting Italians in Rochester, New York. The religious life of Italians before they came to the United States was one of being only nominally Catholic. He adds that Italians were anti-clerical by nature, but hints that they were more anti-clerical toward Protestant clerics than Catholic ones. Moreover, the younger an Italian male the more anticlerical he was likely to be. Older Italian males were a bit more tolerant of clerics. He writes, "The religious conditions in the United States reflect those existing in Italy. No matter how greatly Italians are anticlerical, they are not entirely anti-Catholic." (White, 1940: 18-19).

Additionally, Italians are at a primitive stage of religious development. There are, he notes, "magicians living on the superstitious nature of their countrymen" (White: 19). White offers an example by giving a vivid description of a ritual to cure *mal'occhio* (the evil eye). He addresses the power of the Roman Catholic Church perceptively as well as its attitude toward Italians. "During the early years of Italian immigration not much attention was given to meet their religious needs and other wants. A very large number of Catholic priests were Irish. Many times they would upbraid the Italians, especially at funerals...." (White: 21).

Protestants, especially Baptists in Rochester, used former priests as clergy to minister to Italians. Not surprisingly, Italians who were already anti-clerical "by nature" were especially suspicious of Italian Protestant clergy who had been Catholic priests. The Baptists made the further culturally insensitive mistake of using women missionaries to work with Italian women. Men were jealous of their women in White's terms. They were wary of the modern influences that American women were spreading among their women.

Given the hard work and sincerity of the evangelicals, it is interesting to note that in their own words, their missions were failures. White (1940:-51) reports that "St. Mark's Italian Baptist Mission, Rochester, had the following numbers of members as of his writing baptisms: 10; total membership 85; benevolences $20.00; Sunday School — Teachers and officers

7, pupils 5; one Young People's Society with a total membership of 60."
The Reverend C. R. Simboli "Addresses and Letter Regarding the Italian
Work — the Outlook of the Italian Baptist Work in America" in *L'Aurora*
Feb. 9, 1924 states "If we are honest and frank, we must state that things
are not going so well with our Italian Baptist work in America. To be sure,
there appears here and there a mission that is slowly gaining in number in
the broad reaches of a desert waste" (205). He continues "The majority of
our missions and churches are stagnant and lifeless ... often stimulates the
Catholic Church to great exertion and activity..." The causes for the failure,
he argues, are the lack of proper equipment, old churches controlled by
dying American congregations, and, furthermore, Italians have no liberty
in these churches; their customs, he notes, are interfered with. Nevertheless,
he adds. "The Italians are children of warmth, color, and remarkable sensi-
tiveness" (Simboli: 205).

A file of St. Mark's Baptist Church offers additional insight into how
the Baptists sought to win Italian converts. (Congregational File. St. Mark
Baptist Church, Rochester, New York. American Baptist Samuel Colgate
Historical Library. Rochester, New York).

The Annual Report 1923 of the Home Mission Board records (Baptist
Union of Rochester and Monroe County, Secretary, Alfred E. Isaac) notes
" the Hebard St. Christian Center, among the Italians, The Hebard St.
House has been under the direction of Mr. and Mrs. James W. Herring. The
house has been open 7 days week with gym, sewing, and other classes,
Sunday school on Sunday afternoons. Since February, we have had the aid
of Miss Genevieve H. Pflaum. The building has a gym, shower baths, library, clubroom, etc. (sic.) 100s of boys and girls are found in the most
congested area of the city." Unfortunately, for the cause of evangelism, the
settlement houses provided for the hundreds of kids in Rochester's over-
crowded Italian areas without seeking to change their religions.

It is clear that evangelicals and Pentecostals identified all the traditional American ideals — democracy, individualism, honesty, industrious-
ness, thrift, sobriety, and patriotism — with Protestantism. Therefore, in
converting the foreigner they were serving the nation as well as God. This
motive was stressed almost to the exclusion of all others during and immediately after World War I, when the movement to evangelize the world was
viewed as an extension of Wilson's campaign to make the world safe for

democracy. "If America wants good citizens, let her convert the Italians and all other classes to Christ," suggested Reverend Antonio Mangano... Americanization of the household was critical, and to achieve this the Americanization of the mother and homemaker was essential. As one American evangelist stated, " the greatest problem is not the foreign child but the foreign mother" (Seller, 1978: 125).[2]

The Catholic Response

Evangelicals were not isolated in their recognition of the necessity to reach the mother in order to influence the family. They were, however, mistaken in believing that Italian men were atheistic or even anti-Catholic. Although working-class Italian men did not routinely attend Church, they clearly separated the Catholic Church from the priests whom they distrusted.[3] Evangelicals had company in misunderstanding the nature of southern Italian religiosity. Irish and German Catholics viewed southern Italian Catholicism as barely above the level of folk religion. Jerre Mangione has a wonderful observation in *Mount Allegro* about the difference between Irish and Italian Catholicism.

> Catholicism was so deeply ingrained in their bones that they could violate some of its strictest man-made rules without the slightest feeling of guilt. Unlike their Irish Catholic neighbors, they had almost no fear of God and felt as much at home with him as they did with each other. (Mangione 1981: 68)

This "comfortable" attitude toward God and religion tended to upset Catholics of other ethnic groups, for they viewed it as insolence and disrespect. The history of the accommodation of these viewpoints is, truly, the history of the mutual socialization of Italians to the Catholic Church in Rochester and of that church to the Italians. Simply, the contact of Italians and other ethnic groups within the context of the Catholic Church changed not only the groups but also the Church itself.

As McNamara (1968: 254) states "A certain number of Italians — many of whom left the mother country poorly instructed — accepted not only the philanthropy but the doctrines of Protestant missionaries, as the existence of several small Protestant Italian parishes still bears witness." Under Bishop B.P. McQuaid the Catholic Diocese of Rochester began to address the problems of Italian immigrants more directly. McQuaid, at first taken by surprise at the bitterness of Italian immigrants, quickly realized the necessity for a vigorous policy confronting their needs. Quite consciously, he decided on a strategy of Americanization of children through the schools.

The visit and consequent writings of Abbe Felix Klein supported McQuaid's formulation of a policy of Americanization. Klein came to Rochester in the early 1900s and wrote *Au pays de la vie intense* (*The Land of the Strenuous Life*). Klein loved the democratic vigor of America and sought to democratize the American church. His ideas influenced Bishop McQuaid. McQuaid had already made strides in the direction Klein advocated, and Klein's descriptions of St. Patrick's school where the Sisters of St. Joseph instructed mainly Italian kids who did not speak Italian as a first language. They put them into special classes to learn English and then quickly put them into regular classes. The hope was that these youngsters would teach their family English. It was clearly part of a conscious Americanization policy. McQuaid sought to incorporate Italian immigrants into the Rochester Catholic Church. That policy actively sought to assimilate as best it could Italian, generally southern Italian and Sicilian, religious customs, into the more "orthodox" and genteel practices of the Church in general. Doing so, he realized, was a delicate procedure and required people with empathy and wisdom. McQuaid assigned Father J. Emil Gefell the task of organizing Rochester's first Italian parish, one of a number of national parishes in the diocese. Gefell had received his theological training in Rome and learned to speak Italian fluently. More importantly, perhaps, he liked Italians and worked well with them. Gefell assembled potential Italian parishioners in the basement of St. Patrick's Cathedral. However, plans had been made to ensure that their stay in the basement was not to be for long. Bishop Bernard McQuaid had purchased a parcel of land on Lyell Avenue, Plymouth Avenue and White Street from the City of Rochester. Significantly, the city auctioned the land on the feast of St. Joseph, March 19, 1906. On April 25, McQuaid had transferred the land to the future parish of St. Anthony. The parish borrowed $10,000 to build a church, which was completed in October 1906. Significantly, the affiliated school had opened one month earlier.

On the far east side of the city on the border with the rural town of Irondequoit, Father George Weinmann was sent a few years later after World War I from Mount Carmel parish in the heart of the most crowed Italian ghetto to establish a mission church. He encouraged not only parish societies but also more general Italian religious societies to celebrate their feast days at the mission church of the Annunciation. Members of these societies, scattered throughout Rochester, were delighted to bring their statues to Norton Street to have their special devotions noted. Meanwhile, these regional societies were drawn into the more general Italian atmosphere. People became aware of their fellow Italians while sharing a common institution, the Church.

The purpose of bringing these outside groups to the Annunciation was to use the Catholic Church as a focal point for forging an *Italian* ethnicity, as opposed to numerous regional ethnicities. It also served to tie Italians to the Church as protector of their ethnic practices (Cf. LaRuffa 1988: 110ff. for a parallel situation in the Bronx). That association continued even after Bishop John F. O'Hern banned street processions after an unpleasant incident around 1930. The incident involved a row with Protestants who jeered the parishioners as they marched by. Even after they were confined to parish grounds and made to tone down their fireworks, people still came from various parishes to celebrate their patrons' feast days.

It is easy to understand why these festivals were toned down after reading McNamara's description:

> Before the 1930's these festivals were observed on a big scale. There was a solemn high Mass in the church, and an Italian priest was brought in to deliver the panegirico in honor of the holy person whose feast was being commemorated. Some of the members or their families brought big candles to the church in fulfillment of vows to their patrons. If the orator of the day was a good one, excited listeners might cry out "Evviva"; "Long live Saint So-and-so!" Then, at the moment of the consecration, somebody inside the church would signal a group outside, and the outsiders would set off a round of giant fireworks, while a band struck up the Italian national anthem. (1967: 255)

Each society had its own statue and carried it on members' shoulder while a band played it around the neighborhood. People pinned money to a sash placed around the statue. After the parade, there would be a party of some kind, and, perhaps, there would also be fireworks. The ritual obviously displayed aspects of the interpenetrability of the sacred and the secular, noted by Moore and Myerhoff (1977). It certainly highlighted symbolic proclamations of ethnic identity and a clear drawing of boundary markers.

Although Father Weinmann left the Annunciation in 1938 to concentrate his efforts on turning the St. Philip Neri mission chapel into a real parish, his efforts at the Annunciation bore fruit. He fostered the growth of a viable community that maintained pride in its ethnic background while entering the "American" mainstream. Italians who migrated to Rochester, New York, mainly from the *Mezzogiorno* entered an area had been originally settled by English and Germans. Irish Catholics soon followed. However, even German and Irish Catholics adhered strongly to WASP conformity and to a view of southern Italian religious practices that placed them within pagan ritual. The goal of the Catholic Church in Rochester, no less than civil society, was one of Americanization. Italians had to adapt, somehow, to this ecological niche to survive in Rochesterian society.

Conclusions

Rochester's Italian Catholics had to combat negative stereotypes from both Protestants and their fellow Catholics of other ethnic origins. Protestants, including Italian Protestants, saw them as superstitious people, under the domination of the Pope and his agents. Moreover, these Protestants wondered at a people who persisted in their adherence to the Catholic Church when it was obvious that the Catholic Church in Rochester did not really want that adherence. Italian boys who wanted to become priests were sent to the seminary in Buffalo, New York, rather than be housed with other Rochester boys. German Catholics made it clear that they wanted no part of Italian Catholics in their parishes. Protestants, on the other hand, offered a path to inclusion within American society, jobs, skills and other means to fit and Americanize quickly. In their minds, they offered a way out of superstition and toward enlightenment.

Within the ethnic competition of Rochester's Catholic Churches, Italians had to forge an ethnic identity in order to compete with German and Irish Catholics. In that context, they learned what it was to be Italians, an identity that held no meaning for them within Italy itself. They had to learn to cooperate with fellow Italians from different areas and forge some sort of panethnic identity. Differences among people from various paesi (villages) were acceptable only within the group. They were intolerable and counterproductive in the face of "American" attitudes toward the community, and those attitudes were generally negative.

There were attempts on the part of the Italian community to combat the popular negative image of Italians. The Rochester community, for example, turned to ethnic theater productions. On at least one occasion, an ethnic community used theatre in a deliberate attempt to improve its public image. In 1908 the Sicilians of Rochester, New York, hoped to combat Mafia gangster stereotypes by convincing the non- Italian community that they had "a great reverence and respect for holy things," staged an elaborate drama about the life of Jesus Christ. Author Jerre Mangione described the venture:

> Milkmen, shoemakers, bakers, tailors, factory hands, and ditchdiggers became the actors of the drama. After their day's work and on Sundays these men and women met for... months. The undertaking grew. Hundreds of other Italians in the city pitched in to make the sets and costumes and contributed toward the expenses of the production...
> October 12, the day celebrating Columbus' discovery of America, was the date chosen for the production, perhaps in the hope that on that day the Americans would "discover" the good qualities of the Italians in their midst. (Mangione, 1981: 68)

Although the Italian community came and applauded wildly, no "Americans" were in the audience. A repeat performance drew a large audience from both communities and some favorable comment in the American press, but the impact was short lived. Within a year the press was once again giving disproportionate attention to crime stories involving Italians (Maxine Schwartz Seller, 1983: 11).

The growth of Italian parishes provided a haven for the development of a positive Italian identity while aiding the ultimate assimilation of American values, as Schaeffer (2000: 143) notes in his discussion of the ethnic paradox. Thus, holding onto and exaggerating ethnicity is often a means for entering the mainstream of American life. For Rochester's Italians, Rochester differed from New York City and many other cities with large Italian immigrant populations in having houses, rather than tenements, available for rent. Very quickly, Italians in Rochester moved into these houses and purchased these as soon as possible. The possibility of owning a piece of land where one could grow a garden was irresistible (See Mangione 1981). Moreover, unlike Italians in Buffalo, their Rochester cousins lived in a city noted for light industry (See Yans-MacLaughlan 1977).

Poor Italians found it difficult to adapt to a wealthy city that did not welcome them. In order to do so, they turned to mutual aid societies, political organizations, and the Catholic and Evangelical churches. However imperfectly, they forged a master identity, Italian, to serve in public confrontations with those of other ethnic groups in the city. In so doing, they learned to get their share of the spoils in various competitions. Italians learned to present themselves in various ways, chameleon-like, in order to take advantage of a given situation. It was a survival technique that served them well in an environment so hostile to their basic sensibilities.

In a situation in which, in spite of the efforts of voluntary agencies such as the Italian Mission, landlords and employers joined forces to oppose housing reforms, Italians found it necessary to make choices among various values. Survival was of utmost importance. Employers benefitted from high rents since it made workers desperate for employment of any kind. Undeniably, although the overwhelming majority of southern Italian immigrants were Catholics, their Catholicism appeared quite alien to the Irish and German Roman Catholics who had preceded them. In fact, if it had not been for stiff competition from evangelical missionaries who provided an alternative religion and path to Americanization, the Rochester Catholic Church might have neglected these new immigrants. The evangelicals, however, did provide rivalry for the souls of the immigrants, and their very presence compelled the hierarchy to confront the problems and opportunities the "new immigration" presented.

FRANK A. SALAMONE

---

[1] The image of the "dirty dago" was one Rochester Italians had to fight endlessly.

[2] More sympathetic social workers, such as Florence Manning of Rochester's remarkably successful Lewis Street Center, did in fact provide a bathtub and then showers for Italian immigrants and their children. They, however, knew that the immi-

grants would be delighted to use these resources, lacking in their homes, once they got used to them. They were, of course, correct.

[3] Not all Italian American Protestants in Rochester were disillusioned immigrants. The Lanni family, a very prominent Italian American family, had a long history of protest against the papacy, stretching back to the twelfth century. The Lanni claim to have been Protestants from the Reformation (Lanni Papers Box D 50, University of Rochester). Being Protestants and imbued with the Protestant Ethic, however, did not insulate them from ethnic slurs. Clement Lanni (1931) mentions that he had gone to a Fourth of July parade to see his father march with his fellow Italians. The Italian band was dressed in colorful uniforms. As they approached, someone in the crowd yelled out that the "dagos" were coming. Lanni slipped back into the crowd.

References

Annals of the Sisters of St. Joseph of the Diocese of Rochester.

Anonymous. 1959. *Golden Jubilee: Our Lady of Mt. Carmel Parish.* (Rochester, NY: Mt. Carmel Parish).

Anonymous. 1931. 25o *Anniversario della Chiesa di Sant'Antonio di Padova: 1906-1931.* (Rochester, NY: St. Anthony of Padua Parish).

Baker, James F. 1947. *A History of the Genesee Settlement House.* (Unpublished mss. in Diocese of Rochester, NY, Archives).

Barry, Harriet. *Charles House Book of Minutes.* (Unpublished mss. in Diocese of Rochester, New York, Archives).

Briggs John Walker. 1972. "Italians in Italy and America: A Study of Change Within Continuity for Immigrants to Three American Cities, 1890-1930." Unpublished Ph.D. Dissertation. University of Minnesota.

Cohen, Ronald. 1978. "Ethnicity: Problem and Focus in Anthropology," *Annual Review of Anthropology* 7: 379-403.

DiPietro, Joseph. 1988. *New Rochelle.* New Rochelle Public Library.

Juliani, Richard N. 1986. "The Parish as an Urban Institution: Italian Catholics in Philadelphia," *American Catholic Historical Society of Philadelphia* 96: 49-66.

Kay, Joseph. June 25, 1992. Nazareth College, Rochester, NY.

Keene, Ruth M. 1946 *Acculturation of the First and Second Generation Italians in Rochester, NY.* M.A. Thesis, University of Rochester.

LaRuffa, Anthony York L. 1988. *Monte Carmelo.* (New York: Gordon and Breach Science Publishers).

Mangione, Jerre. 1981. *Mount'Allegro* (New York: Columbia University Press).

McNamara, Robert F. 1967. *Annunciation Parish: The First Fifty Years.* (Rochester, NY: Church of the Annunciation).

Mondello, Salvatore. 1985. "The Italians," In *Ethnic Rochester* edited by James Pula. (Lanham: University Press), 42-62.

Pula, James S., ed. 1985. *Ethnic Rochester.* (Lanham: University Press).

Ray, Charles P. 1961 *American and Italian Women: A Decade of Progressive Reform at the Practical Housekeeping Center of Rochester, NY, 1906-1917*. Senior Thesis, Dept. of History, University of Rochester, January.

Sahlins, Marshall D. 1961. "The Segmentary Lineage: An Organization of Predatory Expansion," *American Anthropologist* 63: 322-345.

Salamone, Frank A. 1993. " Deciding Who 'We' Are: The Order of the Sons of Italy and the Creation of Italian American Ethnicity. Paper Delivered at the Meetings of the International Congress of Anthropological and Ethnological Sciences, August, Mexico City, Mexico.

Schaefer, Richard T. 2000. *Racial and Ethnic Groups*. 8th Edition. (New York: Prentice Hall).

Seller, Maxine Schwartz, editor. 1983 *Ethnic Theatre in the United States* (Westport CT: Greenwood Press).

Simboli, C. R. 1924. "Addresses and Letter Regarding the Italian Work — the Outlook of the Italian Baptist Work in America," *L'Aurora* Feb. 9.

Steward, Julian H. 1955. *Theory of Cultural Change* (Urbana: University of Illinois Press).

Union Advertiser. 1889. "Italian Mission," March 21, 3-4.

White, Alfred Francis. 1940. *A History of the Italian Baptists within the Territory of the Northern Baptist Convention*. M.A. Thesis, Chester, PA.

Yans-McLaughlin, Virginia. 1977. *Family and Community: Italian Immigrants in Buffalo, 1880-1930*. (Ithaca: Cornell University Press).

Archival Material in Rush Rhees Library, University of Rochester

Annual Report Lewis Street Center 1916-1917 Lewis Street Center Files, Box 55 4,4, pp. 104-05.

Annual Reports of the Resident Headworker, 1914-March 1937 1914-1937 Lewis Street Center Papers. D.55 3:7.

Annual Report of the Resident Headworker April 1, 1936-March 31, 1937, 1936-1937 Lewis Street Center Papers. D.55 3:4.

Annual and Monthly Reports of the LSC Dispensary Sept. 1916-April 1921 1916-1921 Lewis Street Center Papers. 3, 7.

Annual Reports of Headworker 1920 -1938 Lewis Street Center Papers, Box D5 4, 5. D.55 3:7LSC Papers.

Annual Reports of the Resident Headworker, 1914-March 1937 April 1, 1936-March 31, 1937.

Anonymous N.d History of Lewis St. Center: 1921-1937. Lewis Street Center Files, Box 55.1:1. University of Rochester, Rochester, NY.

Dispensary Book 1916-1917 Lewis Street Center Papers, Box D 5 4,3.

House Accounts Book 1916-1917 Lewis Street Center Papers. D.55 4:3 Sept.1916-Dec. 1936.

Lanni Papers D50 John Andre Lanni Scrapbook 1 and 2: Scraps East High School
    1906-1910 "Scraps from Here, There, and Everywhere."
Lauterbach, Carl Lewis St. Center: 1937 - 1957. Lewis Street Center Files, Box 55.1:1.
    University of Rochester, Rochester, NY.
Minutes 1914-1950 D.55 2 Lewis Street Center Papers
Record of Baths D.55 4.6 1922-1923 1922-1923 Lewis Street Center Papers D 55 4,
    6. Rogers, Helen Rochester.

Baptist Archives at Colgate Rochester Divinity
Congregational File. St. Mark Baptist Church, Rochester, New York. American Baptist
    Samuel Colgate Historical Library. Rochester, New York.
Annual Report 1923 - Home Mission Board Baptist Union of Rochester and Monroe
    County- Sec., Alfred E. Isaac.

Table One: Growth of Italian Population in Rochester, NY

Date	Number of Italians	Total Population
1865	9	50,940
1870	16	62,000
1875	30	81,722
1890	516	133,896
1910	14,816	218,149
1940	56,329	330,000

Table based on figures in Ray (1961: 4, 11) and Keen (1946: 36).

Chapter 5

♦

JEWISH IMAGES OF CATHOLICISM IN ITALIAN AND ITALIAN AMERICAN LIFE IN THE LATE NINETEENTH AND TWENTIETH CENTURIES: SOME PRELIMINARY REFLECTIONS

This essay suggests that, historically, the relations between Jews and Italian Catholics as individuals and groups on both sides of the Atlantic have been fundamentally respectful. The relationship between Jews and the institutional Catholic Church, however, is more complex. Jewish appreciation of both Italian Catholics and the institutional Catholic Church was at its height especially during World War II, given the assistance the Italian population at large and the Vatican and Pope Pius XII, in the face of the Nazi and Fascist terror, gave to the Italian Jewish population. While relations between both groups remain basically positive, there has occurred, over the past few decades, a certain amount of alienation between the Jewish population and the institutional Catholic Church given the acceptance of the claims of certain revisionist historians that Pius XII and the Vatican either did not sufficiently oppose the Nazi extermination of the Jewish population or actually conspired with Adolph Hitler's designs. My own work and the work of other scholars-both Jewish and non-Jewish-have attempted to refute the falsehoods perpetuated against Pius XII and the role of the Catholic Church during the Second World War. However, as sociologist W.I. Thomas of the University of Chicago put it, "if something is defined as real, it is real in its consequences" (Thomas and Thomas, 1928: 572). The issue that is key to understanding the relationship between Jews and the institutional Catholic Church regarding the subject of this essay, i.e., Jewish images of Catholicism in Italian and Italian American life, is not whether the claims of the revisionists are correct; it is that if they are *believed* to be correct they will have created unnecessary and antagonistic feelings between the Jewish population and an official Catholicism. Regarding the ultimately more important issue of historical *truth*, however, the attack against Pius XII can best be viewed as a cover for a more generic attack against Christianity and, especially, against the Catholic Church and

82

the present pontiff, John Paul II, in the contemporary age. On a more positive note, the activities and program of the pontificate of Pope John Paul II have moved in the direction of countering the effect of the work of the historical revisionists, having attempted to strengthen the relationship between these two ancient religions by highlighting the many bonds that inextricably unite them.

Unification of Italy and the Emancipation of the Jews of Italy

After centuries of foreign domination, Italy was finally united under the House of Savoy in 1870, and the total emancipation of the Jews was achieved. They had contributed to the struggle for freedom; they were educated and had obtained prominent positions in all fields; fifty Jewish generals, for instance, served in the Italian army during World War I. Their contributions were significant in the fields of business, banking, and insurance, as well as in education, the arts, and literature. They participated in government; some joined the Fascist movement. Others were anti-Fascists who understood the danger of an authoritarian government. Italian Jews retained their Jewishness and a deep respect for Jewish ethical concepts, their culture, and their heritage without diminishing in the least their love for Italy.

As the separate Italian states accepted and joined each other politically after centuries of rivalry, war, and foreign domination, and became a single nation in 1870, Jews were emancipated from the ghettos. They were now perceived to be part of the new national group.

The story of the Jewish community in Italy is one of an enduring relationship between the land and its people. The Jewish presence predates Christianity and extends two millenniums before the unification of Italy. Jews and Italians have common values with regard to family, faith, and food. Jews in Italy consider themselves Italians of the Hebrew religion. One can readily understand the respected Jewish presence in the Italian community as well as the willingness and desire of most Italians to protect Jews and defy the Nazis during World War II.

Compared to most European Jewish populations, most Italian Jews were spared the monstrous evil of the Holocaust. *The Italian Jewish Experience* (Di Napoli, 2000) is an excellent survey of the topic.

In June 1940, Benito Mussolini joined Hitler in the war and implemented his anti-Semitic policy with regard to the discrimination and persecution of Jews. This was accomplished through Mussolini's monopoly over the Italian educational system and the mass media. News services in Italy were censored and reports about German atrocities were considered propaganda, as such, by some of the Italian population. Many anti-Nazi and anti-Fascist Italians who helped save the Jews were executed or deported to German concentration camps where they died of starvation, disease, or hard labor. Untold others were compassionate in their daily interactions with the Italian Jewish population and had the courage to take risks in order to save them.

Italian Catholics and Jews in Italy and in the United States

How did Jews view Italian Catholics in Italy and in the United States? And how did Italian Catholics view Jews in Italy and the United States? In both cases, the images were relatively benign, given the relatively respectful treatment of Italy's Jews over the centuries.

They had become well integrated into Italian society. Even during the period when they were forced to live in ghettos, Jews continued to participate in business and other relationships with Gentiles. Italian American Jews include Jews who came from Italy to America and Italians who converted to Judaism or those Jews who married Italians and converted to Christianity.

Italians affirmed that Jews were equal human beings with an equal right to live. This was and is the Italian mentality. It was best described by a Jewish doctor, Dr. Rubin Pick, in a discussion about the safety open to Jews of Italy during the Holocaust:

> I was a Polish national studying medicine in Italy when the racial laws were passed in 1938. Under these laws, foreign Jews — including those who had become citizens after 1919 — had to leave Italy. My father left for Palestine. I applied for permission to stay so I could finish my studies and the authorities let me stay. I was living in Trieste and commuting to Padua. My sister was a student at the university in Trieste, so she applied and again they agreed. The Italian mentality was: How could you let a young girl be by herself, unprotected, in the city? So they told my mother that she could stay too. On hearing this, my brother said, 'You're letting my brother, sister, and my mother stay and making me go alone?!" So they said, "Okay, you can stay too." (Marchione, 1997: 49-50)

When Italy entered the war in 1940, Rubin Pick and his family were sent to the Ferramenti-Tarsia concentration camp in southern Italy. It was a concentration camp Italian-style. If you had a good reason, you could apply for permission to leave. He applied because he still had some requirements to complete — the equivalent of an internship. They let him return to Trieste where he not only completed his internship but remained there free of danger until the Germans arrived in 1943.

Because Italian Jews did not have a history of being severely persecuted, many did not foresee the approaching danger. Dr. Pick, however, stated that two hours after the Germans entered Trieste, he was already outside the city with his mother and sister. For Italian Jews, danger was something new and too often they did not know what they should do. Many Italian Jews who died during the Holocaust might have lived if they has been more attuned to the reality of Nazism. It is ironic and noteworthy to understand that while the nation of Italy was serving Hitler under Mussolini, many Italians were protecting Jews and other refugees.

Never in the past, perhaps, were Italians and Jews so closely connected as during World War II. Italy's humane attitude toward the Jews extended even beyond its own people and national borders. Wherever the Italian army was the occupying power, as it was in parts of Yugoslavia and France, Jews were protected from serious harm. Yugoslav Jews report that they were taken almost from the hands of the approaching German soldiers onto Italian army trains, dressed in Italian military uniforms and brought to Italy where they were concealed.

The case of Captain Joseph Frelinghuysen is also instructive.[1] Frelinghuysen was a lieutenant in the United States Reserves, on active duty in July 1941. He fought with the Allies' heavily armed troops that landed on the shores of northwestern Africa. He was among the American and British soldiers captured by the Nazis to be turned over to the Italians. The prisoners were first sent to Capua near Naples and then to Chieti in the Abruzzi region. While there he realized the importance of learning to speak Italian in order to escape. He also arranged a schedule of calisthenics and followed a routine of strenuous physical exercise.

In September 1943, the Italian Prime Minister, Pietro Badoglio, surrendered to the Allies. After ten months of prison life, Frelinghuysen planned his escape. He crawled through wire fences, under the eye of a

German machine gun, with Captain Richard M. Rossbach. They moved across heavily guarded rail and highway arteries which were carrying supplies and ammunition to the German divisions fighting in the South.

In his book, *Passages to Freedom* (1991), Frelinghuysen explains that what saved them was the compassion and humanity of the people of the Italian mountain ranges. These humble people fed the starving soldiers and risked their lives to protect them. There developed a warm friendship with the people of this remote valley of the Apennines. Years later, when he returned to thank the townspeople, he learned that the German military had murdered many of the Italians who had helped the escapees.

Among many other possible examples of Italian sympathy to the oppressed Jewish people is the account of the family of Sigmund and Minna Jawetz, who lived in Vienna, Austria. Their daughter, Stella Schecter from Cranford, New Jersey, tells the story of how her family was saved. She and her older brother were able to come to the United States, but her younger brother was sent to England. On *Kristallnacht* in 1938, Sigmund Jawetz was taken to Dachau where millions of Jews and non-Jews were murdered. A decorated hero of World War I, he was released after four months and told to leave the country. Her parents then fled to Milan. When Hitler put pressure on Mussolini, they were sent to the Ferramonti-Tarsia concentration camp. Stella Schecter gratefully wrote: "Had it not been for the Italians, our family would not have been reunited" (Marchione, 1977: 44).

Based on cultural similarities of family life, expressiveness, earthiness, and a love of food, Jewish-Italian relations on the American side of the Atlantic in the late nineteenth and through the twentieth century have basically been good and continue, for the most part, to be so. As Jack Nusan Porter puts it, "ethnically these two groups are quite close. They share similar attitudes toward family and children and an indefinable esprit toward life" (2000: 302). (This statement does not deny that relations have not always or completely been harmonious.) Today, there is even a significant amount of Jewish-Italian intermarriage requiring a revision of Will Herberg's triple-melting pot thesis in his classic, *Protestant, Catholic, Jew* (1960) so relevant for the immediate post-World War II era in the United States (Porter, 2000: 203).

The similarities between Jews and Italians in the United States is not merely the product of similar philosophies and life-styles but is also reflec-

tive of certain common historical experiences. Taking advantage of economic opportunities available, they have worked side by side in industry
and have lived in close proximity to each other in the cities and suburbs of
America. On the other hand, migrating away from the inner-city neighborhoods and to the suburbs a bit quicker than Italian Catholics, Jews have
been, relatively speaking, more upwardly mobile in term of occupation,
education, and income (although Italian Catholics have done well in these
categories, nation-wide). Overall, Jews have been more liberal across the
board politically and culturally than has the Italian Catholic population with
the latter taking liberal positions primarily on socio-economic issues and
being more conservative on moral and community oriented concerns (Greeley, 1985).[2]

In 1974, the Seventh Annual Conference of the American Italian Historical Association was co-sponsored by the American Jewish Historical
Society. *The Interaction of Italians and Jews in America* (Scarpaci, 1975),
the product of this joint collaboration, represents an excellent resource.
According to Richard Juliani and Mark Hutter, Italian-Jewish interaction
had, up to the point of the study, passed through three distinct stages:

> 1) the ghetto, with both groups remaining in the initial stages of settle-
> ment and community formation and witnessing the development of a
> nationalistic ethnicity from the original regionalistic ethnicity;
> 2) the period of second generation marginality when both groups were
> immersed into the processes of acculturation and assimilation and wit-
> nessing the transformation and disintegration of immigrant communities
> and the emergence of an ethnic Americanness; and
> 3) the emergence of the new frontiers when contemporary circumstances
> relocate Jews and Italians in largely white-collar and middle-class op-
> portunities and life-styles which are quite remote from their earlier fam-
> ily backgrounds and neighborhoods allowing each to meet each other at
> work, in new communities, and in formal associations. (1975: 54)

The reaction of one of the Italian American Catholic contributors in
The Interaction of Italians and Jews in America, Charles Ferroni, was
based on what he knew about the family in which he grew up:

> My father once told me, there are three things a man should never
> change: his name, his religion, and his wife. There was an element of

permanence, security, and rootedness in our family system which I
sensed also among the Jews. One's name involved honor, respect, and
dignity. There was no conflict in the area of religion. I came to understand that Jesus was a Jew and his Mother Mary (*la Madonna*) was Italian.
My image of the term Jew was and is: Jews are clean, wholesome,
God's people. Finally, I could not see much difference between the Italian wife and mother and the Jewish wife and mother. Dr. Vecoli expressed it best when he wrote that someday some Italian American
should write a *Mario's Complaint*: 'I wonder what we will do when all
of our mothers are gone and there will be no one left to impose her hos-
pitality on us, pack us a lunch, and remind us as we leave for the last
time, be good.' (1975:105)

The Image of Italian American Intellectuals Toward the Jewish Heritage
Given both the generally historically friendly relations between Italians and Jews in the United States and the spectacular success of Jews in
academic life, it should come as little surprise that many Italian American
scholars view the Jewish heritage as liberating and something to be emu-
lated and admired. Author Franco Mulas notes that Jerry Mangione, Pietro
Di Donato, and Mario Puzo are three well known Italian American authors
who depict Jews with an admiration for their intellectual prowess (Krase
and DeSena, 1994: 247-253). Italians and Jews not only have warm relationships but recognize their mutual struggle to free themselves from their
poverty in the early stages of their American experience. Immigrant Jews
relied primarily on education while Italians considered hard work in the
economic realm as their only security.
In an article, "My Experience as an Italian American Writer" (1985),
Jerre Mangione writes with regard to education:

Books nourished my fantasy and drew me close to Jewish friends who,
unlike my Sicilian cousins of my age, shared my fondness for reading,
and discussing what we read. It will be recalled that the Jewish-Amer-
ican immigrants arrived in the United States in massive numbers at
about the same time as did the Italians. Unlike the Italians, many of the
Jews arrived with a traditional respect for learning that made them embrace the educational opportunities offered by the American public
school system. The Italians, on the other hand, often regarded American

schools as a threat to the integrity of the family and to many of the traditions they held to. (Krase and DeSena, 1994: 247)

In Mario Puzo's *The Fortunate Pilgrim* (1985), Octavia rejects all contact with Italian males, loves to read, and deplores the ignorance of Italian Americans. She marries a Jew, Norman Bergeron, who is not only poor but is an aspiring poet. Her mother, Lucia Santa, considers it a scandal that she "picked for a husband the only Jew who does not know how to make money" (Krase and DeSena, 1994:249). Lucia gives a very vivid description: "The macharoni carried a stack of books — a grown man — and with high pompadour, black hair, his horn-rimmed spectacles, thin sliced features curved like a bow" (Krase and DeSena: 1994: 249).

Pietro Di Donato treats the same inter-ethnic themes in his novel, *Christ in Concrete* (1939). Louis Molov is a Jewish boy whose family lives in a tenement with several Italian American families. He establishes a friendship with Little Paul and tries to convince him that education is the only means of survival. In a country portrayed as a land of freedom and opportunity, this novel expresses a radical view of social injustice. Jewish intellectualism helps to free the protagonists from some of the fatalistic values that their Italian-born parents tried to instill in them.

Franco Mulas states that these three "authors resist the common tendency to place undo stress on cultural and religious differences. Although Italian and Jewish values are contrasted, the reader is more often than not directed to the ties and the shared difficulties and goals between the two groups" (Krase and DeSena, 1994:252). What distinguishes the immigrants is the means by which they accomplish their goals: love for family, respect for elders, and fidelity to their respective religions.

Jewish Images of the Church as an Institution

Jews during the years of and surrounding World War II thought of Pope Pius XII and the official Church as a friend of democracy and peace and as an enemy of racism and totalitarianism. It is interesting to note how Jews reacted when Eugenio Pacelli was elected Pope. *The Jerusalem Palestine Post* editorial wrote: "Pius XII has clearly shown that he intends to carry on the late Pope's (Pius XI) work for freedom and peace" (March 6, 1939). The *Jewish Chronicle* in London stated that the Vatican received congratulatory messages from the "Anglo-Jewish Community, the Synagogue Council of America, the Canadian Jewish Congress, and Polish Rabbinical Council" (March 10, 1939).

The Jewish community was also very thankful for the public statements of Pius XII who wrote his first encyclical in October of 1939. The *Jewish Telegraphic Agency* in New York reported that "the unqualified condemnation which Pope Pius XII heaped on totalitarian, racist, and materialistic theories of government in his encyclical, *Summi Pontificatus*, caused a profound stir...Few observers expected so outspoken a document" (October 27, 1939). In yet another instance, the *Jewish Advocate* in Boston reported that "the Vatican Radio this week broadcast an outspoken denunciation of German atrocities in Nazi (occupied) Poland, declaring they affronted the moral conscience of mankind" (January 26, 1940). This broadcast confirmed the media reports about Nazi atrocities, previously dismissed as Allied propaganda.

During the war years and immediately after, there are numerous other documents written by the Jewish community worldwide thanking Pius XII and the Catholic Church for the assistance offered to persecuted Jews. In his lifetime, Pius XII received more praise and expressions of gratitude from the Jewish people than any other Bishop of Rome in history. According to several Jewish historians, including Pinchas Lapide (1967), living in Italy, Pius XII and the Church saved between 740,000 and 860,000 Jews from extermination. In Rome alone, during the Nazi occupation, 4,447 Jews were hidden in over 155 Catholic houses, ecclesiastical institutions, parishes, and schools. In several churches in Rome there are Jewish plaques thanking the Church for saving Jewish lives. The rescue work was done at the express wish of the Pope.

After the War, Moshe Sharrett, former Foreign Affairs Minister and Prime Minister of Israel, went to see Pius XII "to thank the Catholic Church for what it did to save the Jews in all parts of the world" (October 12, 1945). Like Sharrett, Rabbi Herzog of Jerusaleum, as well as the Rabbis of the Italian, United States, Rumanian, and Hungarian Jewish communities came to Rome or sent messages thanking Eugenio Pacelli for the way in which he mobilized the Church in their behalf.

It is the important task of scholars to search out truth in history. Regarding the subject at hand, this means to bring to light the work of rescuers whose extraordinary acts of courage saved many Jews from death and to demonstrate that Pope Pius XII was not "silent" during the war and to explain that his humanitarianism was a new and effective method of fighting anti-semitism. There is an abundance of incontestable evidence to obliterate the myths and lies, propounded by revisionists, that are so widely circulated by a secular and progressive oriented mass media today so unsympathetic to the ideas and positions of the *contemporary* Catholic Church (Marchione, 1997, 2000, 2002a, 200b; Rychlak, 2000; Low, 2002; McInerny, 2001; O'Carroll, 1980; Levai, 1968; Lapide, 1967).

Today Pius XII is falsely maligned and denigrated by the writings of scholars that clearly expose their prejudices against Catholicism and who have an axe to grind against the Church (Goldhagen 2002; Cornwell, 1999; Carroll, 2001; Kertzer, 2001; Wills, 2000; Zuccotti, 2000; Phayer, 2000; Friedlander, 1996; Lewy, 1964; Hochhuth, 1964). These writings have done much harm and have taken their toll on the relationship between many contemporary Jews and the Church as an institution. These writers refuse to grant Pope Pius XII credit for his leadership during World War II. They ignore the testimony of contemporary witnesses and statements in the media that acknowledge his words and actions on behalf of the Jews at the time. They denigrate Pius XII and Catholic leaders who endangered their own lives to protect Jews. Truth and justice demand a strong response to the false claims of the revisionists against Pius XII that he was, variously, "silent," "morally culpable," or "anti-Semitic." Readers are suggested to review Kenneth Whitehead's (2002) recent comprehensive review of the literature on the controversy surrounding Pius XII published in a recent issue of *The Political Science Reviewer.*[3]

The Nazi View of the Catholic Church During the War Years

The revisionists' claim that Pius XII and the institutional Church were silent, morally culpable, and anti-Semitic flies in the face of not only the demonstrated activity of Pius XII and the Church in favor of the Jews, but also that Adolph Hitler despised Christianity and saw the Catholic Church as an opponent. Nazi hatred for the Catholic Church has been documented, for instance, in Konrad Low's new book, *Die Schuld* (The Guilt), published by Resch Press in 2002, with the subtitle, *Jews and Christians in the Opinion of Nazis and in Present Times.* The book is promoted as a response to *Amen* and *The Deputy*, referring to film and theater works that accuse Pope Pius XII of having been too conciliatory to Nazism. Low uses specific historical documents to address aspects of Nazi policy up to now little known, in particular the continuous and systematic persecution of Catholics. He demonstrates that the Catholic Church condemned the racism and nationalism of the Nazi philosophy. He provides ample documentation to show that Catholic bishops condemned Nazi theories. This is why the Nazis persecuted Catholics as well as Communists and Jews. According to the Nazi theory, Christianity's roots in the Old Testament meant that whoever was against the Jews should also be against the Catholic Church. He records Catholic assistance to Jews, which greatly angered the Nazis.

Nazi propaganda constantly portrayed the German bishops and the Pope as traitorous and shameless supporters of the "international Jewish conspiracy." Jewish scholar Jeno Levai, testifying at the Adolph Eichmann Nazi War Crime Trials, stated that "the one person (Pope Pius XII) who did more than anyone else to halt the dreadful crime and alleviate its consequences, is today made the scapegoat for the failures of others" (1968).

The Voice of Pope John Paul II

Despite the setback in Jewish-Catholic relations caused by the revisionists, relations between the Catholic Church and the Jewish people have been helped through the pontificate of John Paul II. No Pope throughout history did more than Pope John Paul II to attempt to create closer ties with the Jewish community, to oppose anti-Semitism, and to try to make certain that the evils of the Holocaust never occur again. Pope John Paul II visited the Chief Rabbi at the Synagogue in Rome and declared that "the Jews are our dearly beloved brothers" and indeed "our elder brothers in faith." He established full diplomatic relations between the Holy See and the State of Israel. He requested forgiveness for past sins by Christians against Jews. For instance, in the 1998 statement of the Vatican Commission for Religious Relations with the Jews entitled *We Remember: A Reflection on the Shoah*, it is stated that:

> We deeply regret the errors and failures of those sons and daughters of the Church. This is an act of repentance, since, as members of the Church, we are linked to the sins as well as the merits of all her children.

At the same time, however, and in the same document, Pope John Paul II acknowledges the important role played by his predecessor during the years of World War II:

During and after the war, Jewish communities and Jewish leaders expressed their thanks for all that had been done for them, including what Pope Pius XII did personally or through his representatives to save hundreds of thousands of Jewish lives. (Rychlak, 2000: xii)

A survivor of both Nazi and Communist oppression himself, John Paul has consistently praised Pope Pius XII for his heroic leadership during World War II. In fact, he defended Pius XII during a meeting with Jewish leaders in 1987, recalling "how deeply he felt about the tragedy of the Jewish people, and how hard and effectively he worked to assist them during the Second World War" (Marchione, 1997:9). Many Catholics from all over the world have petitioned Pope John Paul II to expedite the cause for the beatification of his predecessor, Pope Pius XII.

"Peace" was the clear message John Paul II gave on March 25, 2000, the day of his stay in Jerusalem: "the honor given to the 'Just Gentiles' by the state of Israel at Yad Vashem for having acted heroically to save Jews, sometimes to the point of giving their own lives, is a recognition that not even in the darkest hour is every light extinguished. That is why the Psalms and the entire Bible, though well aware of the human capacity for evil, also proclaims that evil will not have the last word" (Marchione, 2002a: 275).

The Pontiff assured the Jewish people that the Catholic Church was motivated by the Gospel law of truth and love, and deeply saddened by displays of anti-Semitism. The Catholic Church rejects racism in any form as a denial of the image of the Creator inherent in every human being. For the Pontiff,

Jews and Christians share an immense spiritual patrimony, flowing from God's self-revelation. Our religious teachings and our spiritual experience demand that we overcome evil with good. We remember, but not with any desire for vengeance or as an incentive to hatred. For us, to remember is to pray for peace and justice and to commit ourselves to their cause. Only a world at peace, with justice for all, can avoid repeating the mistakes and terrible crimes of the past. (Marchione, 2002a: 275)

SISTER MARGHERITA MARCHIONE

[1] Captain Joseph Frelinghuysen lived in Far Hills, New Jersey, where portraits of his ancestors may be seen. His great-great-grandfather had commanded a company of artillery at the Battle of Princeton in 1776; his grandfather had commanded a detachment of troops in the War of 1812, and his father, a United States Senator from 1917 to 1923, was a Spanish-American War veteran.

[2] For an interesting study of how Jews and Italians in Canarsie, New York, responded to an influx of African Americans into their community, see Jonathan Rieder, 1985, *Canarsie: The Jews and Italians of Brooklyn Against Liberalism* (Cambridge, MA.: Harvard University Press); also see Joseph A. Varacalli's review of the volume in the *International Migration Review*, 1986, 20:678-679.

[3] In the 2002 issue of *The Political Science Reviewer*, Kenneth D. Whitehead reviews much of the literature in his brilliant review essay, "The Pope Pius XII Controversy," 31: 283-387.

References

Blet, Pierre, S.J. 1999. *Pius XII and the Second World War: According to the Archives of the Vatican.* (New York: Paulist Press).

Carroll, James. 2001. *Constantine's Sword: The Church and the Jews-A History.* (New York: Houghton Mifflin).

Cornwall, John. 1999. *Hitler's Pope: The Secret History of Pius XII.* (New York: Penguin Books).

DiDonato, Pietro. 1939. *Christ in Concrete.* (New York: Bobbs Merrill).

DiNapoli, Thomas P. 2000. Editor. *The Italian Jewish Experience.* (Stony Brook, New York: Forum Italicum).

Ferroni, Charles. 1975. "Panelists' Comments on the Discussion of Rudolph Glanz's book, Jew and Italian," pp. 102-105. *The Interaction of Italians and Jews in America.* Jean Scarpaci, Editor. (Staten Island, New York: American Italian Historical Association).

Frelinghuysen, Joseph. 1991. *Passages to Freedom.* (Manhattan, Kansas: Sunflower University Press).

Friedlander, Saul. 1996. *Pius XII and the Third Reich.* (New York: Alfred A. Knopf).

Goldhagen, Daniel. 2002. *A Moral Reckoning: The Role of the Catholic Church in the Holocaust and Its Unfulfilled Duty of Repair.* (New York: Knopf).

Greeley, Rev. Andrew M. 1985. *American Catholics Since the Council: An Unauthorized Report.* (Chicago, Illinois: The Thomas More Press).

Herberg, Will. 1960. *Protestant, Catholic, Jew: An Essay in Religious Sociology.* (Garden City, New York: Doubleday).

Hochhuth, Rolf. 1964. *The Deputy.* (New York: Grove Press).

Juliani, Richard and Mark Hutter, 1975. "Research Problems in the Study of Italian and Jewish Interaction in Community Settings," pp. 42-56. *The Interaction of Italians and Jews in America*. Jean Scarpaci, Editor. (Staten Island, New York: American Italian Historical Association).

Kertzer, David I. 2001. *The Popes Against the Jews: The Vatican's Role in the Rise of Modern Anti-Semitism*. (New York: Alfred A. Knopf).

Lapide, Pinchas. 1967. *The Last Three Popes and the Jews*. (London: Souvenir Press).

Levai, Jeno. 1968. *Hungarian Jewry and the Papacy: Pius XII Was Not Silent*. (London: Sands and Co).

Lewy, Guenter. 1 964. *The Catholic Church and Nazi Germany*. (New York: McGraw-Hill).

Low, Konrad. 2002. *Die Schuld*. (Resch Press).

Mangione, Jerre (1985). "My Experience as an Italian American Writer," *Rivista di Studi Americani*, Anno III, 4 (Albano Terme: Piovan Editore).

Marchione, Margherita. 1997. *Yours is a Precious Witness: Memoirs of Jews and Catholics in Wartime Italy*. (New York: Paulist Press).

_____. 2000. *Pope Pius XII: Architect for Peace*. (New York: Paulist Press).

_____. 2002a *Consensus and Controversy: Defending Pope Pius XII*. (New York: Paulist Press).

_____. 2002b. *Shepherd of Souls: A Pictorial Life of Pope Pius XII*. (New York: Paulist Press).

McInerny, Ralph. 2001. *The Defamation of Pius XII*. (South Bend: St. Augustine's Press).

Mulas, Franco. 1994. "Jews in the Italian American Novel," *Italian Americans in a Multicultural Society*. Editors Jerome Krase and Judith N. DeSena. (Stony Brook, New York: Forum Italicum of Stony Brook University)

O'Carroll, Michael, C.S.S.Sp., 1980. *Pius XII: Greatness Dishonored: A Documented Study*. (Dublin: Laetare Press).

Phayer, Michael. 2000. *The Catholic Church and the Holocaust, 1930-1965*. (Bloomington: Indiana University Press).

Porter, Jack Nusan. 2000. "Italian American Jews," *The Italian American Experience: An Encyclopedia* (Co-editors, Salvatore J. LaGumina, Frank J. Cavaioli, Salvatore Primeggia, and Joseph A. Varacalli) (New York: Garland Press), 302-3.

Puzo, Mario. 1985. *The Fortunate Pilgrim*. (New York: Bantam Books).

Rieder, Jonathan. 1985. *Canarsie: The Jews and Italians of Brooklyn Against Liberalism*. (Cambridge, MA: Harvard University Press).

Rychlak, Ronald. 2000. *Hitler, the War, and the Pope*. (Huntington, Indiana: Our Sunday Visitor).

Thomas, W.I. and Dorothy S. Thomas. 1928. *The Child in America: Behavior Problems and Programs*. (New York: Knopf).

Whitehead, Kenneth D. 2002. "The Pope Pius XII Controversy," *The Political Science Reviewer* 31: 283-387.

Wills, Gary. 2000. *Papal Sin: Structures of Deceit*. (New York: Doubleday).

Zuccotti, Susan. 2000. *Under His Very Windows: The Vatican and the Holocaust in Italy*. (New Haven: Yale University Press).

Chapter 6

✦

MOTHER CABRINI: WOMAN, CITIZEN, SAINT, AND MISSIONARY TO THE POST-MODERN WORLD

Mother Francesca Maria Xavier Cabrini (July 15, 1850-Dec. 22, 1917) never tired of saying that she came to America at Peter's command, the command of Pope Leo XIII — "sent in strict obedience" — she said, to teach the Italian immigrants, their children, and their grandchildren the truth. She came, in the words of the 1950 "Papal Brief" proclaiming her the heavenly Patroness of Immigrants, to protect the immigrants against the "grave spiritual dangers" they were encountering in the New World; against being "shipwrecked in their faith" upon the shores of a strange religion and culture; of forgetting the "holy traditions of their fathers." She came to save souls! (Pope Pius XII, 1891, Sullivan 1992: Appendix I, 285).

As the grandson of southern Italian immigrants, and as a Roman Catholic, I am deeply grateful to Frances Xavier Cabrini, my "Mother" in the Faith, for her mission to America, the evangelizing Apostolate of her Missionaries of the Sacred Heart of Jesus in producing — in Mother Cabrini's own words — "the beautiful fruit of Christ's love in America" (DiDonato, 1960:72).[1] We must never forget that Mother Cabrini believed with all the passion of her Italian heart that she and her Missionary Sisters of the Sacred heart of Jesus were ordered to go to America by the Vicar of Christ and her life-long friend, Pope Leo XIII, and Bishop Giovanni Scalabrini; not just to teach any truth, worthy as that must be to every teacher, but the Truth of the Gospel that Jesus Christ said shall make all men and women free.

"My daughter, your field awaits you not in the East, but in the West," Leo alluded to Cabrini's original intention, indeed her life-long passion, to be a missionary to China. "The house and family of Western civilization must first be put in order. His love must conquer the West before we approach the East" (DiDonato, 1960: 59-60).

The mission of America's first Saint continues today. It continues not just in the Americas but in the West in general. Pope Leo XIII's words ring even more true amid the ruins of the Post- Modern world in which Western man has lost his way; squandering in a "soulless Babylon" — as the Pope feared he would" — the Christian labors of centuries" (DiDonato, 1960: 59-60). Perhaps never more relevant is the terrible Biblical proverb as applied to the West, "Physician, heal thyself!" (Luke 4, 23). Mother Cabrini and the Order that she founded in this way can be seen as Missionaries *par excellence*, Missionaries of Divine Truth and Love, to the millions of disinherited and broken men and women of our times, in the West and throughout the world. "If you continue in My word, then you are my disciple indeed," were the words of her Divine Master that inspired Mother Cabrini's apostolate of evangelization, "and you shall know the truth, and the truth shall make you free" (John 8, 32). Sr. Ursula Infante, MSC, provides the key to understanding the discipleship of the "Saint of the Immigrants" when she describes Mother Cabrini as "a true child of the Church" who was therefore extraordinarily devoted to the Sovereign Pontiff (1970: Foreword).

Like St. Francis of Assisi — who first went to Rome to seek Pope Innocent III's approval for his Order; and like the Jesuit Saint and missionary St. Francis Xavier, her patron, Mother Cabrini — herself a member of the Third Order of St. Francis — took her orders from the man she regarded as the successor of St. Peter. She was absolutely obedient to the Magisterium (the teaching authority) of the Church. "The spiritual always held the primacy in her life," explains Sr. Ursula Infante, MSC, after an exhaustive study of her writings, "and so we find that her directions, her commands and even her reports are interspersed with references to God, to the Sacred Heart, to Mary Immaculate and with heartfelt exhortations to the practice of virtue" (1970: Foreword; and letter to the author, July 10, 1994; Sullivan, 1992:141).

Her zeal was so great to teach the Catholic faith to the poor immigrants and others — in her own words, "to those who as yet do not know Him, and also to those who have forgotten Him" — that the Saint while at sea on April 19, 1890 reproached the steamer that she was taking to New York for its slow progress (*Travels*, 1944: 3). "...I could almost say in all seriousness what I said jocularly a few days ago, that if the Sacred Heart would:

give me the means I would construct a boat called 'The House of Cristoforo' ('Bearer of Christ') to traverse with one Community, little or big,
so as to carry the Name of Christ to all people, to those who as yet do
not know Him, and also to those who have forgotten Him. (*Travels*,
1944: 3)

So many of the Italians in America, poor and illiterate, still did not
know Christ; and this despite the heroic sacrifices of so many priests and
religious in the Old Country. To others the Church was forgotten or misunderstood in the sun-up to sun-down, subhuman working conditions of a
society that exploited the cheap labor of the *paesani* (Maynard, 1945: 96-
98). When Mother Cabrini asked Italian laborers in New Orleans whether
they went to Mass, they answered her with great courtesy that Mass must
wait until they made enough money to return to Italy (Sullivan, 1987: 269-
270). An answer like this from such good but uninstructed men — seemingly unaware of the spiritual consolations of the Church — must have
broken Mother Cabrini's heart and confirmed once again the worst fears of
Bishop Scalabrini and Pope Leo XIII that the Italians in America were
being lost from Catholicism.

Mother Cabrini was a well-educated woman, a teacher, and knew first-
hand the ravages of Marxism, positivism, historicism, agnosticism, panthe-
ism, and the other forms of God-denying immanentism that were tearing
apart Italy and the Western world. In his Encyclical Letter *Pascendi
Dominici Gregis* ("On the Doctrines of the Modernists"), Pope Pius X was
to warn against the Modernists' attempt to accommodate supernatural
Catholic dogma to the mere opinions of ever-changing contemporary philosophy (July 3, 1907; #18). The place of her birth, Sant'Angelo, on the
plains of Lombardy, had seen the centuries-long struggle between the Papacy and members of her own family. Her family during her own life-time
was at war with itself. Francesca's father's first cousin, Agostino Depretis,
a liberal, had even served as prime minister of Italy; while Francesca's
parents stood firmly with the Pope against the anti-clerical radicals (Sullivan, 1992: 3-19; September 4, 1891, *Travels*, 1944: 21).

When she saw Protestant sects growing "like mushrooms" in America;
and so many of her "brethren, purchased by the blood of Jesus Christ...who
through ignorance are losing the inheritance of the children of God and
making themselves unhappy for all eternity," Mother Cabrini remembered

the words of Pope Leo XIII that had turned her and the Missionary Sisters of the Sacred Heart away from their first love, China, and on to the New World:

> The house and family of western civilization must first be put in order. His love must conquer the West before we approach the East. There are sad truths you will learn by seeing with your own eyes. America, growing with titanic strides, will soon achieve world influence. If she becomes another soulless Babylon, she will topple, and with her fall she will drag down lesser nations, and the Christian labors of centuries. (DiDonato, 1960: 59-60)

Then the Holy Father addressed the problem of the Italians:

> Hundreds of thousands of our Italian souls in America have become lost and battered sheep, isolated from Christ, understanding, and ordinary decency. The New World cries for the warmth and the compassion of a mother's heart, a heart tempered by love and sacrifice, the heart of the apostle. Francesca Cabrini, you have that very heart. My daughter, your field awaits you not in the East, but in the West. I desire very much a great missionary expansion in America. Francesca Cabrini, go to America. Plant there, and cultivate the beautiful fruit of Christ. (DiDonato, 1960:59-60; Borden, 1945: 81-82; Sullivan, 1992: 54-55, 297 and note; Sullivan, 1987:274)

Sr. Mary Louise Sullivan, MSC, in a quotation from the Foundress' spiritual journal for September 1897, makes dramatically clear the impact of this and other audiences of Mother Cabrini with Pope Leo XIII. And in so doing she provides the capstone to Mother Cabrini's "Life and Legacy." "What joy comes to the soul at the words of that holy old man," she writes, "I would resist an angel for fear of an illusion, but I would believe the Pope. His words fill my soul with tranquility, comfort and assurance" (Sullivan, 1992: 45). DiDonato adds an even more revealing quotation from the American Saint: "His mouth is the mouth of Our Lord, his words the words of our Spouse. He is the illumination of Divine Wisdom, and therefore his words and benedictions, for me, is the true column of fire that guides me through every difficulty and danger" (1960: 87).

Mother Cabrini's supernatural gift of faith in the Catholic Church, her uncompromising belief that the "Roman Pontiff and the bishops in communion with him" are "authentic teachers, that is, teachers endowed with the authority of Christ," and that they exercise "the ordinary and universal Magisterium" and teach those "entrusted to them" "the truth to believe, the charity to practice, the beatitude to hope for" — in the words of the *Catechism of the Catholic Church* — this is what Mother Cabrini believed and came to America to share with all men and women in the Sacred Heart of Jesus (1994: #2034).

In her papal mission as a Christ-bearer to the New World, one cannot help but compare Mother Cabrini to Christopher Columbus. The comparison is important. Mother Cabrini seems to suggest it herself, as we have seen, when she muses in her letters about constructing the great ship "The House of Cristoforo" (the "Bearer of Christ"). Both Columbus and Mother Cabrini were members of the Third Order of St. Francis, and as followers of the "Poor Man of Assisi," or what are called today Secular Franciscans, recognized without reservation the primacy of Peter. Columbus, as is well-known, even had a personal monograph on his stationary showing St. Christopher fording a river carrying the Christ-child on his shoulders, so serious was he about the mission he believed providence had assigned to him of evangelizing the New World. Columbus assisted at Mass often, and with his crew "the Admiral of the Ocean Sea" was present for the Divine Liturgy on August 2, 1492, the Franciscan Feast Day of Our Lady of the Angels. He named his flagship the Santa Maria in honor of Our Lady and every evening he and his crew sang the Salve Regina, prayed together the Lord's Prayer, the Hail Mary, and the Credo of the Church.

Mother Cabrini also in her many Atlantic crossings — when the steamer "rolled like St. Peter's boat" — often invoked St. Maria in traditional prayers. She would light a candle to "Our Lady of Loretto, so efficacious against sea storms, and" — she wrote — "our Most Holy Mother did really come to our aid, delivering us from the extreme danger which surrounded us" (October 15, 1891, September 18, 1891, *Travels*, 1944: 30). Even when faced with drowning at sea, it seems that Mother Cabrini's humor never failed her. "Here we are," she wrote on another occasion of near shipwreck in April 1890, "the *see-saw* has started, moved by the hand of God: willy nilly we have to play the game" (*Travels*, 1944: 5).

October 12, 1492, the day of Columbus' landfall and the "discovery" of America, was the Marian Feast Day of Our Lady of the Pillar. The great Captain offered the New World first to Our Lady and only then to Ferdinand and Isabella.

Mother Cabrini's respect for Christopher Columbus, like her a "Christbearer" to America, led her in August 1892 to name the Missionary Sisters' new hospital on East 12th Street "Columbus Hospital" (Sullivan, 1992: 93-94). She could easily identify with Christopher Columbus as an Italian and fellow-evangelist. But Mother Cabrini, as always loyal to the Holy Father and opposed to the occupation of Rome by the nationalists in 1870, would have nothing to do with an offer made to her Missionary Sisters to staff the existing Italian Hospital in New York; run, she said, by Freemasons and Garibaldini (Sullivan, 1992: 92).

Francesca Maria Cabrini! This woman is a saint! "Here was a life lived for God alone." No one has said it better than Sister Ursula Infante in describing Mother Cabrini's heroic sanctity. "...No task was too great. No labor was too hard, no journey too long and fatiguing, no sufferings were unbearable when the saving souls and succoring of suffering humanity were in question" (Sullivan, 1992: 245).

Obedience, humility, unconditional love, reparation to the Sacred Heart of Jesus, our childhood with Jesus Christ in God's merciful Fatherhood, and absolute submission and trust in Providence (which the Church has proclaimed anew in our day as the Divine Mercy in the canonization of another great lady and saint, Sister Kowalska Faustina), this is the "needful" thing, the "Pearl of Great Price," "the Kingdom of our inheritance" as men and women, which Mother Cabrini dedicated her life to making known in America (Luke 10, 42; Matt 13, 46). This remarkable woman, like Mother Teresa of Calcutta of recent memory, took seriously her Divine Master's command to love God and neighbor. She understood and obeyed Jesus Christ's injunction to "take no thought, saying what shall we eat! Or, what shall we drink? Or, wherewithal shall we be clothed?" (Matt 6, 25)

> For after all these things do the Gentiles seek. For your heavenly Father knoweth that you have need of all these things. But seek you first the Kingdom of God and His righteousness; and all these things shall be added unto you. (Matt 6, 31-33)

This is the practical, everyday Catholic mysticism of Mother Cabrini!

Mother Cabrini's obedience, we must remember, was obedience to Truth, the saving Truth of the Gospel. Her act of obedience to Pope Leo XIII, which all her biographers emphasize as critical to Mother Cabrini's life and legacy, remind us of the story of the Holy Roman Emperor Saint Henry, surnamed the Good, who lived from 972 to 1024. Mother Cabrini, providentially enough, was born on his Feast Day, July 15th. The Emperor in his great wisdom, a gift of the Holy Spirit, came to know what Mother Cabrini knew, i.e., that "obedience is better than sacrifices" (I Kings 15, 22).

The story goes that Henry, an exemplary Christian King in very troubled times like ours, wanted to abdicate his throne and spend the rest of his days as a simple monk in the cloister seeking holiness in the Kingdom of God. As sincere and great-hearted as he was, writes Fr. Hyacinth Blocker, O.F.M., (1950: 2), Henry had not yet learned what Mother Cabrini already knew in her Christ-likeness as seen in the decisive audience with Pope Leo XIII many centuries later; and what the Italian American saint was heroically to consume herself trying to teach her generation and ours.

The Emperor did not yet know, until he met with the now nameless and wiser holy abbot of the monastery of St. Vanne at Verdun, that "holiness is not a matter of who you are or where you are but of what you are and how you live" (Blocker, 1950: 2). "But your majesty," the abbot tried to convince Henry that he should stay where God had placed him, "being a monk demands obedience, submission to your superiors. Do you think that you, accustomed to issue edicts and being served by others, can bow your neck to the yoke?" "I promise," Henry said without hesitancy to the abbot, "unqualified obedience to you as my lawful religious superior. Whatsoever you say, that I shall do." "In that case," the abbot replied, "since you are now my subject, I command you under obedience to continue as king. Your people need you. As king you can accomplish more good for others, and for yourself, than as a monk. Make the entire Holy Roman Empire your monastery" (Blocker, 1950: 2).

Mother Cabrini, aspiring to become a "true daughter of the Church," herself always submitted to "lawful" authority, that of the Magisterium. Without the authority of Christ's Church founded on Peter, she believed, we lose our bearings and like the "foolish virgins" of the parable miss the coming of the Bridegroom (Matt 25, 1-13). The Magisterium was to Mother Cabrini like the bracing and invigorating "pure air" of the sea that she so often writes about in her letters (e.g., August 20, 1890, *Travels*, 1944: 17).

All around us today, Mother Cabrini would say, we see people desperately looking for legitimate authority, as the popular bumper sticker: "Question all authority" reveals. "If you are obedient, you will be true Missionaries....submit as so many lambs. This is the secret of obtaining peace, and of obtaining great graces and blessings for the Institute." "Be obedient and your sacrifice will be entire, you will be true Spouses of Christ, you will enjoy Heaven in anticipation." "The sweet yoke of obedience," Mother Cabrini told her Sisters and tells all of us today, will make us a "haven of Christ, a haven of peace" (September 8, 1891, *Travels*, 1944: 27-28).

But while the American saint was always obedient to the "servants of the servants" of God, notably the bishops; while it is true that she was never less than whole-hearted in her submission to lawful authority in the Church, she could be creative in obeying while at the same time gently calling the superior and herself to a higher standard of Faith. A story, too characteristic of Mother Cabrini's trust in God not to be true, has had wide currency and must be told here.

Upon first arriving in New York, she met with Archbishop Michael Corrigan to ask for permission to open an orphanage for the poor Italian children of the City (DiGiovanni, 1991: 56-77). When his Excellency, a man of practical good sense, learned how little money Mother Cabrini actually had to purchase and maintain the building she wanted, Archbishop Corrigan exclaimed that it was a "mere drop in the bucket." "How long do you think five thousand dollars will last? Why, it won't last a year!" Mother Cabrini shot back: "Your Grace, in the Our Father we pray, 'Give us this day our daily bread.' We don't ask God to give us bread to last a year. Don't you honestly think that God will look out for our future?" (Blocker, 1950:233)

"Faith produces hope," Mother Cabrini reminds us today as she did Archbishop Corrigan, "and prayer is at once the supplicating hope." "No, our prayers are never in vain," she repeated again and again to her Sisters, "but everything is disposed of by the wisdom of the Omniscient God. Confide in God above all, hope, and you will not be confounded" (September 9, 1891, *Travels*, 1944:226). "I can do all things in Him who strengthens me," was Mother Cabrini's motto which she took from St. Paul (Philippians

4, 13). Our hope is in Christ, Who is present among us as He promised in the Blessed Tabernacle of every Catholic Church, the Italian American foundress profoundly believed. He has not left us orphans! "Go often, my dearest daughters," Mother Cabrini wrote on one of her trans-Atlantic crossings, "place yourselves at the feet of Jesus in the Blessed Sacrament. Behold that Divine Heart! He is our comfort, our way and our life. Listen to Him with great faith and devotion" (August 21, 1890, *Travels*, 1944: 19).

Christ is no absent God. This was Mother Cabrini's deep conviction. The God-Man has not left us orphans. He is present, as He said He would be, in His Church and in the Sacraments, especially in the Eucharist; pre-eminently the Sacrament of the union of God and the human person and of all humankind. So important for Mother Cabrini was assisting at Holy Mass and visits of adoration to the Blessed Sacrament that when there was no priest on board ship she would scan the horizon of the coastline for the steeples of Catholic churches so she could assemble the sisters travelling with her on deck to make a spiritual communion together (Turchi, Introduction, *Travels*, 1944: xv).

Mother Frances Xavier Cabrini and her Missionary Sisters of the Sacred Heart of Jesus are still here among us, sent by Peter and the Church Jesus Christ founded. They are still on the same God-given mission to make the ever-growing numbers of "orphans," the religiously disoriented, alienated and neglected men, women, and children of the Post-Modern World, whole — spiritually and physically. They are still on their mission to make us authentic human persons, real men, women, and children; to make us "holy" in Christ, to make us saints! The words of Pope Leo XIII still reverberate with truth: "The house and family of western civilization must first be put in order!" Not to the East, but to the West!

There is no doubt in my mind that if she were alive today, if she were still with us in the Church Militant, Mother Cabrini would agree with Msgr. Romano Guardini's analysis of the plight of Mankind in the Post-Modern World. In his classic work *The End of the Modern World*, written in Nazi Germany during World War II, the Italian-born Catholic philosopher posed the question of what has gone wrong with things in the time in which we live? What has gone wrong is the question that we still pose. And the answer was already clear in Mother Cabrini's day that Western man — no matter what he professes — is really living as though God does not exist, as though there is no Revelation and no Church. In fact Modern man has succeeded in abusing his God-given freedom to the incredible degree of exiling his Creator and Sustainer from human life, from nature, from culture — and from God's master handiwork, Creation itself! "With the denial of Christian Revelation genuine personality," Guardini wrote, "had disappeared from the human consciousness. With it had gone that realm of attitudes and values which only it can subsume" (Guardini, 1968:123).

And then these powerful words in which Romano Guardini prophecied our own day, the Post-Modern World after the collapse of Christian civilization:

> The coming era will bring a frightful yet salutary preciseness to these conditions ...as the benefits of Revelation disappear even more from the coming world, man will truly learn what it means to be cut off from Revelation. (Guardini, 1968:123)

"Man will truly learn what it means to be cut off from Revelation!"

This era, this day has come! In the minds of most people today, as Guardini and others foresaw, man, nature, and culture have become autonomous! Created to have meaning only in relation to God, Who gives them existence and sustains them, man, nature, and culture stand alone as "absolutes," or pseudo-absolutes — when in reality there is only one Absolute, God. Vainly, absurdly and tragically, man, who has meaning only as created by God and in the Image and likeness of God, tries to explain himself, nature, and culture without God. And in trying to explain these things exclusively in relation to himself as Godless, denying God and placing himself outside his created relationship to God (atheism, naturalism, relativism) man, nature, and culture must become absurd or worse.

Once we understand what Guardini has described as Post-Modern radical individualism, which is also a kind of *anomie* (Durkheim), of standardlessness and autonomy in man, nature, and culture, we should be and are in fact today finally disposed to listen to Mother Cabrini and the Church she personified. That is, if we have any sanity left! The hopelessness of Post-Modern man's predicament, without Divine Revelation and himself choosing to be completely on his own in revolt against God, presents to us — as Guardini predicted it would — that stark clarification of the situation which in Mother Cabrini's times was still only in outline. Today, living among the ruins of Christian civilization, we no longer have the luxury of seeing the situation as anything other than practical atheism.

Mother Cabrini, mediated through her writings, speaks to our present situation and demonstrates again, as she did in her own historical moment, the path that Catholics must follow. It is the Way, the Truth, and the Life of Jesus Christ in the teachings of His Church. It is, for example, the truth about man and woman — that they are created in the Image and likeness of God and called to perfection in Jesus Christ, to divinization — called in the words of St. Peter to be "partakers of the divine nature" (2 Peter 1, 4).

Specifically, we learn from Mother Cabrini for this reason — that God says so! — to cherish every child, born and unborn. We are "persons," as she so deeply believed, in some mysterious way like the three Persons of the Blessed Trinity, not mere individuals (cf. Pope John Paul II's philosophy of Personalism). We see this today in the work of the Missionary Sisters of the Sacred Heart of Jesus at West Park — Mother Cabrini's beloved "Manresa" — not far from New Paltz, New York. Mother Cabrini said of this holy place, where her body was interred just after her death, that she would always be there "in spirit." She was so dedicated to the children of this orphanage and all children, each of whom she believed to be the Christ-child in Our Lady's arms, that she wanted her daughters not to be called Sisters" but "Mothers" — because she knew that women by nature are mothers, some in the physical sense but all in the sense of being spiritual mothers (Sullivan, 1992:84). No genderless creature but a whole person, woman for Mother Cabrini was like Our Blessed Mother, a mediator between God and man, a tabernacle of Life (September 8, 1891, *Travels*, 1944: 24-25).

And what does this great lady, this "Saint of the Immigrants" who became a naturalized American citizen, teach us about love of country? "I am made all things to all men, that I might by all means save some," Mother Cabrini said with St. Paul. "And this I do for the Gospel's sake, that I might be a partaker thereof with you" (1 Corinthians 3). Her motive is clear. No narrow nationalist in any ideological sense, Mother Cabrini was a citizen and patriot and loved her new country — "one Nation under God." She was a true, authentic multiculturalist who worked along with other Italian immigrants to enrich the United States with Italian culture — but especially with Catholic Italian culture (Varacalli, 1999).

In an age of computer icons and other cold, immanentistic symbols, advertising automobiles and breakfast cereals; in an age when the stupendous Fact of the God-Man on the Cross is hardly noticed and there is a pervasive "opaqueness" which blinds Post-Modern man to the Transcendent, Mother Cabrini's Italian "Spirit of the Festa," her Catholic sacramentalism and incarnationalism — so beautifully and mystically expressed in her Order's Adoration of the Sacred Heart of Jesus — must be our "Spirit" too (D'Elia, 1999; Varacalli, 1999).

And as a Missionary of the Sacred Heart — "Love is not loved!" — Mother Cabrini urges her daughters and each of us, believer or non-believer, spiritually disinherited and orphaned as we are in the Post-Modern world, "to burn with love" for God, "Love Himself," and for one another (August 21, 1890, *Travels*, 1944: 18-19). "If, then, we do not burn with love, we do not deserve to bear the beautiful title which ennobles us, elevates us, makes us great, and even a spectacle to the angels in Heaven" (August 21,1890, *Travels*, 1944: 18). "Go often," Mother Cabrini lovingly counsels us, and "place yourselves at the feet of Jesus in the Blessed Sacrament," the Center of Life. "Behold the Divine Heart! He is our comfort, our way, and our Life" (August 21, 1890, *Travels*, 1944: 19).

Because of Mother Frances Xavier Cabrini, broken men, women, and children of the Post-Modern World — especially Italian Americans — need not be disinherited, spiritual orphans any longer. They have returned, at the urging of Mother Cabrini, from the "far country" of the Post-Modern World to the Eternal Father's House.

DONALD J. D'ELIA

---

[1] I would like to thank Sr. Ursula Infante, MSC, the translator of Mother Cabrini's letters, and Sr. Mary Louise Sullivan, MSC, Ph.D., Mother Cabrini's biographer, for their support and encouragement. I am also indebted to Sr. Regina Peterson, MSC of the St. Cabrini Home in West Park, New York; Mr. and Mrs. Eugene Cusatis and Mr. and Mrs. Leonard Piccoli of St. Joseph's R.C. Parish, New Paltz, New York; and Mr. and Mrs. John Mazzetti of St. Augustine's R.C. Parish, Highland, New York for their support of the Conference on "Mother Frances Xavier Cabrini, Bishop John Scalabrini, and Italian Immigration" held at SUNY - New Paltz in April 1999.

References

Blocker, Hyacinth O.F.M. 1950. *Walk with the Wise* (Cincinnati, Ohio: Francis Book Shop).

Borden, Lucille Papin. 1945. *Francesca Cabrini: Without Staff or Scrip* (New York: Macmillan).

Catechism of the Catholic Church. 1994 (New York: Image/Doubleday).

D'Elia, Donald J. 1999. "The People of the Festa: The Incarnational Realism of the Italian-Americans," Joseph A. Varacalli, Salvatore Primeggia, Salvatore J. LaGumina, and Donald J. D' Elia's (editors) *The Saints in the Lives of Italian-Americans: An Interdisciplinary Investigation* (Stony Brook, NY: Forum Italicum, 203-227).

DiDonato, Pietro. 1960. *Immigrant Saint: The Life of the Mother Cabrini* (New York: Dell Publishing).

Di Giovanni, Stephen Michael. 1991. "Mother Cabrini: Early Years in New York," *The Catholic Historical Review* 77: 56-77.

Guardini, Romano.1968, 1998. *The End of the Modern World* (1968 ed.; trans. Joseph Theman and Herbert Burke; edited with an Introduction by Frederick D. Wilhelmsen; 1998 ed., New York: Sheed and Ward; with a new Introduction by Rev. Richard Neuhaus, Wilmington, Del.: Intercollegiate Studies Institute, 1998).

Infante, Ursula, MSC. 1970. *Letters of Saint Frances Xavier Cabrini* (trans. from the Italian by Sr. Ursula Infante, MSC (Chicago, Ill.: Archdiocese of Chicago).

Maynard, Theodore. 1945. *Francesca Cabrini: Too Small a World* (Milwaukee, Wis.: Bruce Publishing).

Missionary Sisters of the Sacred Heart of Jesus. 1944. *Travels of Mother Frances Xavier Cabrini, Foundress of the Missionary Sisters of the Sacred Heart of Jesus* (with a biographical sketch by the Most Reverend Amleto Giovanni Cicognani, D.D. and an Introduction by Ottavio 18 Turchi, S.J., Chicago, Ill.).

Pope Pius X, *Pascendi Dominici Gregis* ("On the Doctrines of the Modernists"), July 3, 1907.

Sullivan, Mary Louise, MSC, Ph.D. 1987. "Mother Cabrini: Missionary to the Italian Immigrants," *U.S. Catholic Historian* 6: 265-279.

Sullivan, Mary Louise, MSC, Ph.D., 1992. *Mother Cabrini; Italian Immigrant of the Century* (New York: Center for Migration Studies).

Varacalli, Joseph A. 1999. "The Saints in the Lives of Italian American Catholics; Toward a Realistic Multiculturalism," Joseph A. Varacalli, Salvatore Primeggia, Salvatore J. LaGumina, and Donald J. D' Elia's (editors) *The Saints in the Lives of Italian-Americans: An Interdisciplinary Investigation* (Stony Brook, NY: Forum Italicum, 231-249).

Chapter 7

✦

CATHOLICISM, ITALIAN STYLE: A REFLECTION ON THE RELATIONSHIP BETWEEN THE CATHOLIC AND ITALIAN WORLDVIEWS

Building on academically respected and reliable descriptions of Italian Catholicism and Italian American Catholicism[1], the primary purpose of this essay is to move in the direction of articulating formally Italian Catholic multicultural or pluralistic versions of the Catholic faith. Put another way, what is intended is to provide coherent philosophical visions of the historic manner in which the majority of Italians and Italian Americans have appropriated the religion of Catholicism.

A secondary purpose of this essay is to critique such Italian style versions of the Catholic faith from the viewpoint of an "official," i.e., Magisterially-defined Catholicism, noting the ways in which the former is and isn't consistent with the latter.[2] Simply put, not all historical Italian cultural and Italian American ethnic appropriations of the Catholic religion have fallen within the parameters of the Catholic faith as defined officially by legitimate religious authority.

The key issue in this attempt is what is ultimately transformative: an official Catholicism or an historic Italianness? Put another way is the Catholic faith to be cut down to the key principles of an Italianità? Or is the Italian religious and cultural heritage to be incorporated into a Catholic faith that, in essence, is the in the driver's seat? Examples of the former would fall outside the parameters of an "official Catholic" Italian Catholic multiculturalism. Examples of the latter would be consistent with the philosophical and methodological approach of Pope John Paul II which is that one can best evangelize the faith through the process of "inculturation," that is, building the Catholic faith on whatever is true, holy, beautiful, and useful in the various cultures and subcultures of the world, Italy and the Italian American community included.

110

Themes for an Italian Catholic Multicultural Perspective

Offered here is a series of slightly overlapping themes, with corresponding intellectual justifications, that would be part of the conceptual scaffolding necessary for designing any specific Italian Catholic multicultural perspective. No claim is made that the following set of themes is exhaustive and adequately portrays the Italian Catholic religious experience totally and completely. Each theme is followed by a brief critique from a Magisterially-defined Catholic worldview. The end result, hopefully, will provide useful ideas for the construction of various Italian Catholic multicultural perspectives, some of which could, and others could not, claim to be legitimate interpretations of Catholicism.

Theme One: Rejection of any sharp distinction between the "sacred" and the "profane."

This theme is clearly enunciated in Richard Gambino's essay, "Italian American Religious Experience: An Evaluation" (1991). As Gambino puts it, "The Italian experience of life was a unity without dualism. The comic and the sober, the everyday and the sacred were experienced as a continuity..." (Gambino, 1991:41). He continues, "...in the old Italian religious experience, any event in inner life, as for example, in dreams, fantasies, visions, and prayers (that were often meditative in nature) could awaken one to the Divine Presence. Any outer experience, such as work, religious, social or family rituals, making love, the death of a beloved, music or the other arts, the beauty of nature or the sublimity of the sea or the night sky, could concentrate itself into a laser-like illumination that pierced the individual's heart with the eternal, with that which is beyond all time and space" (Gambino, 1991: 46). Robert A. Orsi supports Gambino in this contention, specifically in the case of *la festa*: For Orsi, the Italians "did not make a distinction between the 'religious' aspects of the *festa* — the praying and penitential devotions, the religious sacrifice — and what outside observers felt were the inappropriate 'profane' characteristics of the celebration — the food, noise, dancing, partying" (Orsi, 1985: xviii). Gambino relatedly notes that given that "there was no dualism of earthly love and heavenly love in the deeper levels of the experience...this caused much consternation to those Catholic and Protestant clergy who preached this type of dualism. It fell on deaf Italian ears, leading to the frequent charge that the Italians could not 'rise above their earthly nature'" (Gambino, 1991: 56-7). All of this is put more simply by Gambino when he simply states that, for the Italian, "life is the religious adventure" (Gambino, 1991: 48).

Critique: Collapsing the "sacred" and "profane" distinction is going to the other extreme.
The Italian Catholic belief that much legitimate religious activity occurs outside of the institutional structures of the Catholic Church is not only a useful corrective to the erroneous theological beliefs held by some Catholics but is actually, with qualifications, held by the Church herself. The Catholic religion not only believes that there are "rays of truth" found in other religions, but, more to the point, advocates the universal reality of the "natural law" written into the human heart, into healthy social institutions like that of the intact nuclear family, and into nature itself. However, the Italian tendency to elevate merely cultural institutions and practices which are temporally and spatially bound to religious status is a form of idolatry. On the one hand, there is nothing wrong in manifesting, for instance, a strong attachment to one's family traditions, neighborhood, and forms of music and food. They often are useful and life-affirming ethnocentrisms. In the final analysis, however, they are but "dust and ash before the Lord" and will ultimately vanish much like sand castles at the beach front. They should not be confused with the permanent and eternal truths about the relationship between God and the individual and society. Put crudely and following the spirit of Freud, it can be the case, alas, that a sausage and pepper sandwich is only a sausage and pepper sandwich.
Another related point that should be mentioned here is that not all transcendent, supernatural phenomena are necessarily of God and hence should not be recognized as such. And sometimes such phenomena as dreams, fantasies, and visions are things that both start and end with the human; every human ejaculation is not necessarily imbued with a transcendent sacredness.

Theme Two: At base, religion is "existential," a matter of "experience;" conversely, it is not, at base, "doctrinal."
This theme overlaps to a significant degree with the previous theme and is developed by Gambino who, in talking about the nature of religion, states that "it was existential; that is, the meaning or essence of life was

carved out of life's experience. To be sure, this was within a Catholic Christian structure, but the structure itself was interpreted and understood via living experience. There was little use for abstractions, even those carrying official church doctrine. Stories about the Holy Father and the saints were more vital than any theologian's scheme or papal encyclical" (Gambino, 1991: 49). The fact that religion is intrinsically existential or experiential, for Gambino, presents problems for any religious institution like the Catholic Church that relies heavily on doctrine and other aspects of religious tradition. As he claims, "a real danger for many Italian Americans who remain in the church is de facto spiritual indigence, coupled with empty traditionalism. The danger is losing living contact with the old experience and instead taking a religious stance which merely defends the 'old ways' with which they have, in fact, no lived connection" (Gambino, 1991:44).

Critique: All religion, indeed, all culture, is at least in one sense "doctrinal" in that "experience" requires for purposes of meaningful interpretation the existence of some prior set of concepts, whether those concepts are derived from religious or non-religious sources.

Fact and experience never speak for themselves. Indeed, — and following the Durkheimian cultural tradition in sociology and the religious studies perspective of George Lindbeck (1984) who opts for a "cultural linquistic" alternative to the an "experiential expressive" model — they are necessarily shaped and given meaning by human language, symbols, norms, and values. The conceptual apparatus needed to make sense out of otherwise meaningless experience, "a big, buzzing confusion," in the phrase of psychologist William James, certainly need not be highly articulated, formalized, and abstract as in the cases of Catholic encyclicals, or for that matter, Marxist philosophy, Freudian psychology, or feminist writings. All cultural formations, whether religious or not, can be found in both their "high" and "low" traditions. In the case of Italian Catholicism, the cultural concepts most often employed to interpret the Catholic faith inordinately have historically come from the "low tradition" associated with the social worlds of peasants, immigrants, and non-intellectual ethnics. It is correct to note that the average Italian Catholic may have had or has little use for official church doctrine but it is, I submit, wrong to assert that "the meaning or essence of life was carved out of life's experiences...(albeit)...within a Catholic Christian structure." To the contrary, what carved out the meaning of life, for many Italian Catholics, has been, variously, either southern Italian, Italian American, and now progressively more so, American culture. Simply put, "the meaning or essence of life" for the Italian Catholic — as with any other individual — has been carved out of "culture," not "experience." In other words, it is emphatically not the case that pristine experience leads the way; dreams, fantasies, prayers, and outer experience all require an interpretive framework. The issue is the nature of the religious or cultural worldview that provides the individual with a conceptual lens from which to interpret everyday life. Put another way, the claim that religion is, at base, "experience" masks the reality that the ultimate concern of many Italian Catholics is their Italianità or, now, "Americanness" and not the Catholic religion.

Finally, it should be pointed out that spiritual poverty and empty traditionalism is not always or mostly a function of sticking obstinately to outdated ideas and rituals but, more often, is a reflection of an insufficient and inadequate presentation of religious tradition. Tradition, following the thought of someone like Russell Kirk, can be a very dynamic and life-affirming reality.

Theme Three: Religion, Catholicism included, is contoured to the demands of the family.

Speaking of Italian American religion, Robert A. Orsi states that many observers, even the most hostile, admitted that Italians were faithful attendants at baptisms, weddings, and funerals, and no one denied that they went to church on the days of their feste.... The popularity of baptisms, weddings, and funerals begins to point us toward the meaning of Italian popular religion.... Why be so faithful about baptisms and not about Catholic schools? Why weddings and not Sunday mass?" (Orsi, 1985: xvi). For Orsi, in order to penetrate the meaning of Italian religion, one needs "to identify and study what it was that the people themselves claimed, implicitly or explicitly, as the foundation of their understanding of the good and the basis of the moral judgment" (Orsi, 1985: xix). "Without exception, the Italians...(replies Orsi in his study of East Harlem, New York from 1880 to 1950)...identifies this as the 'domus'" (Orsi, 1985:xix).

For Orsi, then, one must analyze the relationship of Italian Americans to Catholicism by including their more inclusive attachments to culture, more specifically to the family and, by extension, to the local community. The selective use that Italians have made of an institutional Catholicism cannot be understood apart from the reality that Italians have mediated their religion through the family. They were serious Catholics, again, when it came to baptisms, weddings, funerals, and festas; they were less serious Catholics when it came to attending Sunday Mass, going to Catholic schools, participating in the sacrament of penance, making financial contributions, and encouraging their sons and daughters to enter religious life.

Critique: The family — or, for that matter, any other social institution or practice — should never be viewed as one's ultimate concern or allegiance; that honor belongs to God alone.
There is, perhaps, no single theme more closely associated with the historic Italian worldview than a central attachment to the family. Familism can be defined as the philosophy that all other involvements and considerations are subordinate to the family; what is perceived to be in the interests of the health and welfare of the family is primary (Varacalli, 1992). It should be immediately noted here that familism is progressively being rejected in contemporary American society, even on the part of many among the younger generations of Italian Americans as the associated philosophies of individualism and statism become more entrenched facts of life. Part and parcel of this movement is the idea that it is the alleged "autonomous self" that is the ultimate locus of decision-making with the state supposedly guaranteeing legally the final authority of the individual.
Catholicism, with important qualifications, is more sympathetic to historic Italian familism than is the contemporary American and even contemporary Italian attitude. Catholicism sees the family as a more "natural" institution than the state and as one that existed prior to the development of any governmental power. The right of the family, — "a little Church," if you will — to exist and the proper authority of parents must be respected. On the other hand, Catholicism insists that *all* social institutions — from the family to the state — as well as all the activities of individuals, must be consistent with, and subordinate to, God's law. An official Catholicism, therefore, stands opposed to the typically historic Italian idolatry of the

family. For instance, resort to the use of birth control and abortion have typically been justified in the pre-modern Italian vision as necessary concessions made in order to protect the interests of *la famiglia* and the *domus*. These activities, from a Magisterially defined Catholic point of view, are just as morally wrong as the modern American and Italian defense of birth control, abortion, and divorce on the grounds of *individual* rights. Regarding the modern Italian exercise of these latter activities, contemporary Italian society now has the lowest birth rate of all the nations of Europe. An official Catholicism is seen here simultaneously criticizing what might be referred to as, respectively, "pre-Christian" and "post-Christian" forms of paganism.

Theme Four: For the Italian, religion, Catholicism included, is mediated through vehicles that are, respectively, 1) personal, concrete, immediate, and situational, 2) affective, spiritual, and "of the heart," 3) communal/parochial, and 4) lay centered.

Italian religiosity, especially southern Italian religiosity, is characterized, among other things, as "mystical" and "passionate" (Vecoli, 1977:42) and "warm," "personal," and "parochial" (Russo, 1977: 196,198). As such, the southern Italian immigrant to American shores had no natural attraction to either American Protestantism or to Irish and American Catholicism. In the first case, and despite the occasional instances in which the Italian immigrants used and "played off" Protestant missionaries in their battle with the Irish, the Italian religious sensibility was profoundly at odds with the individualistic ethos of Protestantism. The Protestant Reformation, at least in many cases and to a significant degree, had eliminated — referring to sociologist Peter L. Berger's phrase — the "cosmological baggage" between Christ and the individual characteristic of the Catholic faith and interpreted by Protestants as a retrograde paganizing movement within Christianity. Indeed, the barren Protestant cosmology, stressing the direct and unmediated relationship between the individual Christian and Christ was precisely the antithesis of the "saint drenched" world that the southern Italian peasant and first generation immigrant inhabited, a world characterized by D'Elia for its "incarnational realism" (1999). While the Irish Catholic and American Catholic world at the time was anything but barren, it was clearly perceived by the southern Italian as too abstract and doctrinal, cold and distant, legalistic and nationalistic. As Rudolph Vecoli concludes in his essay, "Prelates and Peasants," the "Italian problem was many things to many people, but to the Italian immigrants themselves it may have been that the Church in the United States was more American and Irish than Catholic" (1969:268).

Critique: Italian Catholic religiosity contains many valid, indeed indispensable, elements of the universal religious impulse; however, by themselves these elements are incomplete.

The official Catholic worldview heartily affirms all the characteristics of Italian religion. And as Italian American Catholic thinkers such as Father Nicholas J. Russo (1977) and Francis X. Femminella (1961) have argued, they have served as a necessary corrective to Irish Catholic interpretations of the faith. A universal Catholicism, however, balances Italian, Irish, and other particularistic interpretations of the Faith and completes them by setting them against their dyadic opposites which produces a religious worldview that is far more universal and comprehensive than any specific multicultural option standing alone as an entity by itself (Varacalli, 2001). Likewise, Protestantism, from the official Catholic perspective, is depicted as a splinter of the fuller truth of the Catholic religion.

Italian personalism encourages a religious orientation stressing responsiveness to the concrete, immediate, and situational needs of the individual and group, whether it be the family, village, or neighborhood. The typical southern Italian appropriation of the saints is one in which the latter have represented palpable, warm, and accessible intercessors in a cosmos otherwise seen as too alien, distant, and abstract. While meeting the needs of all of his people, however, the Church teaches that God answers the individual or group in his own way and in his own time. The God of the Catholic Church operates "with the big picture" in mind, a picture many times opaque to human understanding and one not necessarily responsive in any immediate sense. "The ways of God," in short, are not necessarily "the ways of man." Referring to the terminology of the Harvard psychologist Gordon Allport (1960), Italian religiosity has historically been more "extrinsic" and "immature" than "intrinsic" and "mature."

Italian religiosity is spiritual, affective/emotional and "of the heart." Italian religiosity emphasizes an all-merciful and therapeutic God; it is opposed to the Jansenistic tendencies of some variations of Irish Catholicism. This is balanced off in the more comprehensive Catholic worldview by including a vision of a God that understands the necessity of absolute moral norms of conduct, Church law, the implementation of justice in the form of the reparation of sins, and of strong intellectual traditions capable of self-criticism and continual reform.

Italian religion is parochial and communal. Italian American religious sensibilities extend primarily to a concern for the health and welfare of, primarily, one's family and, secondarily, to friends and local neighbors. This limited vision is externally reflected in the typical Italian emphasis on the worship/veneration of the Holy Family consisting of Mary, Joseph, baby Jesus, and grandmother Ann as well as to the local village or regional patron saint. Charity, for many Italians, begins in the home and one's duty is to tend to the needs of one's own garden. Obviously, if the whole world were successful in this approach, all would be well. Empirically, however, this isn't the case. On this count, the official Catholic worldview necessarily extends the range of concern to all fellow human beings all of whom are made in the image of God and also adds the issue of "justice" to "charity" (Varacalli, 1999). For an official Catholicism, a concern for justice is a constitutive element of the preaching of the Gospel.

Given the rejection of a sharp distinction between the sacred and the profane, it should not be surprising that the Italian religious sentiment is heavily lay-centered. Honor is paid to those who fulfil their religious and moral commitments to the family, hence the special role of Mary, "mother of mothers," and Joseph, role model for fathers and to the local neighborhood, hence the veneration of local patron saints and folk heroes. Catholicism, while absorbing the Lutheran emphasis on the apostolate for the layperson, nonetheless still reserves a special role for those who are called to holy orders. The Catholic vision is one in which, ideally, priest and laity both perform essential and complementary, not opposing, missions in the Church and world, respectively. For the priest, the focus is on the ministerial administration of the sacraments; for the lay person, it is to serve as a leaven in various worldly apostolates.

Theme Five: Italian religiosity is both pragmatic and this-worldly and lacks any "surrender" to the transcendent.

Southern Italian religiosity, whether peasant or immigrant, was used and manipulated in a practical, world-affirming way as a vehicle to defend the individual from an otherwise hostile, poverty-ridden, and politically oppressive social order referred to as the state of *la miseria*. As Rudolph Vecoli put it in his "Cult and Occult in Italian American Culture": "the life of the *contadini* was hard, mean, and cruel... religion and magic merged into an elaborate ensemble of rituals, invocations, and charms by which they sought to invoke, placate, and thwart the supernatural" (1977:26). Speaking of the various entreaties of the peasants to the saints, Vecoli follows: "the faithful were not simply making prayerful appeals; rather they regarded these supernatural beings as personalities who could be enlisted in their cause by the performance of certain acts" (1977:28). Vecoli then quotes A. L. Maraspini in regard to the peasant's belief and practice of magic: "for the peasant does not pray to the saint in the pious hope that the latter may take pity on him or that by any meritorious act he may deserve the saint's sympathy; he believes that the saying of the prayer, the lighting of the candle, and the offering of the ex-voto, are in themselves sufficient to enforce the saint's interest of his behalf" (1977:28). As Vecoli concludes, "should the saint fail his petitioner, he stood in danger of having his statue or image cast out or destroyed in retribution" (1977:28).

Critique: Any religious worldview that claims to be authentically Catholic necessarily requires a significant acceptance of an attitude of "world rejection."

As sociologists of religion from Max Weber to Robert Bellah have noted, the great world religions, in one way or another and to some significant degree, entail a rejection of the idea that salvation can be achieved by accepting and making paramount things of this world. Put another way, in all the great world religions, the spiritual must trump in importance the material. Within a Catholic framework, the spiritual must *also* transform things of the world into action and service to God. The Church teaches that the ultimate intended destiny of each human being is not to acquire worldly "success" but to reach heaven. However the Church simultaneously believes that without good works, faith is dead. In the final analysis, from a

Catholic perspective, individuals are seen stumbling to the cross in a this worldly "vale of tears." But as we stumble, the Church teaches, we are also commissioned by Christ to do so in solidarity with, and concern for, all of His children and for all of His creation, the political and civil order included. Put another way, the Catholic Church claims perennially to offer mankind a bright promise, that of eternal salvation resting with the Lord in the next life and a balanced, purposeful life in this world serving God, and through God, oneself, one's family, one's local community, and one's society (Varacalli, 2001).

The ideal vision of the purpose of life as depicted by an official Catholicism can be contrasted to that of the more secular Italian and Italian American worldview. Historically, for the peasant of *il Mezzogiorno* and for the overwhelming majority of the immigrant and second generation working class Italian American population, the most sacred or ultimate concern was to fulfill the quite worldly requirements of *la via vecchia* (the old way) as mediated through, first, the family, and second, the village or local neighborhood. For most contemporary Italian American Catholics, the most sacred or ultimate concern seems to be what Norman Mailer termed the "American Dream," although, perhaps, celebrated *la dolce vita* style. This is especially true for those Italian Americans born during and after the Second Vatican Council and hence twice removed from the highly integrated and religiously successful immediate post World War II American Catholic scene.

Regarding the issue of the typical turn of the twentieth-century appropriation of the saints on the part of Italian peasants and immigrants, it is clear that it was used, in their mindset qua a magic-infused folk religion, in a utilitarian way that underplayed the Catholic emphasis on the necessary transvaluation of the world and a selfless surrender and service to God. From a Catholic perspective, the saints are seen as "intercessors to Christ" who should be venerated. Put another way, they are not to be seen as demigods who are to be either worshiped, bargained with, or manipulated magically for purely worldly considerations. The latter approach of many Italians and Italian Americans to the idea and practice of the saints was but a small part of the generally recognized, from the Church's point of view, "Italian problem" of the Catholic Church in the United States (Browne, 1946).

Theme Six: Italian religiosity accepts an understanding of human nature and social institutions, including that of the Catholic Church, that ranges from an anti-utopian realism to, more often than not, an overt cynicism.

Typical of Italian, especially southern Italian, thought is the propensity to "debunk" the oftentimes pretentious claims of those who occupy the official authority positions in a society or an organization (Varacalli, 1992). This debunking, realistic, or cynical attitude is quite consistent with the famous claim of the English Catholic historian, Lord Acton, who posited that "power tends to corrupt and absolute power corrupts absolutely." Many traditional southern Italian *contadini* and first generation Italian Americans held attitudes, as such, that ranged from "indifference" to an institutional Catholicism to an overt anticlericalism. These Italians were quite content to practice their version of a folk Catholicism on the borders, at very best, of the official Church and inside their homes, neighborhoods, and local ethnic-religious associations. Interestingly enough, some modern-day Italian American thinkers formulate in a more articulate, intellectualized manner their opposition to an institutional Catholicism through their selective use of the Vatican II statement that the Church be defined "as the people of God." Such Italian American intellectuals, as such, have now posited a (presumably, true) "people's religion" over and against a (presumably, false) Church hierarchy.

Critique: Italian religiosity does not admit of the special graces that God bestows on the Church Magisterium, i.e., on the Pope as the Vicar of Christ and those Bishops who stand in loyal communion with him.

The relationship between Italian realism/cynicism and an official Catholicism is complex. To the degree that Catholicism sees human creation as basically good and social institutions as necessary to implement God's will, it does not share a fundamental compatibility with Italian, especially southern Italian, thought. To the degree, however, that the Church recognizes the impact of original sin on both human nature and individual persons, there is room for a healthy dialogue with southern Italian cynicism.

Where Italianness and Catholicism especially come to loggerheads in this area is on the issue of the authority of the Magisterium. The Church teaches that the Holy Spirit bestows certain graces on a Bishop who, if willingly open to such graces, protects the Bishop from his imperfect nature in his religious duty, qua Bishop. "Grace works through nature," the Church says. Sometimes, however, as is evident through today's disgraceful scandals involving not a few Bishops and priests, it is palpably evident that some are too weak to integrate successfully the life of grace with the human condition. Put another way, nature shows itself capable of resisting what God offers. Indeed, from the most cynical Italian attitude, *any* attempt to transform a less than perfect human nature is full of hubris and constitutes the task of the fool. As the Church believes, however, all individuals — Bishops, priests, and lay people included — are free to either accept or reject the invitation from God to pursue holiness. Holiness, in short, from the Catholic perspective, is a realistic ideal that all can and should strive after.

The complex and constantly changing relationship between the Catholic and Italian worldview is a central concern in the field of Italian Catholic and Italian American Catholic scholarship. Hopefully, this brief reflection is suggestive of future lines of research.

JOSEPH A. VARACALLI

---

[1] In the construction of an Italian Catholic multiculturalism, I rely mainly, but not exclusively, on the work of three major contemporary scholars in Italian American scholarship on religion. They are, respectively, Richard Gambino (1991); Robert A. Orsi (1985); and Rudolph Vecoli (1969, 1977). Key, but not exclusively so, to the former's perspective is his understanding of the inextricably intertwined relationship between the "sacred" and the "profane" typical of Italian Catholicism. Key, but not exclusively so, to the second thinker is his claim that the Italian culture is "domus-centered" and therefore that Italian Catholic religiosity is contoured to the needs of that historically most central institution to the Italian, i.e., the family. Key, but not exclusively so, is the latter's claim that the Italian religious sensibility rejected any cosmology that was viewed as either abstract and/or barren, including an Irish version of Catholicism and even more so, a Protestant viewpoint of the relationship of the individual to God. The work of other scholars of Italian and Italian American religion have also influenced this essay. Most prominently these would include essays by Browne (1946), Femminella (1961), Russo (1977), and D'Elia (1999).

[2] It is important to point out two issues immediately. The first is that "Magisterially-defined" is not identical with "Irish or Irish American Catholicism;" the Irish have their own particular interpretation of an official Catholicism, parts of which are variously inside and outside of the bounds of an official Catholicism. The second is that the Catholic religion, "officially understood," and perhaps especially

exemplified in the worldview of John Paul II, is one that argues that the faith can and must build on whatever is true, holy, beautiful, and useful in the various cultures and subcultures of the world, the Italian and Italian American culture and subculture, respectively, most definitely included. The process of inculturation, however, if it is to be implemented correctly from a Catholic frame of reference, is not indiscriminate and must follow a basic principle, to wit, that the cultural idea or practice to be baptized and religiously converted must, minimally, not be intrinsically incompatible with the authentic Catholic heritage.

References

Allport, Gordon. 1960. *The Individual and His Religion* (New York: MacMillan).

Browne, Henry J. 1946. "The 'Italian Problem' in the Catholic Church of the United States, 1880-1900," *Historical Records and Studies* (New York: The United States Catholic Historical Society) 35:46-72.

D'Elia, Donald J. 1999. "People of the Festa: The Incarnational Realism of the Italian Americans," in Joseph A. Varacalli, Salvatore Primeggia, Salvatore J. LaGumina, and Donald J. D' Elia's (editors) *The Saints in the Lives of Italian-Americans: An Interdisciplinary Investigation* (Stony Brook, NY: Forum Italicum), pp. 203-207.

Femminella, Francis X. 1961. "The Impact of Italian Migration and American Catholicism," *The American Catholic Sociological Review*, 22:223-241.

Gambino, Richard. 1991. "Italian American Religious Experience: An Evaluation," *Italian Americana*, 10:38-60.

Lindbeck, George A. 1984. *The Nature of Doctrine: Religion and Theology in a Post-Liberal Age* (Philadelphia: The Westminster Press).

Orsi, Robert Anthony. 1985. *The Madonna of 115th Street: Faith and Community in Italian Harlem, 1880-1950*. (New Haven and London: Yale UP).

Russo, Rev. Nicholas J. 1977. "Three Generations of Italians in New York City: Their Religious Acculturation," pp. 195-213, *The Italian Experience in the United States* (edited by Bishop Silvano M. Tomasi and Madeline H. Engel) (Staten Island: Center for Migration Studies).

Varacalli, Joseph A. 1986. "The Changing Nature of the 'Italian Problem' in the Catholic Church of the United States," *Faith and Reason* 12:38-72.

Varacalli, Joseph A. 1992. "Italian American Catholic: How Compatible?," *Social Justice Review* 83:82-85.

Varacalli, Joseph A. 1994. "Multiculturalism, Catholicism, and American Civilization," *Homiletic and Pastoral Review*, 94:47-55.

Varacalli, Joseph A. 1999. "The Saints in the Lives of Italian American Catholics: Toward a Realistic Multiculturalism," in Joseph A. Varacalli, Salvatore Primeggia, Salvatore J. LaGumina, and Donald J. D' Elia's (editors) *The Saints in the Lives of Italian-Americans: An Interdisciplinary Investigation* (Stony Brook, NY: Forum Italicum), pp. 231-249.

Varacalli, Joseph A. 2001. *Bright Promise, Failed Community: Catholics and the American Public Order* (Lanham, Maryland: Lexington Books).

Vecoli, Rudolph. 1969. "Prelates and Peasants: Italian Immigrants and the Catholic Church," *Journal of Social History* 2:217-268.

Vecoli, Rudolph. 1977. "Cult and Occult in Italian American Culture," pp. 25-47, *Immigrants And Their Religion in Urban America* (edited by Randall Miller and Thomas D. Marzik) (Philadelphia: Temple UP).

Part III
Feminist Models and Images

Chapter 8

✦

ITALIAN AMERICAN WOMEN AND THE ROLE OF RELIGION

The role of religion in the lives of Italian American women is a subject of interest and complexity limited to neither a singular nor definitive interpretation. History, culture, politics, economics, society, and gender are among the multitude of factors that may be said to inform our understanding of this largely "neglected" topic (Brown, 2000: 681). Important contributions in current scholarship notwithstanding, changing and emergent realities in the relationship between the Church and Italian Americans in general and Italian American women in particular, necessitate ongoing reassessment. Nor is our historical knowledge of the subject without gaps and puzzling questions. A critical approach would necessarily require researching any number of cultural and historical dimensions. In the writing of this essay, bibliographic sources spanning a wide variety of subject areas were highly informative. One text aiming at a comprehensive approach to Italian American subjects proved particularly invaluable: *The Italian American Experience: An Encyclopedia* (LaGumina, Cavaioli, Primeggia and Varacalli, eds., 2000).

The aim of this study is, first, to provide an historical overview of the correspondence between Italian and Italian American women and religion, and second, to raise issues pertinent to this connection, particularly in the context of contemporary Italian American women. Some important questions are ultimately raised, ones which may well become the basis of further study.[1] One such question is given extensive treatment near the end of this essay: "Despite the professional and educational advancements of Italian American women and their increasing contributions to scholarship and literature, within the Church and other public venues and spheres of influence, what explains their comparative silence?" This question prompts others, many of which transcend gender-based issues, becoming more broadly relevant to Italian Americans and any persons with an interest in cultural and/or religious studies.

Historical Overview

Elizabeth Messina's article, "Women in Transition," opens with the following thought-provoking lines:

> Cross-cultural psychologists and ethnographers believe that history and individual development are linked and that historical events affect personality development, emotional functioning, and psychopathology of the immigrant group and subsequent generations. In their country of origin and in the United States Italian immigrant women exemplify this dynamic by the difficulties they have endured. (2000:687)

For the present study, if one agrees with Messina that history and individual development are linked, then one should want to know what difficulties Messina has in mind here and to what extent, if any, they contributed to, or were influenced by, the role religion would play in the lives of these immigrant women. To address this issue, one should begin by examining aspects of Italy's history before the period of mass migration to the United States. This is because, with minor exception, southern Italians experienced an "Italy" culturally and qualitatively different from that of northern Italians. This difference would, by the turn of the century, be a primary catalyst in the mass exodus to the United States from the regions south of Rome, the Mezzogiorno.

What would cause a people to leave in droves from their homeland? What contributing factors fed the difference between north and south? One social condition permeating the latter was that of "pervasive racial prejudice" (Messina, 2000:687). Centuries-old bias by northerners, largely made up of landowners, the middle class, the nobility, and the clergy, resulted in southerners regarding with suspicion virtually all economic, political, educational, and religious institutions and spheres of influence.

In part because of an oppressive, exploitative, and antagonistic, if not hostile, social environment, southern Italians turned their trust inward toward the family (Gambino, 1974b:429; Messina, 2000:688). Within this most basic of social structures, the mother typically held a key position. As the traditional southern Italian expression goes, "the man stands at the head of the family, the mother stands at the center."[2] Precisely because of the importance of family "over any other Italian institution" (Barolini, 1985:9), the mother would indeed hold a "central" position.

Within this familial scheme of equity and difference, the peasant women of southern Italy played an active, if not crucial, role. Mangione and Morreale describe some of their everyday activities:

> They had to fetch water from the well, wash the family clothes in a public water trough, bake bread, prepare food, shop and bargain, and, weather and time permitting, roam the open fields with their children for dandelions or wild fennel. In the fall they gathered *rusticia* (twigs and dried wheat stalks) and returned with a giant bundle on their heads, children beside them. They helped harvest olives, almonds, and walnuts, and hired themselves out to gather grapes for the making of wine. Women were also *levatrici* — midwives (doctor, pharmacist all in one), who did cupping, set leeches, brought infants into the world, and helped bury the dead. (1992:235)

Women likewise contributed to the social life and protection of their communities. Mangione and Morreale tell of one dramatic example:

> During the Fasci Siciliani, in many Sicilian towns it was the women who stormed the police barracks or city halls to release prisoners or help throw out bureaucratic files that were burned in the streets. Often, when the men were in need of help, the cry was heard: 'Go get the women.' (1993:236)

The Church, also integral to the social fabric of the community, offered women a carefully measured part within its sphere of influence. Church decoration and cleaning were typical tasks for women, as was teaching catechism. Unmarried women served children, the sick, and the elderly. Many became nuns. A select number became midwives, a role sanctioned by the Church, and a subject taken up later in this essay.

The relationship between southern Italians and the Church was nevertheless oddly dichotomous. On the one hand, Catholicism was recognized as perhaps the sole "unifier" of the many and diverse regions of the Mezzogiorno and northern Italy. Yet northerners, for their part, had little knowledge of the comparatively difficult circumstances of southern Italian Catholics. Moreover, while southern Italian women often found a kind of stature and purpose within the Church, becoming instrumental in many of its activities, the men were characteristically distrustful of it and not inclined to participate.

Saints and *La Madonna* were venerated, particularly by southern Italian women. Their prayers generally centered around petitions for comfort from personal troubles and the difficulties of peasant life. Yet the Church itself would be a refuge more symbolic than real as centuries of oppression, some of it church-inspired, had left its mark upon the psyches and emotions of a people whose responses ranged from apathy and indifference to suspicion and cynicism.

Italian Unification

Centered in northern Italy, the unification movement[3] issued laws most economically advantageous to the North. Thus as the North experienced economic growth, the South was left impoverished and socially marginalized. Hearder notes that, "The peasants were burdened with crushing taxes, which led to riots, and a violent insurrection in Sicily in 1866" (2001:204).

The Unification was, in one sense, more fiction than fact, having been described as "one of the great cosmetic myths of all time" (Cornelisen,[4] 2001:172). One year before his death in 1882, Garibaldi himself is quoted as saying, "It is a different Italy than I had dreamed of all my life, not this miserable, poverty-stricken, humiliated Italy we see now, governed by the dregs of the nation" (Morreale and Carola, 2000:57). This sentiment is directly linked to "the sturdiest myth of them all" (Cornelisen, 2001:173): the "Southern Myth," otherwise known as the "Southern Question." As Cornelisen explains:

> To generations of governments it was the excuse for a policy of indifference with equal taxation, softened slightly by graveside injections of emergency funds, not intended to cure the patient, but to keep him from disturbing the neighbors. Mussolini was frightened by it and chose to banish it with rhetoric. Poverty was a disgrace to the new Roman Empire. Southerners must try harder, produce more. More wheat, more children, more emigrants for the African colonies! His greatest contribution to the South may have been an enormous tax on goats that reduced the number of the beasts gnawing at what little shrubbery still held the

earth to the hillsides — of course, at the expense of the local diet. (2001:173)

"There are many ironies in Italian history," say Mangione and Morreale, "but none greater perhaps than the phenomenon of Italians leaving Italy so soon after it became a united nation in 1871" (1993:xiv). Of the Unification, Cancilla Martinelli explains that it was "essentially a civil war between the new, upwardly mobile middle class of the Industrial Revolution and the entrenched aristocracy." She says, moreover, that the "'blessings' of unification came to peasants in the form of increased taxes, military conscription, loss of feudal privileges, and increasing amounts of deforestation and land fragmentation" (2000:498). When the government did not respond to a parliamentary report on "rural conditions," which exposed conditions such as malnourishment, child labor, problems with housing and illiteracy, and plagues of malaria in the south and pellagra in the north, mass emigration would begin (2000:498). While great numbers left the Mezzogiorno region, northern emigrants to the U.S. were sharply declining. In fact, not less than four fifths of Italian emigrants were from the seven regions south of Rome: Abruzzo, Molise, Campania, Lucania (now Basilicata), Apulia, Calabria, and the island of Sicily. In 1860, only 2.5 percent of the "Italian" population could speak Italian. Most Italians were peasants for whom nationalism had little meaning. Nor did they actively endorse a unified Italy. Most of the people spoke the individual dialects or, in some cases, languages of their respective regions. In fact, despite over four hundred years of collective subjugation under Spanish and Bourbon rule, each region of the Mezzogiorno was, in most respects, culturally distinct, with its own traditions, folklore, patron saints, and language or dialect. This cultural variety would find its way to America via the immigrating peasantry.

Another irony of the Unification pertains to the Catholic Church which, despite its part in bias that added to social discord, "did not recognize the new nation of Italy" (Morreale and Carola, 2000:26), becoming, in a sense, the de facto unifier of Italy's many regions. Thus persons from Naples or Sicily could find common ground in their Christianity. Each was otherwise wholly identified members of their respective regions. Moreover, no southerner considered him/herself "Italian" — that is, unless he/she migrated to America, where officers of immigration and others would identify them as such. To this day, many southern Italians and Italian Americans, particularly those with ancestral origins in the Mezzogiorno, identify primarily with their towns, villages, and regions as opposed to Italy as a nation. Scholar James T. Fisher says of Italian emigrants that "they all shared an abiding attachment to their village community. Many, if not most, hoped one day to return to Italy" (2002:73).

Immigration to America
Northern Italians were first to migrate to America. From the late 1700s to the mid-1800s some twelve thousand Italians, mainly well-educated males from northern Italy, migrated to America. With but few exceptions, little is known about them.

Southern Italians began immigrating in the late 1870s. They initially settled in other European countries and then in Argentina and Brazil. The number of Italians migrating to South America would be drastically and permanently reduced, however, by the devastation of Brazil's yellow fever epidemic, which claimed the lives of some nine thousand Italians.

Regions in North America, where cheap labor was in demand, would now become a more viable destination than South America. From 1870 to 1880 over fifty-five thousand Italians left the largely poverty stricken agricultural regions of the Mezzogiorno for the U.S. This initial migratory surge escalated to over two million within the first decade of the twentieth century. Over 4.5 million Italians had entered the U.S. by 1930.

American immigration laws, including the Immigration Act of 1917, the Quota Act of 1921, and the Johnson-Reed Act of 1924, effectively ended the mass migration of Italians. Additionally, deportations and voluntary repatriation to Italy increased due to the 1929 economic depression. Also, Italian Fascist policy placed a prohibition on emigrating altogether.5

As other Catholic immigrant groups had done before them, Italians tended to settle in industrial cities (Fisher, 2002:73). Yet if there was a harmonious link between Italian individuals and their greater external world, it would be found in nature. In his article, "From Urban to Suburban: Italian Americans in Transition," LaGumina speaks of Italian Americans settling in the suburbs of Long Island due to its favorable, more natural environs:

Long Island, New York suburban communities had an early attraction for people who loved being close to the land, people who yearned to raise their families in hospitable physical environments in their own homes. Among these people were small groups of turn of the century Italian Americans who in the mid-twentieth-century were joined by large numbers of their ethnic group and currently form the largest single nationality bloc (approximately 700,000) in proportion to the area's population. (1990:27)

Luciano J. Iorizzo observes that Italian American history is commonly understood in "urban" terms and sometimes from a "rural" perspective (1985:360). Of LaGumina's observations about Italian Americans and suburbia, he concludes:

What LaGumina is suggesting is that Italian immigrants took part in the vanguard of a significant movement in American history, that is, the move to suburbia, a movement usually associated with middle class America. If this can be widely demonstrated, it would suggest, from still another vantage point, that the role of Italian immigrants in American history is not yet fully appreciated. (1985:360)

Religion and nature appeared to find a kind of unity in "LaMerica." Decades prior to the exodus, while southern Italians of the "Kingdom of the Two Sicilies" were subject to Bourbon rule, the "dream" of America was flourishing to the extent that it has been described as a "quasi-religious vision of a paradise on earth" (Mangione and Morreale, 1993). Inspiratory stories from Columbus to Garibaldi fueled the ever-growing LaMerica myth. Italians returning to Italy from the States "Americanized" the myth to such a degree that, as has been observed, the Statue of Liberty became "the Madonna of liberation," and the American dollar bill, something sacred to attach to the garments of religious statues (Mangione and Morreale, 1993).

With purpose and resolve, southern Italians left the stagnation of their hopeless lots behind and, for the most part, never looked back. They would soon discover, however, that the myth of L'Merica was just that, a myth, replaced by realities far less grand-indeed, rather stark. In *The Italian American Family Album*, the publishers quote an Italian immigrant as saying, "I came to America because I heard the streets were paved with gold.

When I got here, I found out three things: First, the streets weren't paved with gold; second, they weren't paved at all; and third, I was expected to pave them" (Hoobler and Cuomo, 1998).

Their new-found lot did carry with it one crucial difference from the past: the potential for a measure of upward economic mobility. "Of all developed societies," says Rocco Caporale, "[Italian immigrants] came from a country with one of the lowest [records] of upward mobility" (2000:598). According to LaGumina:

> The fabled "gold coast" of the north shore that was the setting for one of the country's major concentrations of mansions for the wealthy, Nassau's north shore was also an attraction for Italian immigrants, albeit of a proletarian social class. (2000a:31)

Italian immigrant men, competing for jobs against more well-established ethnic groups,[6] found work largely as unskilled laborers. Women would also become wage earners, a significant change from their wholly domestic conditions. Moreover, as Oteri Robinson puts it, "Immigrant women of the early 1900s built a cornerstone for future generations by overcoming oppressive work conditions" (2000:685). Sources of income for Italian immigrant women included home piecework, renting to boarders, and various manufacturing trades, including the garment, articificial flowers, lace, tobacco, paper products, and candy industries. Oteri Robinson notes, however, that "because of their strong work ethic and frugality, Italian American women were more easily exploited, often receiving the lowest wages" (2000:685). Until the garment strikes of 1913 and 1919, Italian American women had not participated in the labor movement as had other women. With these strikes, however, their voices were heard and would make a difference. From the strike of 1919, with key organizers Angela and Maria Bambace, Local 89, a separate Italian-speaking local, of the International Ladies' Garment Workers Union, was established.[7]

Midwives and the Church

One role for select women that linked them to the Catholic Church was that of midwife. Angela D. Danzi, in her informative article, "Midwives and Childbirth," names midwives "the most integral and familiar figures in Italian American communities" (2000:378). Furthermore, she says that, though medical doctors aimed at discrediting the foreign tradition of midwifery, many Italian American women employed their services into the 1930s.

As for the use of midwives in Italy, the tradition appears to have lasted longer there than in other parts of Europe. Danzi believes it probable that this was because "the Catholic Church and later the state, in varying degrees, sanctioned and regulated her role, and because of the longstanding aversion of the populace to the involvement of men in obstetrics" (2000:378).

The link between midwifery and the Catholic Church should not be underestimated. Dating back to the sixteenth century, the Church accorded midwives official status for the right to baptize.[8] Interestingly, in 1746 a Catholic priest wrote a popular instructional manual for midwives. According to Danzi, the manual "combined scientific and religious instruction, emphasizing midwives' solemn responsibility for the soul of the child" (2000:378).

Sanctioned by the Church, the midwife, in her capacities to deliver babies, baptize them, and nurse and heal the sick, was an important figure. Among her varied roles within the Church, she often served "as witness and confidante in intimate family matters and female sexuality." She thus "provided a way for the male clergy to reform and control the local populace" (Danzi, 2000:378).

The strong bond between midwives and the Church would nevertheless fade in the U.S. Though data into the 1930s indicated that, compared to the medical specialists, midwives had better survival rates for mother and infant, the medical establishment did eventually succeed in eliminating essentially all midwives from practice. This initiative found even greater force due to the Johnson Acts of 1921 and 1924, which caused a huge drop in the number of immigrants, including Italians and thus Italian midwives and those women who would have used her services.

Midwifery also fell out of fashion due to, as Danzi explains, better education and employment opportunities for women. Thus the midwife would essentially be relegated to the memory of their mother's generation. Danzi says, furthermore, that "...the preference for a physician and hospital birth was another way of demonstrating that they were 'modern, Americanized women'" (2000:380). The long-held tradition of midwifery had become neither desirable nor viable and the once respected midwife, sanctioned by the Church, would permanently lose a measure of influence and status, as did, in a sense, the Church itself.

La Madonna

Italian Catholics typically filled gender-specific roles. Men interfaced publicly to organize feast days, or *feste* that honored a local patron saint. Women, on the other hand, represented the family in devotional prayer. Feast day rituals included prayerful petitions and gratitude for blessings received. In the U.S., immigrants would continue practices of many such feste.

As with Catholicism in Italy, certain observances and practices of Italian American Catholicism may be said to extend beyond the boundaries of official Catholic doctrine. The celebration of the *Giglio* Feast[9] in Williamsburg, Brooklyn is one such example. In a recent radio interview, noted scholar Salvatore Primeggia[10] explained that such deviation from strict doctrine stems from the merging of Roman Catholicism with southern Italian folk religion, with its mixture of Mohammedan,[11] and early Greek and Roman belief structures. Primeggia and Varacalli note that "Research on the *Giglio* Feast...forces a reconsideration between the relationship between the sacred and the profane" (1996:423).

The *Giglio* Feast also shows clearly delineated, gender-specific roles:

> There are three ideal-typical relationships between men and women that, theoretically at least, could be applied to the feast under study. The first would be androgyny, or the movement claiming to witness the merging of male and female characteristics into a generic humanity. Such a movement would expect to document progressive stages by which social-role differentiation between men and women continually becomes smaller and less significant. Translated into our study, the indication of such a move would be the virtual interchangeability of men and women in the various roles and functions performed during the feast. Evidence of such a movement is almost nil. (Primeggia and Varacalli, 1996:431)

Pagan and folk culture in southern Italy was influential in shaping archetypal aspects of womanhood. Within this cultural framework women would also venerate various patron saints and La Madonna, the latter also strongly associated with notions of womanhood. Lucia Chiavola Birnbaum, writing on Italian folklore, explains the blending of Catholic and folk aspects by distinguishing between "the all-male trinity of papal catholicism" and "the cosmological center of popular catholicism, a center visibly inhabited by the madonna and saints" (1990:106). In southern Italy, the Virgin Mary was, in fact, one in a number of venerated madonnas, or goddesses. These archetypal figures of womanhood have their origins in prechristian, Graeco-Roman times. One such example is the Black Madonna, which dates anywhere from 20,000 to 10,000 BCE. Three goddesses sometimes portrayed as black figures include Isis, Cybele, and [Artemis (Diana)] of Ephesus. As Chiavola Birnbaum has observed, "early church images of the madonna and child bear a striking resemblance to the dark pagan goddess Isis holding Horus" (1990:107). Furthermore, she cites former Dominican cleric, Ean Begg, who asks, "Is the Black Virgin a symbol of the hidden Church and of the underground stream?" (1990:107). What did she, in fact, symbolize? What is the Virgin's significance in America? Robert Orsi puts it that the madonna in America has "inspired rich devotion, but she is not the madonna of popular culture elsewhere in the world who symbolizes resistance and revolution" (Chiavola Birnbaum, 1990:107; Orsi, 1985). Moreover, while the resurrection of Christ is the central theme of Christianity in the U.S., in southern Italy, "the observer of Easter folk processions is struck by the significance of the mother, and the open-endedness of popular Italian religious beliefs" (Chiavola Birnbaum, 1990:106).

The Virgin Mary, Mother of God, remains powerfully attractive to southern Italians. As Cornelisen puts it:

Mary, the Earth Mother figure, can be loved, trusted, and prayed to, while God and His son, Jesus, remain cold symbols. The Holy Ghost, so elusive anyway, is quite literally the white plaster bird that hovers uncertainly over all altars. Men find it hard to be humble before other men, even harder to lose face, so praying to a woman for her intercession is less abrasive to the ego. Women can identify immediately with the all-suffering Mother and perhaps take consolation in her importance to all men. Much as the Vatican may deplore it, in the South Christ is on the altar, but the people pray to and worship the Virgin Mary. (2001:23)

Among the Italian laity, Mary's maternity made her "a strong interces-sor" (Brown, 2000:681). Yet Mary is also said to be an "ambiguous model" for Italian and Italian American women. Brown cites Orsi's opinion that her veneration placed all women in a "double bind." On the one hand, they were honored for "maintaining family life." On the other, they were "held responsible" when problems arose within the family (Brown, 2000, 681; Orsi, 1985).

Veneeta-Marie D'Andrea argues that, for some women, the Italian Catholic image of womanhood fostered the belief that they could manage responsibilities, developing in them a "greater sense of identity." Character-istics of such an identity would include a serious-minded, assertive, protec-tive, and practical nature. According to D'Andrea, this, in turn, would increase one's self-esteem and reinforce skills needed for greater "social" and "economic" advancement (Brown, 2000:681; D'Andrea, 1983).

The Madonna or patron saints were often venerated for past blessings by a processional show of women walking barefoot and carrying lighted candles. They would also light candles or place upon an altar a waxen image of a particular body part that required healing. Religious veneration and petition rituals were also practiced by women at home, where they would keep a lighted candle or recite the rosary before a statue of a patron saint or the madonna. Through these practices, women would be fulfilling their cultural obligation to the family (Brown, 2000:682).

La Madonna, as Chiavola Birnbaum puts it, "has presided over Italian beliefs since the beginning. Her powerful valence in Italy—and her many dimensions—are visible in folklore rituals and in local political practices. A question that Italian American scholars might well ponder is why...the many multi-dimensioned madonnas of Italy continue to thrive while popu-lar devotion to the madonna in the United States is giving way to devotion to the male Saint Jude" (1990:110).

Immigrant Women and the Church

Immigrant Italian women, like Italian women, were the essential link between home and Church. They represented the family in fulfilling such religious duties as hearing Mass on Sundays and holy days of obligation, going to confession, receiving communion, and seeing that children re-ceived the sacraments and a religious education. An Italian American woman's position in this regard was so influential that Protestant denominations would appeal to her if they were to secure a possible conversion by other members of the family (Seller, 1978).

Yet, in caring for their families, women often consulted their pastor. Local parishes offered various family services. Societies, parochial schools, and other Catholic organizations, offered assistance in the form of childcare and children's activities as well as money, food, fuel, and clothing.

There are instances of individual women assuming leadership roles in public religious life. Brown cites Rudolph J. Vecoli who has documented that of Emmanuella De Stefano. Her husband, as Brown explains, "migrated from Laurenzana Potenza to Chicago in 1873. He was a pioneer of the migration of Laurenzanesi to Melrose Park, a Chicago suburb. In gratitude for his recovery from an illness, Emmanuella vowed to establish an annual feast day celebration of Our Lady of Mount Carmel and erected a chapel to house it. In 1903 the devotees became the basis for the parish of Our Lady of Mount Carmel in Melrose Park" (2000:682).

Pastors nevertheless preferred that women act "in an organized manner under pastoral direction" (Brown, 2000:682). One organizational initiative under such direction was the sodalities, or religious societies. A common purpose of sodalities was to join women in prayer. They also raised money for their parishes. Parishioners became members based upon their age and marital status. For unmarried women, the primary sodality was the Children of Mary. Married women of the late nineteenth and early twentieth centuries might join the Rosary Society of their parish (Brown, 2000:682). Women entering the church in procession, with special dress and the sodality banner, added a formality and solemnity to a Sunday or special-occasion Mass.

Earlier devotional practices in Italy were often brought to the United States. Brown mentions women from near Genoa who had a devotion to Our Lady of Guadalupe. Upon arrival in New York City, these women became members of Our Lady of Pompeii in Greenwich Village and organized an annual festa to Our Lady of Guadalupe there (2000:682).

With comparatively fewer studies of women in history, little information exists regarding the impact Italian women would have on Italian religious vocations. We know, for example, that, compared to men of other ethnicities, Italian and Italian American men have not joined the priesthood and religious life in nearly the same numbers. Vocational commitments of Italian American women is less certain. Interestingly, Italian culture seems to have been unopposed to the idea of women becoming nuns, while at the same time exhibiting an ill-defined, anticlerical stance. According to Brown, families might nevertheless have preferred that their daughters remain at home to care for their elderly parents (2000:683).

Notwithstanding the availability of educational and professional training in certain circumstances for some women, Brown puts it that "the Catholic Church shaped traditional conceptions of womanhood everywhere on the Italian peninsula" (2000:681). Their influence, including their role in family and parish life, as well as life in religious vocations, also presents itself to Italian American women. Yet within American Catholicism, Italians were considered a "problem" (Brown, 2000:681; Browne, 1946). This irreligious image of Italians was due to the fact that many Italians were anticlerical (Brown, 2000:681; Vecoli, 1983), an attitude that may be explained by both southern Italian folk culture and the unfavorable social and political climate of nineteenth-century southern Italy. One problem was thought to be that the majority of early migrants were male — indeed four out of five from the 1880s through the 1910s (Brown, 2000:681). However, the arrival of significant numbers of women and children after the first World War, as Brown explains, did little to solve the "Italian problem." It did, however, make Italian American Catholicism a greater reality.

Women, Religious Institutes, and the Church
Research on religious institutes in Italy and the United States reveals the dedication of numerous Italian and Italian American men and women.[12] In nineteenth-century Italy, a new sort of religious institute was emerging. This was due to a wave of secularism set in motion by Italy's unification. Catholicism was now subject to criticism by politicians who considered it an impediment to national unity. Consequently, as Brown explains, there was acquisition of Church property and usurpation of religious communities. In addition, free-market capitalism developed alongside a diminishing sense of state responsibility for the welfare of citizens. From these secular developments emerged new, more self-sufficient religious institutes, institutes more inclined towards current social issues related to social services (2000:683).

Italian Catholic women who initially joined religious vocations in America found that institutes lacked a "strong Italian ethnic identity" (Brown, 2000:683). This is likewise true of parishes, which continue to be relatively few in number. The Church of St. Mary Magdalen da Pazzi, founded in Philadelphia in 1857, constitutes the first permanent Italian American parish (Fisher, 2002:72). LaGumina notes, "Although there were Catholic parishes in Glen Cove, [New York], Italian Americans in the 1930s petitioned and built their own- St. Rocco." It remains an exception for the region, constituting "the only ongoing Italian national parish on Long Island" (2000a:37).

Newer ethnic groups desiring their own "ethnic" parish no doubt were inspired, at least in part, by the bias they encountered from more established American Catholics, including some clergy. Italian immigrants, already more than familiar with bias from the Church and elsewhere in their homeland, would again encounter it in the United States According to Fisher:

In New York, Philadelphia, and other eastern cities, the overwhelmingly Catholic Italians were often scorned by Irish and German American priests in whose territorial parishes they had settled; Italians were sometimes even relegated to attending separate masses in church basements. (2002:73)

To put such difficulties in perspective, the American Catholic Church seems early on to have had issues of bias from every conceivable vantage point. According to Morris' account:

Italian priests refused to baptize Lithuanians, Poles detested Czechs, Germans contended with Irish. Immigrant Irish priests attacked the laxity of Irish-American priests. Ethnic pastors shamelessly competed with one another for parishioners, and territorial pastors protested bitterly when ethnic parishes siphoned off their revenues. By 1916, half the Catholic parishes used a foreign language at least some of the time. European bishops often treated America as a kind of Australia for wayward priests, a dumping ground for clergy of the lowest quality... (1997:128-9)

A number of Italian Americans displeased with Catholicism converted to Protestant, often evangelical, sects. Social reformers also seized the opportunity to appeal to Italian dissatisfaction and anticlericalism. As Morris tells it, Chicago priests were outraged to learn of the Giordano Bruno Anticlerical Society, which was sponsored by Hull House, the original American settlement house. Bruno, of course, was Galileo's contemporary and the last in a long line of heretics to be burned at the stake. "Even Pastors who disliked Italians were embarrassed to be seen losing souls to socialists and holy rollers. Most bishops did their best to recruit qualified Italian priests, and eager young curates who were willing to help solve 'the Italian problem' found that promotions came quickly" (Morris, 1997:129-30).

As can be said of the many Catholic immigrant groups in America, Italian immigrants were neither disuaded in their faith, nor discouraged in their resolve to make their way in America. As immigrant German and Irish Catholics had done before them and others would do after, Italian American Catholics had become largely assimilated into American Catholicism. In fact, despite past pervasive, widespread bias, third-generation Catholics were more inclined to choose marriage partners from an ethnic group different than their own. According to the New York survey nearly one in four third-generation Italians married an Irish Catholic (Morris, 1997:131). Currently, American Catholicism is more multiethnic than ever. Under fourteen percent of American Catholics are identified as Irish American, and even less are Italian, German, or Polish Americans (Morris, 1997:301).

What is perhaps most noteworthy is the enduring benefits eventually accrued to some Italian immigrants because of the dedication of other Italians and Italian Americans — particularly women — to the Church. Mother Tommasini (1827-1913), who joined the Society of the Sacred Heart, is one such example. Maria Luigia Angelica Cipriana Stanislas Tommasini was born in Parma and came from an educated family. Due to the liberal revolutions of 1848, she left for France and then, the United States. She taught there at Manhattanville and then served in various administrative positions within her institute (Brown, 2000:683; Schiavo, 1949).

Italian women also established contemplative institutes. In 1875, for example, Bolognese blood sisters, Mother Mary Magdalen Bentivoglio and

Mother Mary Constance Bentivoglio, members of the Order of Saint Clare, established a monastery in the United States.[13]

In the nineteenth century, a number of institutes newly formed in Italy were also established in the United States. Though this was a period of Italian mass migration, and though they operated within migrant populations, Italian immigrants were often ignored. The first such organization, the Franciscan Missionary Sisters of the Sacred Heart, was founded near the Italo-Austrian border by a French countess and a Roman priest in 1861. It was then established in the United States in 1865 and its first United States mission was to the German parish of Saint Francis of Assisi in New York City (Brown, 2000:683).

Ultimately, numerous Italian institutes would be established in the United States.[14] The most famous one working directly with Italian immigrants was that of Saint Frances Xavier Cabrini (b. Lombardy, Italy, 1850; d. Chicago, 1917), whose Missionaries of the Sacred Heart, founded in Italy in 1880, came to New York in 1889. Naturalized in Washington State in 1909, Mother Cabrini, originally from Lombardy, Italy, was the first American citizen to be proclaimed a saint by the Roman Catholic Church. Canonized in 1946, Saint Frances Xavier Cabrini had been deeply devoted to Italian immigrants, making numerous visits to Italian American centers throughout America and establishing schools, orphanages, hospitals, and social service programs.

Institutes, such as The Pallottine Sisters of Charity (New York, 1889) and The Daughters of Saint Mary of Providence (Chicago, 1913), also worked with Italian immigrants. Many others such as, Sisters of Our Lady of Sorrows (1947) and The Camboni Sisters (1952), came to the United States after the period of Italian mass migration.

Contemporary Italian American Women

Each generation of Italian Americans experience challenges unique to their time. Of the second generation, Gambino remarks:

> Part of [their] compromise was the rejection of Italian ways which were not felt vital to the family code. They resisted learning the Italian culture and language as well, and were ill-equipped to teach it to the third generation. (1974b:430)

The third generation of Italian American young women are described by Gambino as "confused" by conflicting messages from their parents: "Get an education, but don't change"; "go out into the larger world but don't become part of it"; "grow, but remain within the image of the 'house-plant' Sicilian girl"; "go to church, although we are lacking in religious enthusiasm." In short, maintain that difficult balance of conflicts which is the second-generation's lifestyle" (1974b:431)

In her article, "Women and the Church," Brown concludes:

> Perhaps ethnic people assimilate when their ethnicity no longer predicts their position in American life. This seems to have happened to Italian American Catholic women, but not without cost. In 1984 vice-presidential candidate Geraldine Ferraro was considered a groundbreaker for women and Italian Americans. Among the critics of Ferraro's positions was the principal broadcaster on the Eternal Word Television Network, Mother Angelica, born Rita Rizzo. (2000:684)

The role of religion in the lives of contemporary Italian American women is complex and, in a sense, altogether divergent from that of previous generations. Of the modern Italian American woman, writer Helen Barolini says:

> [Her] traditional power (based upon selflessness and sacrifice) has been lost in the name of autonomy and self-awareness. She cannot be co-ruler in a family, perhaps, in the old way, but she can have power of choice in her own life and, as upholder of a democratic family style, benefit herself as well. The Old World family style and mother role that developed in response to *la miseria* are no longer relevant in a democratic society nor tolerated in affluence. (2000:13)

Immigrant Italian women provided a bridge between home and Church. They were the ones who represented the family in such religious duties as hearing Mass on Sundays and holy days of obligation, going to confession, receiving communion and seeing that children received the sacraments and a religious education. The relevancy of religion in the lives of today's Italian American women, however, cannot be generalized. As with many American Catholics in general, the new reality of its significance would have much more to do with personal choice. Of southern

Italian American Catholics, Primeggia and Varacalli explain that they "do not recognize the legitimacy of magisterial authority to normatively define the standards of the Catholic faith. Rather than accepting an objective understanding of an institutional Catholicism, these Italian Americans embrace a subjective understanding, in that they believe their form of religious adherence to be a legitimate variation of authentic Catholicism" (1996:437).

Some Italian American women may be said to have dismissed religiosity altogether. Others have converted from Catholicism to other Christian denominations or non-Christian religions. A large majority remain Catholic. Yet within this majority, some remain so essentially in name only, making ceremonious appearances perhaps a few times a year for the major religious holidays. This reflects a growing trend within the greater American Catholic community. According to Fisher, "more and more American Catholics were choosing for themselves which aspects of church doctrine they would accept while rejecting those in which they did not believe... Forty-nine percent did not believe that a good Catholic must attend Mass on a weekly basis, despite clear Church teaching to the contrary" (2002:160).

Again reflecting a segment of the greater American Catholic community, some Italian Americans, and in particular the women, find the patriarchal and hierarchical Church structure, with its implicit and explicit challenges to modern sensibilities, problematic. Riane Eisler observes that "a male-dominated and generally hierarchic social structure has historically been reflected and maintained by a male-dominated religious pantheon and by religious doctrines in which the subordination of women is said to be divinely ordained" (1987:24).

Hans Küng,[15] himself a Catholic and author of a book on the history of the Catholic Church, says, "Unfortunately the establishment of hierarchical structures, in particular, prevented the true emancipation of women and still does" (2003:28). He says, furthermore, that "scarcely any of the great institutions in our democratic age...discriminate so much against women-by prohibiting contraceptives, the marriage of priests, and the ordination of women" (2003:xxi).

Of the last of these prohibitions, on the other hand, Catholic theologian George Weigel argues, "the Catholic tradition of ordaining only men to the priesthood is an expression of Catholic sacramental imagination. It is not a matter of misogyny. It is not a question of rights. It is not a question of power. It is a question of sacramentality" (2001:69). He says, furthermore, that "the equality of men and women, made in the image of God and redeemed by Christ, does not mean that men and women are interchangeable as icons of God's presence to the world" (2001:65).

John Shelby Spong, an Episcopal bishop, believes that before long all churches will have women pastors, priests, and bishops. On ordination to the priesthood, he argues:

> No less a person than Pope John Paul II has supported a document and an attitude that proclaims, 'Women will never be priests in the Roman Catholic Church because Jesus did not choose any women to be his disciples.' I submit that this is a literal misuse of the Holy Scriptures. In the social order and mores of the first century, a woman as a member of a disciple band of an itinerant rabbi or teacher was inconceivable. The female role was too clearly circumscribed for that even to be imagined. Here, however, biblical literalism is eclectic rather than thoroughgoing. Perhaps it has not yet occurred to the bishop of Rome that Jesus did not choose any Polish males to be disciples either, but this did not exclude from the priesthood the Polish boy Karol Jozef Wojtyla, who became John Paul II. (1992:6-7)

Of further interest, in the clerical hierarchy, it appears that representative Italian American men are virtually non-existent:

> With the notable exception of Cardinals Bernardin and Bevilacqua, and several bishops, Italian Americans have been slow to gain even a proportionate share of the American Catholic Church hierarchy, where they are notably underrepresented in proportion to their number as parishioners in the Roman Catholic Church. Their presence in the "Social Register" is insignificant. (Caporale, 2000:600)

This fact is included here because, perhaps, with regard to the Church, gender-related issues can be expanded to include ethnic ones as well. Of course, there may be any number of reasons accounting for why Italian American men are virtually absent from the religious hierarchy, reasons of possible interest, but, at the same time, beyond the scope of this paper.

Among other Church-related issues, a number of American Catholics, including Italian Americans, have chosen to selectively ignore the Church's position on such matters as birth control and divorce, and likewise certain religious obligations such as confession, while otherwise considering themselves "good Catholics." Survey data indicates that "...30 percent [of American Catholics]...believed that it was not necessary to accept the Church's position on abortion to remain a 'good Catholic'..." (Fisher, 2002:160).

The Silent Factor

Italian Americans can be said to have largely assimilated into American culture. They also continue to achieve greater professional and educational advancements. Italian American women, moreover, are increasingly more visible in the areas of scholarship and literature, as a glance at the reference list in this essay attests. Yet what explains the relative silence of many Italian American women within the Church and certain other public venues and spheres of influence? Again, the issue of gender in general must be considered. In the opening to his book, *Born of a Woman*, Episcopal Bishop John Shelby Spong declares, "For most of the two thousand years of history since the birth of our Lord, the Christian church has participated in and supported the oppression of women" (1992:1). The question may nevertheless apply, albeit to a lesser extent, to Italian American men. Caporale explains that "...even though some individuals have managed to reach high positions in the political and social arena, the collective prestige of the group remains disproportionately low in relation to its accomplishments." This is explained in part, according to Caporale, by "the tardiness with which the group has succeeded in creating powerful national organizations and institutions that could represent the group as a whole in proportion to its numerical and economic strength..." (2000:600).

No single answer would satisfactorily address the comparative silence of Italian American men and women in both certain public and religious forums. According to Brown, "Catholicism has shaped Italian and Italian American understanding of womanhood, the role Italian family women played in Italian American parish life, and Italian women in religious vocations" (2000:681). Thus if one is to better appreciate what role religion might play in the lives of Italian American women today, it would be important to know something of the Church's position on women. Nor could we necessarily discount factors as seemingly distant as the early and medieval Church, to say nothing of the early days of the Old Testament. Consider the following commentary by author H.W. Crocker III, a Catholic convert from Anglicanism:

> Despite the prominence of women among the converts to Christianity — and despite the role that priestesses played in pagan religion — Paul no more raised the status of women than did Jesus who chose twelve male Apostles. On the contrary, Paul's message was one of submission. He told the Corinthians,[16] 'Let your women keep silence in the churches: for it is not permitted unto them to speak; but they are commanded to be under obedience...And if they will learn anything, let them ask their husbands at home: for it is a shame for women to speak in the Church.' (2001:17-18)

In Eisner's view:

> The Church's condemnation of women to subordinate and 'silent' status can be seen not as a minor historical mystery but a primary expression of the Church's possession by the androcratic/dominator model. It was essential to subordinate and silence women — along with the 'feminine' values originally preached by Jesus — if androcratic norms, and with them, the medieval Church's power, were to be maintained. (1987:140)

One may also speak of the so-called "madonna complex." In what way might this, too, be a dimension of, what I call, the "silent factor" in Italian and Italian American women? Messina puts it that a southern Italian wife and mother was expected to "emulate" La Madonna. The madonna complex assumes women to be "spiritually superior" to men. Perhaps this seems generous. However, her superiority is based upon,

> ...an ideal and ethic of suffering and self-sacrifice of a woman's needs as an individual, in favor of her children and family. The cultural respect for maternal authority entails several behavioral norms: in return for her children's and husband's respect, the mother is expected to develop and endure hardship and suffering without complaint...Family solidarity appears to be maintained at considerable psychological cost to Italian American woman. Although there is no empirical data available that demonstrates the impact of the madonna complex on Italian

American women's psychological functioning, or the interaction effects of the machismo complex on their sexual functioning, clinical and anecdotal evidence suggests that they have internalized the madonna complex and have psychologically subordinated their needs. This, in turn, discourages direct expression of their feelings or assertion of their needs. (Messina, 2000:690-91)

This depthful silence would sometimes take remarkable forms. Writer Ann Cornelisen recalls how southern Italian women would remain silent during childbirth! Cornelisen's vivid and insightful recollections convey that this "expression" of silence has been misunderstood by "outsiders" as "submission":

The young midwife assigned to the hospital...told me once that peasant women feel less pain and suffer less from what they do feel than "other women" — a nice bit of medical snobbery that assumes that either peasant sensitivity is blunted by work, or there is some mystical ratio between nerve ends and socioeconomic class heretofore unrecognized. I have always wondered if she understood anything of the peasant woman's strict code of behavior or of the shame she brings on herself if she flails about in pain...And too the young midwife probably never thought of the delivery "thrones," facing each other in one communal delivery room, as the final indignity a modern world could force upon innately modest women... There, in silence, they wait,...their teeth clamped firmly on those clean white towels they brought from home and their heads turned aside that they may neither see nor be seen. Down the way in separate cubicles their more sensitive sisters groan and cry out for painkillers. (2001:110-11)

Of Sicilians, historian Denis Mack Smith, says:

Whether one accepts or denies the existence of a Sicilian nation is a matter of words. What is hard to deny is the pervasive presence of what Lampedusa calls 'a terrifying insularity of mind.' No doubt the origins of such an attitude must be sought in the reaction of a much-conquered and misruled people to one government after another. Its effects, however, are more difficult to ascertain. (1968:xvi)

Indeed varieties of conspicuous silence are not attributable solely to the female gender of Italians and Italian Americans. Italian immigrants and subsequent generations have exhibited notably repressed, silent behavior. As Morreale and Carola explain:

> Unlike other immigrant groups, Italian Americans kept silent when newspapers reported about them inaccurately. This attitude was a carry-over from their impotent political position in Italy. Also, Italian Americans stayed within very small social circles, and had no interest in mixing with people in high places. In fact, Italian Americans distrusted all social, political, and cultural contacts, an old Italian tradition called campanalismo: Anything beyond the sound of the campanile (parish belltower) was not to be trusted. This was probably one of the most difficult self-defeating practices Italian Americans had to overcome. (2000:147)

Had Italians who migrated to the United States finally rid themselves of bias? Are Italian Americans rid of it now? In the case of Italian American women, how much of the "silent factor" might be attributable to the effects of ethnic bias and how much to gender discrimination? In the United States, both subtle and crude forms of ethnic bias continue to target both Italian American men and w omen, particularly in the form of media-sensationalized ethnic and mafia stereotypes. LaGumina notes that "parodies and caricatures of Italian Americans as silly, boisterous and igno-rant...[lend] credence to an unflattering image of people who best serve society as the butt of jokes, completely devoid of positive features that deserve comment let alone emulation" (2000b:1-2). The most prominent stereotype links Italian Americans with criminality. "Given a spate of popular commercial and television movies such as Godfather I, Godfather II, Godfather III, The untouchables, and Goodfellas," says LaGumina, "it could almost be guaranteed they would reinforce Italian American criminality in the public mind" (2000b:2). This all-too-familiar form of "entertainment" is met with enthusiasm or characteristic ambivalence by some Italian Americans and a kind of twisted pride by others. Yet mass proliferation of such media stereotypes ensure the continuation of bias against Italian Americans. LaGumina notes that "Against this background it becomes possible, even acceptable, to describe Tony Barone, a respected Creighton University basketball coach, 'with a face out of Goodfellas.' One wonders

whether the press would have been so heedless in describing other ethnic groups with disapproving connotations" (2000b:2).
Is it possible to know the long-term effects of this type of pervasive media exploitation? Gambino, writing in 1974, says:

> Whether the [mafia] "image" or reputation of Italian Americans was damaged by this popularization is difficult to assess. But there is no doubt that many Italian Americans felt that the adventures of the Corleone family reflected badly on them. Italian American civic organizations disparaged the film and its popular reception as well as other stereotypes shown in the mass media. They desired the public to regard Italian Americans not in terms of any "mafia myth" but simply as...fellow citizens...(1974b:428)

Nearly a generation later, we read:

> The story of Italians and their American offspring must of course include the relatively few (nearly all of them American-born) whose criminality encouraged Hollywood, ambitious politicians, and the national media to exploit the dramatic nuances of the term "mafia," engendering a stereotype that slanders the great majority of Italian Americans. This, too, is a feature of Americanization. (Mangione and Morreale, 1993: xvii)

Some Italian Americans choose to deny, side-step, or dismiss this decades-old, "tired" bias out of hand. After all, "it's only a movie." Yet "guilt by association," as LaGumina defines it in one of his recent books, is "another insidious anti-Italian pattern." Indeed, "the mere suggestion that one bearing an Italian name is connected with mobsters or illicit activity is enough to consign the unfortunate individual to censure and disapproval" (2000b:3).

Often the more vocal proponents against such discrimination are criticized for their insecurity, piling more layers of stigma onto the dignity of all Italian Americans. Fortunately, as evident from some of the voices included in the present essay and other venues, there are those who speak on behalf of the integrity of Italian and Italian American men and women. Author Ann Cornelisen, having lived among the peasantry of southern Italy for decades, takes up the question of women and silence in her own way, saying, "...before [these women] disappear to become the ghostly shadows behind a myth, I think they should have their say" (2001:9). Her book, *The Women of the Shadows: Wives and Mothers of Southern Italy*, ensures that some of these otherwise forgotten women-little known in the first place-are brought to life.

Had Italian women who migrated to the U.S. become any more visible than those they had left behind? As Barolini observes, since they could not "Americanize on the spot," they "suffered instant obsolescence (an American invention), and became an anachronism, a displaced person, a relic of a remote rural village culture" (1985:13).

What can be said of succeeding generations of Italian Americans? Recall Messina's words quoted at the beginning of this essay. Consistent with the findings of psychologists and ethnographers, she believes that the "personality development," "emotional functioning," and "psychopathology" of Italian immigrant women and subsequent generations are, and have been, affected by historical events (2000:687).

Several important points emerge from this study. First, the role of religion in the lives of Italian American women is a subject which must be placed within broader and deeper historical contexts if any measure of appreciation is to be attained. Second, in exploring the relationship of these women to the Church, a more broadly applicable gender bias presents itself as an issue. Third, to study Italian Americans in the context of public, societal, or religious institutions, one cannot exclude the question of ethnic bias, even if, given its often subtle nature, comparatively little research on the subject has been done to date. Finally, in women's studies, one cannot exclude or ignore the relevant issues or difficulties experienced by men. Nor should one forget the male voices that ring out in support of equity and fairness towards women. If one probes deeply enough, issues thought to pertain solely to a particular ethnicity or gender become relevant to all. For some Italian Americans, forgetting a past they perhaps do not quite remember may afford a superficial comfort in an "Americanized" sort of way. However, as Gambino puts it:

> The dilemma of the young Italian American is a lonely, quiet crisis, so it has escaped public attention. But it is a major ethnic group crisis. As it grows, it will be more readily recognized as such, and not merely as the personal problem of individuals. If this is to be realized sooner rath-

er than later, then these young people must learn whence they came and why they are as they are. (1974b:432)

Nearly thirty years later, these words remain relevant, particularly for Italian American women. "More than for men," says Barolini in 1985, " the displacement from one culture to another has represented a real crisis of identity for the Italian woman, and she has left a heritage of conflict to her children; ...even third and fourth generations feel the remnants of it" (1985: 13).

I thought it fitting to conclude this essay with a poem titled "Mafioso," by Sandra M. Gilbert (neé Mortola),[17] author of numerous feminist literary pieces. I believe it conveys a powerful message to Italian American women and men: when the path to self-discovery leads beyond a shallow past, one's individuality comes into greater focus as does the identity and dignity of a people.

Mafioso[18]

Frank Costello eating spaghetti in a cell at San Quentin,
Lucky Luciano mixing up a mess of bullets and
calling for parmesan cheese,
Al Capone baking a sawed-off shotgun into
huge lasagna —
are you my uncles, my
only uncles?

Mafiosi,
bad uncles of the barren
cliffs of Sicily — was it only you
that they transported in barrels
like pure olive oil
across the Atlantic?

Was it only you
who got out at Ellis Island with
black scarves on your heads and cheap cigars
and no English and a dozen children?

No carts were waiting, gallant with paint,
no little donkeys plumed like the dreams of peacocks.
Only the evil eyes of a thousand buildings
stared across at the echoing debarcation center,
making it seem so much smaller than a piazza,

only a half dozen Puritan millionaires stood on the wharf,
in the wind colder than the impossible snows of the Abruzzi,
ready with country clubs and dynamos
to grind the organs out of you.
–Sandra M. Gilbert (neé Mortola)

LINDA ARDITO

---

[1] For readers who may indeed wish to further explore the theme of this essay, I have included a more extensive reference list.

[2] Primeggia and Varacalli, in their article "The Sacred and Profane Among Italian American Catholics: The *Giglio* Feast," allude to this traditional saying as quoted in Gambino's book, *Blood of My Blood* (Primeggia and Varacalli, 1996; Gambino, 1974).

[3] For concise and authoritative accounts of Italy's unification, see Duggan, Christopher. 1984. *A Concise History of Italy* (Cambridge, UK: Cambridge University Press), 117-42; Hearder, Harry. 2001. *Italy: A Short History*, 2nd ed, revised and updated by Jonathan Morris (Cambridge, UK: Cambridge University Press), 198-218.

[4] Ann Cornelisen, author of *Women of the Shadows: Wives and Mothers of Southern Italy*, lived for decades among the southern Italian peasantry in the villages of Lucania after World War II.

[5] For more information on Italian migration patterns, see Phyllis Cancilla Martinelli (2000:496-500).

[6] Fisher alludes to the Irish and such newcomers as Jews and Polish Catholics, (2002:73).

[7] For an overview of women in the workforce, see Oteri Robinson (2000: 685-7).

[8] Danzi dates this sanctioning granted to midwives from at least the time of the Council of Trent, in 1546 (2000:378).

[9] For a scholarly and informative account of the *Giglio* Feast, see Primeggia and Varacalli (1996).

[10] Salvatore Primeggia, Professor of Sociology at Adelphi University, is nationally known for his work in the area of Italian American Studies. Primeggia's

interview on the topic of "Southern Italian Folk Religion," aired March 3rd at 6:00 PM and March 5th, 2003 at 11:30 AM on the radio program "The Catholic Alternative," hosted by Joseph A. Varacalli (radio station WHPC, 90.3 FM at Nassau Community College).

[11] Kreutz speculates that in tenth-century Campania "...no one was likely to have objected to putting Islamic wares to Christian liturgical use-for example, employing a Fatimid perfume censer in an ecclesiastical setting. Islamic textiles, sometimes adorned with cufic inscriptions signifying their Muslim origin, might even be used to wrap relics" (Kreutz, 1991:143).

[12] For an extensive accounting of religious institutes and sodalities both in Italy and the U.S., see Brown (2000, 680-85).

[13] They settled in Omaha, Nebraska. Mother Mary Constance died at Omaha on Jan. 29, 1902 and Mother Mary Magdalen at Evansville, Indiana, Aug. 18, 1905.

[14] For a detailed listing of such institutes, see Brown (2000:683-4). The institutes listed include: The Adorers of the Blood of Christ, The Sisters of the Holy Family of Nazareth, The Sisters of the Sorrowful Mother (3rd Order of Saint Francis), The Daughters of Charity of the Most Precious Blood, The Daughters of the Most Holy Cross, The Sisters of the Resurrection, The Institute of the Sisters of Saint Dorothy, The Mantellate Sisters, Servants of Mary of Blue Island, The Daughters of Our lady of Mercy, The Franciscan Sisters of Saint Elizabeth, The Nuns of the Perpetual Adoration of the Blessed Sacrament, and The Pious Daughters of Saint Paul.

[15] Küng, a Catholic, has a doctorate in theology from the Sorbonne (1957). He was named a theological consultant for the Second Vatican Council by Pope John XXIII (1962). He played a primary role in the writing of Vatican II documents which modernized aspects of Catholic teaching. He lives and teaches in Tübingen, Germany.

[16] I Corinthians 14:34-35.

[17] Gilbert was born in 1936 and raised in Jackson Heights, Queens, New York. She has been professor at the University of California at Davis.

[18] Gilbert's poem Mafioso appears in Barolini, Helen, ed. 1985. *The Dream Book: An Anthology of Writing by Italian American Women* (New York: Schocken Books), 348-9.

References

Allen, Beverly, Muriel Kittel and Keala Jane Jewell, eds. 1986. *The Defiant Muse: Italian Feminist Poems from the Middle Ages to the Present* (New York, NY: The Feminist Press at the City University of New York).

Azen Krause, Corinne. 1991. *Grandmothers, Mothers and Daughters: Oral Histories of Three Generations of Ethnic American Women* (London: Macmillan Publishing).

Barolini, Helen, ed. 1985. *The Dream Book: An Anthology of Writing by Italian American Women* (New York: Schocken Books).

_____, et al. 1986. *Images: A Pictorial History of Italian Americans*, editorial and research consultants, S.M. Tomasi, E. Are, A. Brizzolara, A. Bujatti, B.B. Caroli, O.D. Cava, S.J. LaGumina, T.J. Marino, R. Pane, G. Rosoli, and L.F. Tomasi. (Staten Island, NY: Center for Migration Studies).

Barreca, Regina 2002. *Don't Tell Mama: The Penguin Book of Italian American Writing* (New York: Penguin).

Battista-Lanzalli, Geraldine. 1990. "Family vs. Career: A Dilemma for the Italian American Woman of the 80s," *Italian Americans in Transition*, eds., Joseph Scelsa, S. LaGumina, and Lydio Tomasi, 115-26 (Staten Island, NY: American Italian Historical Association).

Bona, Mary Jo. 1999. *Claiming a Tradition: Italian American Women Writers* (Southern Illinois UP).

Boscia-Mule, Patricia. 1999. *Authentic Ethnicities: The Interaction of Ideology, Gender, Power, and Class in the Italian American Experience*, Vol. 124 (Greenwood Publishing).

Brown, Mary Elizabeth. 2000. "Women and the Church,"*The Italian American Experience: An Encyclopedia,*eds. Salvatore J. LaGumina, et al., (New York: Garland Publishing), 680-85.

_____. 2000. "Religion," *The Italian American Experience: An Encyclopedia*, eds. Salvatore J. LaGumina, et al., (New York: Garland Publishing), 538-542.

Browne, Henry J. 1946. "The 'Italian Problem' and the Catholic Church in the United States, 1880-1900," *Records and Studies* 25 (New York: United States Catholic Historical Society).

Cancilla Martinelli, Phyllis. 2000. *The Italian American Experience: An Encyclopedia*, eds. Salvatore J. LaGumina, et al., (New York: Garland Publishing), 496-500.

Caporale, Rocco. 2000. "Social Class Characteristics," *The Italian American Experience: An Encyclopedia*, eds. Salvatore J. LaGumina, et al., (New York: Garland Publishing), 595-600.

Capozzoli, Mary Jane. 1990. *Three Generations of Italian American Women in Nassau County, 1925-1981* (New York: Garland Publishing).

_____. 1987. Nassau County's Italian American Women: A Comparative View," *The Melting Pot and Beyond: Italian Americans in the Year 2000*, Proceedings of the XVIII Annual Conference of the American Italian Historical Association, eds., Jerome Krase and William Egelman (Staten Island, NY: American Italian Historical Association), 285-96.

Chiavola Birnbaum, Lucia. 1993. *Black Madonna: Feminism, Religion, and Politics in Italy* (Boston: Northeastern UP).

_____. 1990. "On the Significance of Italian Folklore for Italian Americans: The Case of Easter Rituals," *Italian Americans in Transition*, eds., Joseph Scelsa, S. LaGumina, and Lydio Tomasi (Staten Island, NY: American Italian Historical Association), 105-13.

Cornelisen, Ann. 2001. *The Women of the Shadows: Wives and Mothers of Southern Italy* (South Royalton, VT: Steerforth Press; originally published by Little Brown and Company, 1976).

Crocker, H.W. 2001. *Triumph: The Power and the Glory of the Catholic Church-A 2,000-Year History* (Roseville, CA: Prima Publishing).

D'Andrea, Veneeta-Marie. 1983. "The Social Role Identity of Italian American Women: An Analysis and Comparison of Families and Religious Expectations, *The Family and Community Life of Italian Americans*, ed. Richard N. Juliani, 61-8 (Staten Island, NY: American Italian Historical Association).

Danzi, Angela D. 2000. "Midwives and Childbirth," *The Italian American Experience*, eds. Salvatore J. LaGumina, et al., (New York: Garland Publishing), 378-80.

DeSalvo, Louise A. and Edvige Giunta, eds. 2002. *The Milk of Almonds*, Vol. 1. (New York, NY: The Feminist Press at CUNY).

DeSena, Judith N. "Involved and 'There': The Activities of Italian American Women in Urban Neighborhoods," *The Melting Pot and Beyond: Italian Americans in the Year 2000*, Proceedings of the XVIII Annual Conference of the American Italian Historical Association, eds., Jerome Krase and William Egelman (Staten Island, NY: American Italian Historical Association), 239-47.

_____. 1990. "New Images: An Examination of Italian American Women," *Italian Americans in Transition*, eds., Joseph Scelsa, S. LaGumina, and Lydio Tomasi (Staten Island, NY: American Italian Historical Association) 99-103.

DiLeo, Domenica, Patrizia Tavourminoa, and Gabriella Micallef, eds. 1999. *Curraggia: Writing by Women of Italian Descent* (Toronto: Women's Press).

Di Prima, Diane. 2002. *Recollections of My Life as a Woman: The New York Years* (New York: Penguin Publishers).

Duggan, Christopher. 1984. *A Concise History of Italy* (Cambridge, UK: Cambridge University Press).

Eisler, Riane. 1987. *The Chalice and the Blade: Our History, Our Future* (New York, NY: Harper & Row).

Eula, Michael J. 1993. *Between Peasant and Urban Villager: Italian Americans of New Jersey and New York, 1880-1980* (New York: Peter Lang Publishing).

Ewen, Elizabeth. 1990. *Immigrant Women in the Land of Dollars: Life and Culture on the Lower East Side, 1890-1925* (New York: Monthly Review Press).

Fisher, James T. 2002. *Communion of Immigrants: A History of Catholics in America* (New York, NY: Oxford University Press).

Gabaccia, Donna R. 1984. *From Sicily to Elizabeth Street: Housing and Social Change Among Italian Immigrants, 1880-1930* (Albany, NY: State University of New York Press).

Gallo, Patrick J. 1974. *Ethnic Alienation: The Italian Americans* (Rutherford, NJ: Fairleigh Dickinson University Press).

Gambino, Richard. 1974. *Blood of My Blood: The Dilemma of Italian Americans* (Garden City, New York: Doubleday).

_____. 1974b. "Italian Americans Today," *A Documentary History of the Italian Americans* (New York, NY: Praeger Publishers), 428-32.

Giunta, Edvige. 2002. *Writing with an Accent: Contemporary Italian American Women Authors* (New York: St. Martin's Press).

Hearder, Harry. 2001. *Italy: A Short History*, 2nd ed, revised and updated by Jonathan Morris (Cambridge, UK: Cambridge University Press).

Hoobler, Dorothy and Thomas, and Mario M. Cuomo. 1998. *The Italian American Family Album* (Oxford: Oxford University Press).

Iorizzo, Luciano J. 1985. "Religion and Community Life Among Italian Americans: Some Comments and Observations," *Italian Americans: New Perspectives in Italian Immigration and Ethnicity*, ed. Lydio F. Tomasi (Staten Island, NY: Center for Migration Studies), 358-64.

Johnson, Coleen Leahy. 1985. *Growing Up and Growing Old in Italian American Families* (New Brunswick, NJ: Rutgers University Press).

_____. 1982. "The Maternal Role in the Contemporary Italian American Family," *Italian Immigrant Women in North America*, (Toronto, Canada: The Multi-Cultural History Society of Ontario), 234-44.

Juliani, Richard N. 1990. "Family Life,"*The Italian American Experience: An Encyclopedia*, eds. Salvatore J. LaGumina, et al., (New York: Garland Publishing), 209-12.

Krase, Jerome and William Egelman, eds. 1987. *The Melting Pot and Beyond: Italian Americans in the Year 2000*, Proceedings of the XVIII Annual Conference of the American Italian Historical Association (Staten Island, NY: American Italian Historical Association).

Kreutz, Barbara M. 1991. *Before the Normans: Southern Italy in the Ninth and Tenth Centuries* (Philadelphia: University of Pennsylvania Press).

Küng, Hans. 2003. *The Catholic Church: A Short History* (New York, NY: Modern Library).

LaGumina, Salvatore J. 2000. *The Italian American Experience: An Encyclopedia*, eds. Frank J. Cavaioli, Salvatore Primeggia, and Joseph A. Varacalli, (New York: Garland Publishing).

_____. 2000a. *Long Island Italians* (Charleston, SC: Arcadia Publishing).

_____. 2000b. *Wop!: A Documentary History of Anti-Italian Discrimination*, Vol. 1 (Toronto: Guernica Editions, Inc.).

_____. 1992. *New York at Mid-Century: The Impellitteri Years*, Vol 147 (Westport, Ct: Greenwood Publishing).

_____. 1990. "From Urban to Suburban: Italian Americans in Transition," *Italian Americans in Transition*, Proceedings of the XXI Annual Conference of the American Italian Historical Association, eds., Joseph V. Scelsa, Salvatore LaGumina, Lydio Tomasi (Staten Island, NY: American Italian Historical Association), 27-37.

_____. 1989. *From Steerage to Suburb: Long Island Italians* (Staten Island, NY: Center for Migration Studies).

_____, ed. 1979. *Immigrants Speak: The Italian Americans Tell Their Story* (Staten Island, NY: Center for Migration Studies).

Lapomarda, Vincent A. 2000. "Rita Antoinette Rizzo (Mother Angelica)," *The Italian American Experience: An Encyclopedia*, eds. Salvatore J. LaGumina, et al., (New York: Garland Publishing), 553.

Laurino, Maria. 2000. *Were You Always an Italian?: Ancestors and Other Icons of Italian America* (New York, NY: W.W. Norton).

Leckey, Dave. 2002. *Boundless Lives: Italian Americans of Western Pennsylvania*, Mary Brignano, ed. (Pittsburgh, PA: Historical Society of Western Pennsylvania).

Mack Smith, Denis. 1968. *A History of Sicily: Medieval Sicily, 800-1713* (New York: Dorset Press).

Mangione, Jerre and Ben Morreale. 1993. *La Storia: Five Centuries of the Italian American Experience* (New York, NY: Harper Collins).

Mannino, Mary Ann and Justin Vitiello, eds. 2002. *Reflections on Italian American Women Writers* (Lafayette, IN: Purdue University Press).

Mansueto, Anthony. 1987. "Blessed are the Meek...Religion and Socialism in Italian American History," *The Melting Pot and Beyond: Italian Americans in the Year 2000*, Proceedings of the XVIII Annual Conference of the American Italian Historical Association, eds., Jerome Krase and William Egelman (Staten Island, NY: American Italian Historical Association), 117-36.

Marchione, Margherita. 1995. *Americans of Italian Heritage* (Lanham, MD: University Press of America).

Messina, Elizabeth G. 2000. "Women in Transition," *The Italian American Experi-ence: An Encyclopedia*, eds. Salvatore J. LaGumina, et al., (New York: Gar-land Publishing), 687-94.

Morreale, Ben and Robert Carola. 2000. *Italian Americans: The Immigrant Experi-ence* (Beaux Arts Editions).

Morris, Charles R. 1997. *American Catholic: The Saints and Sinners who Built America's Most Powerful Church* (New York: Vintage Books).

Noonan, Daniel P. 1990. *The Catholic Communicators* (New York: Xavier Society for the Blind).

O'Neill, Dan. 1986. *Mother Angelica: Her Life Story* (New York: Crossroads).

Orsi, Robert Anthony. 1985. *The Madonna of 115th Street: Faith and Community in Italian Harlem, 1880-1950* (New Haven, CT: Yale University Press).

Oteri Robinson, Bridget. 2000. "Women in the Workforce," *The Italian American Experience: An Encyclopedia*, eds. Salvatore J. LaGumina, et al., (New York: Garland Publishing), 685-7.

Parca, A. 1963. *Italian Women Confess*, tran. Carolyn Gaiser (New York: Farrar, Straus).

Primeggia, Salvatore and Joseph A. Varacalli. 1996. "The Sacred and Profane Among Italian American Catholics: The *Giglio* Feast," *International Journal of Politics, Culture and Society*, Vol.9, No. 3, 423-49.

Pucelli, Rodolfo, trans. 1955. *Anthology of Italian and Italo-American Poetry* (Boston: Bruce Humphries, Inc.).

Quinn Caro, Edythe. 1987. "Celebration, Conflict and Reconciliation at Saint An-thony's," *The Melting Pot and Beyond: Italian Americans in the Year 2000*, Proceedings of the XVIII Annual Conference of the American Italian Histori-cal Association, eds., Jerome Krase and William Egelman (Staten Island, NY: American Italian Historical Association), 249-58.

Raptosh, Diane. "Italian/American Women on the Frontier: Sister Blandina Segale on the Santa Fe Trail," *Italian Ethnics: Their Languages, Literature and Lives*, eds., Dominic Candeloro, Red L. Gardaphé, and Paolo A. Giordano, (Staten Island, NY: American Italian Historical Association) 91-107.

Russo, Nicholas John. 1977. "Three Generations of Italians in New York City: Their Religious Acculturation," *The Italian Experience in the United States*, eds., Silvano M. Tomasi and Madeline H. Engel (Staten Island, NY: Center for Migration Studies) 195-213.

Schiavo, Giovanni [Ermenigildo]. 1949. *The Italian Contribution to the Catholic Church in America* (New York:: Vigo Press).

Seller, Maxine Schwartz. 1978. "Protestant Evangelism and the Italian Immigrant Woman," *The Italian Immigrant Woman in North America*, eds., Betty Boyd

Caroli, Robert F. Harney, and Lydio F. Tomasi, 124-136 (Toronto: The Multi-Cultural History Society of Ontario).

Spong, John Shelby. 1992. *Born of a Woman: A Bishop Rethinks the Birth of Jesus* (New York: Harper Collins Publishers).

Sullivan, Mary Louise. 2000. "Mother Frances Xavier Cabrini (1850-1917)," *The Italian American Experience: An Encyclopedia*, eds. Salvatore J. LaGumina, et al., (New York:: Garland Publishing) 88-9.

_____. 1992. *Mother Cabrini: Italian Immigrant of the Century* (Staten Island, NY: Center for Migration Studies).

Tirabassi, Maddalena. 1990. "Emancipation Through Americanization? The International Institutes and Italian Immigrant Women," *Italian Americans in Transition*, op. cit., 81-98.

Tomasi, Silvano M. "The Ethnic Church and the Integration of Italian Immigrants in the United States,"*The Italian Experience in the United States*, eds., Silvano M. Tomasi and Madeline H. Engel (Staten Island, NY: Center for Migration Studies) 163-93.

Torgovnick, Marianna. 1994. *Crossing Ocean Parkway: Readings by an Italian American Daughter* (Chicago: University of Chicago Press).

Vecoli, Rudolph J. 1983. "The Formation of Chicago's Little Italies," *Journal of American Ethnic History* 2 (Spring), 5-20.

Vigilante, Mary Ann and Ann Mannino. 2000. *Revisionary Identities: Strategies of Empowerment in the Writing of Italian/American Women* (New York: Peter Lang Publishing).

Weigel, George. 2001. *Truth of Catholicism: Ten Controversies Explored* (New York: Harper Collins).

Chapter 9

✦

CONFESSING THE SELF: ITALIAN AMERICAN WOMEN AUTHORS AND THE CATHOLIC CHURCH

> *You were always irish, god/in a church where*
> *I confessed to being Italian.. . . /O god,/god,*
> *I confess nothing.*
> —Elaine Romaine

Elaine Romaine's poem encapsulates the experience of many Italian Catholics in early and mid-twentieth-century America. Capitalizing only two words in the entire poem, "Italian," and "St. Anthony," Romaine effectively diminishes the power of Irish domination and the conventional ritual of the Catholic sacrament of communion. Even Romaine's description in the middle of her poem of the annual festa illustrates its association with patriarchy as only male figures — priest, father, uncles, and brother — are involved in the procession. But at the end of the poem, in an apostrophe, Romaine makes her own confession: "O god, /god, I confess nothing."

For many Italian American writers, American Catholicism is treated with the same Old-World skepticism and distrust as the peasants held toward the church in nineteenth-century pre-industrial Italy. Italian American women writers in particular have achieved independence from conventional Catholicism, but Italian Catholic iconography nonetheless pervades their work as they negotiate, challenge, and recreate religious rituals and beliefs. Their refusal at times to engage in the sacraments of penance and communion demonstrates their realization that the church has traditionally refused to assist in female development.

Despite their decision to minimize outward forms of piety in their creative works, many Italian American women have responded in writing to their Catholic upbringing. Writers such as Octavia Waldo, Susan Leonardi, Mary Gordon, Helen Barolini, and Mary Cappello for example, have directly confronted what they have observed are the limitations of the traditional Catholic Church, especially for its women. Carole Maso, Nancy Sa-

162

voca, Tina De Rosa, and Louisa Ermelino, on the other hand, covertly defy traditional Catholicism through their portrayals of highly singular women. In both cases — direct confrontation and quiet subversion — women writers have had much to confess. Refusing to abide by the cultural injunction of *omertà* (silence), Italian American women use the religious beliefs of their ancestors in the service of empowering their female characters. An inherited tradition of anti-clericalism enables these writers to deepen an understanding of the abiding influences of old-world beliefs in a new-world setting.[1]

In his analysis of Italian Catholicism, Rudolph Vecoli (1969: 228) explains that "the peasants [of the *Mezzogiorno*] were intensely parochial and traditional. While nominally Roman Catholics, theirs was a folk religion, a fusion of Christian and pre-Christian elements, of animism, polytheism, and sorcery with the sacraments of the Church." The anticlericalism that survived the transatlantic crossing caused serious conflict between Italian immigrants and the Irish-American Catholics, who dominated the American Church hierarchy and who did not forgive the Italians their anti-institutional attitudes. For Italians, the Church hierarchy had been historically linked in the *Mezzogiorno* with the exploitive landowners, developing as a result, an "intense distrust and scorn for the Church as an institution" (Gambino, 1974: 229).

Italians nonetheless were deeply Catholic, believing in the "local presence or immanence of the divine, often through intercessors, within everyday life" (Russo, 1). As a case in point, Italians maintained a fluid consanguinity between religion and magic, best epitomized by their engagement in Italian processions and *feste*, elaborately ritualistic. Carlo Levi's apt description of southern Italians is echoed in the religious rituals, and later, the writings of southern Italians in America: "Even the ceremonies of the Church become pagan rites celebrating the existence of inanimate things, which the peasants endow with a soul, and the innumerable earthly divinities of the village" (117). While the Italian immigrants may have been turned away from the church, seated with Negroes, and denounced as "Dagos" from the pulpit (Vecoli, 230), they continued to engage in feast days, despite the fact that such expressions did not accord with the standards of religious conduct prescribed by the official Catholic American church. As anthropologist Kay Turner explains, the "unorthodox desire to bring the sacred into the streets, thereby collapsing rigid boundaries between sacred and secular realms," has its roots in ancient fertility ceremonies in which circular movement parallels the progress of reaping to sowing and "the cyclic movement of women's menstrual-fertile period" (7). Italian American women have responded to both conventional forms of Catholicism and to the unorthodox rituals Italians sustained in America.

Octavia Waldo and Susan Leonardi are two such writers who create female characters to challenge the rules and obligations of formal Catholicism. These characters maintain a vexed spirituality, partly caused by the doctrinal orientation of the American Catholic Church. Octavia Waldo's 1961 *A Cup of the Sun* explores the limitations of conventional Catholicism on one of its central female characters, Niobe Bartoli. Set in the outskirts of Philadelphia during World War II, *A Cup of the Sun* revolves around a neighborhood of Italian immigrant families who struggle to maintain old-world customs in a rapidly changing milieu. The novel traces the painful childhoods of second-generation Italian Americans. Their growth into adulthood is as much influenced by *l'ordine della famiglia* (the strict Italian family code of responsibility) as it is by outward influences of the American Catholic church, public schooling and the looming threat of a world war.

Waldo incorporates two confessional scenes in her novel to explore the limitations of traditional Catholicism and the lack of understanding between first and second-generation Italian American families. The fourteen-year old Niobe has committed incest with her brother, though she neither initiated nor desired sex with him. This one-time incident, though confusing and damaging to Niobe, triggers her desire to understand herself as a woman capable of achieving autonomy outside the rigid strictures of her family and church. Niobe realizes that screaming for help during the incident would have incurred the wrath of her authoritarian father. Believing that confessing their act to the family would get them killed by Mr. Bartoli, Niobe keeps silent during and after intercourse with her brother. Niobe also remains silent in the confessional, for which she must suffer the vitriolic response from the priest, who censors her: "It is a grievous sin to withhold your sins in the confessional.... You must not receive communion after a confession of lies. If you should die afterwards, God will abandon you to the rodents of the soil... His disgust will be worse than any torment you can now imagine. Worse than having your eyes gouged from their sockets to roll their blood over the driest sand" (64). Sounding more like a Calvinist than a Catholic, the priest's condemnation of Niobe does not move her to confess any quicker; in fact, Niobe holds off confessing her act until she has processed it further.[2]

Before Waldo incorporates a second confession scene, she inserts a description of a neighborhood procession, in which a plaster statue of the Madonna is paraded through the streets. That the statue is modeled after the local sculptor's mistress parallels the relationship between sacred and secular in Italian American Catholicism; the movement of the statue from sacred space (the church) to secular space (the street) nurtures observers like Niobe, who remains outwardly unmoved by the procession. Recognizing the deeply religious motivation of this kind of procession, Turner explains that its primary purpose is "to release the holy, to extend the 'power field' of the sacred to encompass profane realms" (7). Later in the narrative, the very artist who sculpted the icon of the Virgin confers healing absolution on Niobe when he intuits her secret suffering and reassures her with these words: "'It doesn't matter-not to anyone-not to anyone who loves you'" (203).

While Niobe's first-generation parents do not rigidly follow the rules of Catholicism, they expect their children to go to confession and to receive communion. What remains imperative for Niobe is religious confession, an obligation to utter the truth about illicit sex with her brother. Before the second confession scene, however, Waldo clarifies that Niobe has lost her faith in the power of doctrinal Catholicism to heal its practitioners: "The Church offered confession for comfort and nothing to grasp hold of afterwards except fear" (69). The confession scene itself "unfolds within a power relationship" (Foucault,1980: 61). As such, the priest meticulously follows rules governing "confession of the flesh," including "insinuations of the flesh: thoughts, desires, combined movements of the body and soul" (Foucault 19). Accordingly, the priest involves the young Niobe in a question-and-answer format about the details of her sexual encounter, effectively intensifying her memories of the act. The confession of Catholicism parallels what Foucault describes as an "institutional incitement to speak about [sex]... a determination on the part of agencies of power to hear it spoken about" (18), which occasions from the confessor another oratory of condemnation. Niobe Bartoli will enter adulthood without the limited ministrations of the Catholic Church.

Susan Leonardi's short story, "Bernie Becomes a Nun," also explores the inability of American Catholicism to nurture its women. Told from the perspective of a poor Italian American woman, whose route toward independence is limited in the 1950s to entering a convent, this story also examines the ways in which class status and gender circumscribe and determine a young girl's life. The oldest of ten, Bernadette may very well love God, but she becomes a nun "to get out of the house" (217). While the humorous tone of Leonardi's story is reminiscent of Regina Barreca's and Nancy Maniscalco's works,[3] like them, Leonardi also uses humor to expose the fragility and difficulty of women's lives.

By the age of eleven, Bernadette Frances Palermo inchoately realizes that she does not want to repeat her mother's life of multiple pregnancies and bad health. An exemplary student, Bernadette's first avenue of independence is through education. Leonardi introduces sixth-grader Bernadette as class librarian appointed by Sister Mary Ascension, which reinforces her desire to be distinguished outside the role of first-born caregiver to her mother's many children. The initial setting Leonardi places Bernadette in is the cloakroom (where the books are stacked), positioning her as closeted. Bernadette's desire to control the checkout of one of the books in the 154-book library reveals her recognition that her life's script may very well be written for her already. *Bernie Becomes a Nun* is the book that becomes the very "Hound of Heaven" barking at Bernadette's heals (205). A story about a privileged girl who becomes a Maryknoll nun, Bernie takes the name "Sister Joseph Marie," the name Bernadette cynically takes years later when she enters the novitiate herself. However, the order Bernadette enters, The Sisters of St. Clare, recalls the first Franciscan order of nuns, the Poor Clares, whose vows of impoverishment and seclusion do not initially improve Bernadette's life.

In time, however, Bernadette receives a room of her own. Never before had she private time for personal reflection to cry, to keep a diary, to "sleep for eight hours straight without listening for a waking babe or hearing a sister toss and turn" (222). Leonardi establishes a parallel between sixth-grade cloakroom and private room because both places are spaces for Bernadette to experience a second avenue of independence: through lesbian sexuality. Throughout the narrative, however, only fleeting references are made to Bernadette's sexual feelings for women, which began in grammar school. The narrative's limited omniscience to Bernadette's thoughts proscribe such a confession, a full telling of the sexual desire influencing her decision to enter a convent.

During the climax of the story, after twenty-seven years of service, Bernadette reexamines the limitations set by her social class, gender, and sexuality. In her forties now and reassessing her limited life choices, Bernadette describes the sixth-grade book, *Bernie Becomes a Nun* as "pure propaganda," humorously describing the main character in Protestant terms: "white, thin, Anglo-Saxon, pretty, popular Bernie-the kind of girl God likes best" (229). As a child of working-class Italian parents, Bernadette describes in contrast her own dire situation, which occasions her own personal confession: "And I was brown and fat and poor and ugly and there didn't seem to be many alternatives. I've been a nun for twenty-seven years because I found that fucking book in the sixth-grade library" (229). Refusing any longer to deny either the socioeconomic or the sexual forces that informed her decision to enter the convent, Bernadette requests an extended leave of absence from the order.

Like Niobe of *A Cup of the Sun*, Bernadette cannot be healed by her family of origin or the traditional Catholic Church. Her medical degree in psychiatry may have unlocked her desire for self-analysis, but a non-traditional nun her mother's age becomes the key to Bernadette's emotional healing. Retreating to a monastery of The Sisters of St. Clare allows Bernadette to observe doctrinal church work performed by another woman, who says Mass and hears confessions (227). By excluding the voices and precepts of traditional Catholic ideology, Leonardi challenges the institutional structure and rules of the Church, which continues to limit the autonomy of women and prevent their public participation in the sacraments.

Tina De Rosa and director Nancy Savoca elide direct representation of American Catholic traditions in order to explore the power of female characters who transcend the secular and join the sacred. Both De Rosa's *Paper Fish* and Savoca's *Household Saints* examine how women go about achieving singularity within traditional Italian American Catholic family culture. While not overtly about Catholicism and its effects on women, both works highlight female empowerment and the possession of authority over a young woman's identity with help from a female mentor figure. De Rosa's *Paper Fish*, a novel of Little Italy set in Chicago and spanning the years 1949-1958, portrays the paternal grandmother as the singular individual in the novel. Grandma Doria is the matriarch of the Bellacasa family and, as such, she inevitably limits the power of her non-Italian daughter-in-law, Sarah. Sarah's Lithuanian background is entirely muted in *Paper Fish*, illustrating her own feeling of being overwhelmed by her husband's Italian American family and, in particular, by the illness of first-born daughter, Doriana.

Much of Sarah's maternal energy is spent nursing Doriana, whose unnamed illness profoundly affects each member of the Bellacasa family. Doriana's fevers and fits, and her inability to speak or develop mentally causes an unresolved crisis in the family. When Sarah thinks "[Y]ou look into your pockets, and there is only a spool of thread and some needles" (107), she accepts her role as a mother who mends without proper materials. With no financial resources or access to medical help, Sarah futilely offers sustenance to her sick daughter. Carmolina, second-born daughter, and a primary focus of De Rosa's narrative, is not so much estranged from her mother in childhood as she is actively nurtured by her Grandma Doria. Through storytelling, Grandma Doria extends Carmolina's lineage by connecting her to Neapolitan great- grandparents and to healing memories of the booted country. Such stories save Carmolina emotionally, succored as she is by Grandma Doria's allegiance to oral traditions, to stories that explain as they heal.

Storytelling is Carmolina's most cherished and necessary plaything, since her sister is unavailable and her parents fearfully observe their second-daughter's behavior, hoping that Doriana's disease will not spread to her younger sister (36). When she is eight years old, Carmolina runs away from home for several days. During this period, Carmolina clarifies her membership in the family, employing the story-telling methods learned from her grandmother to understand her sister's illness. Grandma Doria uses an image of the forest to explain her granddaughter's illness; Carmolina recalls this story when she is away from home: "One day Doriana go into the forest. She forget the key. She get lost in the forest" (112). Learning from her grandmother that the forest is a metaphor of the mind, Carmolina embraces her sister's illness and, through language, paradoxically gains necessary distance from her. Healed by her grandmother's poetic explanation, Carmolina understands that her sister's physical beauty is co-extensive with her illness: Doriana's "face it turn hot like a little peach and she scream and try to get out of the forest.... When you fight to come home, you beautiful" (112).

Grandma Doria gives her granddaughter a legacy of words. In adulthood, Carmolina must accept her calling as a literary artist, mentored by her storytelling grandmother. As though in celebration of Grandma Doria's singularity as an artist of oral traditions, De Rosa concludes her narrative with a procession. Unlike the Italian *festa*, culminating with a high mass and procession of the statue of the saint or Madonna through the streets, De Rosa's depiction of the procession is understated but nonetheless more unorthodox than the folk ceremonies of immigrant *contadini*. Aware that she is dying and must pass the torch of inspiration onto her granddaughter, Grandma Doria requests to see Carmolina dressed in bridal garb. The ensuing ceremony between grandmother and granddaughter poignantly depicts Carmolina's difficult acceptance of her grandmother's imminent death and her own future vocation as a writer.[4]

As Kay Turner explains, the primary purpose of a "*procession* is to affirm sacred membership in community ... [it is] a form of public religious ritual ceremony performed by members of a community" (6). In *Paper Fish*, the procession begins outdoors as Doria's two sons carry her in her mahogany chair because of the pain in her feet. They take her up the stairs to Carmolina, who awaits her in her parent's apartment. The neighborhood people witness the ceremony and validate its significance by their very presence, the phenomenon of *companilismo* (village-mindedness) fully intact. In this scene, De Rosa reverses the traditional movement of the patron saint from indoors to outdoors in order to privatize Carmolina's entrance into adulthood. As Fred Gardaphé explains, the ceremony in *Paper Fish* "recreates the traditional procession of the Madonna, a staple of Italian culture, through which the people receive the blessing of God's mother" (1996: 137). De Rosa affirms that Grandma Doria is the sacred icon, the word made flesh, and her movement through the neighborhood "transforms secular into sacred territory" (Turner, 1980: 7).

In an effort to heighten Grandma Doria's singularity, De Rosa not only equates her with the Madonna but also with the village priest. Unlike the young priest in *Paper Fish* whose ineffectuality is highlighted by the fact that he is dying, the elderly Grandma Doria functions like a priest who has the power of transubstantiation: she is able to convert secular into sacred. She does this through the procession itself, sanctifying the community through her spiritual presence and later through her capacity to ignite the flame of creative inspiration for her disciple granddaughter. In addition, like the father in a wedding ceremony, Grandma Doria gives Carmolina away, not to someone else, but to herself. Doria compels her granddaughter to look at herself in the mirror with the final injunction "'Now it you turn. You keep the fire inside you'" (130). De Rosa has equated creative inspiration with the divine; bestowing sacred status upon the storytelling Grandma Doria encapsulates her artistic belief.[5]

Equally singular in her spirituality is third-generation Teresa Santangelo of Nancy Savoca's *Household Saints*. Based on Francine Prose's 1981 novel of the same name, the film *Household Saints* focuses on the lives of three generations of women in New York's Little Italy of the 1950s through the 1970s. Like Grandma Doria, these Italian American women do not fully abide by religious conventions, either rejecting formal liturgical precepts outright or refusing to adopt the formalities of an Irish-dominated Catholic hierarchy. Carmela Santangelo, the widowed matriarch and first-generation Italian woman, is old-world Catholic, a hearty blend of superstition, including spells and incantations, direct access to the saints, the Madonna and Christ. In this sense, Carmela is considered by formal Catholicism to be unorthodox because she directly worships the saints and maintains a mystical relationship with her patron saint, the Madonna, and her dead husband. Her daughter-in-law, Catherine, new-world Italian American, eventually assumes the role of making the Santangelo "Miracle Sausages," and even though she has rejected her mother-in-law's access to the divine, she supports her husband's butcher business through this activity.

Catherine's child, Teresa, resumes her now-deceased grandmother's mystical piety *in extremis*, modeling herself on her namesake, the young French saint who served God by performing routine chores. Both female mentors — Carmela Santangelo and Teresa of Lisieux — enjoyed a mystical relationship with the unseen and both sought divinity in the ordinary. As Baker and Vitullo have illustrated, Teresa Santangelo's activities parallel those of female mystics of the late medieval period, religious "women [who] searched for transcendence through the physical as well as the contemplative, especially in their service to others" (59). Seen through the lens of generational enumeration, moreover, *Household Saints* works typologically, with the grandchild — Teresa — functioning as the antitype of the foundational character, Carmela.[6] As such, Teresa fulfills individually the role that her grandmother fulfilled through marriage under the traditional female domestic role, within the household. By early adulthood, however, Teresa Santangelo becomes the quintessential antinomian. Antinomians in the Puritan era rejected the orthodox notion of visible sainthood, offering instead a "perfectionist theology wherein election is witnessed and sealed by the spirit and cannot be tested by outward means" (Lang, 1987: 7).

Refusing to allow his daughter to enter the Carmelite order (which is an orthodox response to serving God), Joseph Santangelo unwittingly enters into a power struggle with a woman fully committed to Jesus Christ. After experiencing a vision of Christ while ironing her boyfriend's checkerboard shirt, Teresa is soon exiled and committed to a Catholic psychiatric institution. She dies shortly thereafter of unknown causes. She is nineteen. Of the traditional roles open to women in normative Italian culture, those of wife and mother remain preeminent. Teresa's "transgression" is to choose sainthood as a path she feels fated to follow, but within the domestic space traditionally reserved for wives and mothers-the home. Like nuns and other cloistered religious, Teresa is able to free herself from the bodily constraints of motherhood, but preserve the sanctity of domestic space by residing inside and spending entire days focused on the minutia of household chores.

Teresa becomes the ultimate antinomian — she manages to avoid the notion of "visible sainthood," that is, the idea of grace emanating from good works in the community, and instead believes in an individual identity subsumed in divinity — and which cannot be tested by outward means. Thus the confusion (and multiple interpretations) at her death. Teresa Santangelo comes to mean different things to different people in her Italian American community. Like a saint, she is prayed to and revered after her death. Her otherworldliness not only frees her from patriarchal family control but also functions as her ticket to immortality. Like a good Puritan, Teresa's brand of mystical Catholicism quenches all endeavor. She is free to indulge the further extremes of self-assertion under the seeming auspices of liturgical belief in sainthood.

Breaking codes arms Italian American women with the means to assert themselves within structures that would otherwise diminish them. Italian American authors have regularly developed characters whose brand of Catholicism-unorthodox at best-profoundly affects the way they live in the world. For most writers, coming out of the confessional box and writing about it is an act of liberation. Italian American writers as different as Jerre Mangione and Mary Cappello have shared their confessional stories, some humorous, some scathingly critical, of the way orthodox Catholicism has mistreated its practitioners.[7] Maintaining independence from conventional Catholicism does not prevent Italian American women writers from recreating in their works distinctively Italian-based Catholic iconography and beliefs. What they continue to confess outside the box reminds us that spiritual belief transcends institutional authority.

MARY JO BONA

---

[1] Many Italian American writers have examined the effects of Catholicism on their own lives or the lives of their created characters. The following list provides a short sampling. In nonfiction, see Mary Gordon's "More Catholic than the Pope" in *Good Boys and Dead Girls*; Helen Barolini's "Another Convent Story" in *Chiaroscuro: Essays of Identity*; Susan Caperna Lloyd's *No Pictures in My Grave: A Spiritual Journey in Sicily*, Mary Cappello's "Nothing to Confess" in *Night Bloom* and Beverly Donofrio's *Looking for Mary: Or, the Blessed Mother and Me*. In fiction, see Mary Tomasi's *Like Lesser Gods*, Carole Maso's *Ghost Dance*, and Louisa Ermelino's *The Black Madonna*.

[2] Denunciatory rhetoric informs the most famous of all sermons preached in America, Jonathan Edwards' "Sinners in the Hands of an Angry God." In 1741 Edwards warned his Enfield congregation that unconverted men "walk over the pit of hell on a rotten covering" (477), attempting to restore their commitment to a stern God. American Catholicism, influenced by the Irish hierarchy, followed a more doctrinal-based ideology, focusing on reverence and fear of the Godhead in contrast to the intimate association Italian Catholics felt toward the divine.

[3] See Nancy Maniscalco's novel, *Lesser Sins* and Regina Barreca's critical study, *They Used to Call Me Snow White But I Drifted: Women's Strategic Use of Humor*.

[4] For a more extensive analysis of Carmolina as a flowering artist, see my *Claiming a Tradition: Italian American Women Writers*: 145-162.

[5] Personal conversation with the author. De Rosa has said: "Creation is close to priesthood. It's taking the ordinary and turning it into the extraordinary" ("An Interview," 23). See De Rosa's biography of John Bishop Scalabrini: *Bishop John Baptist Scalabrini: Father to the Migrants* and "My Father's Lesson." For a more recent interview see Lisa Meyer's "Breaking the Silence: An Interview with Tina De Rosa."

[6] On generational imagery in America, see Marcus Lee Hansen, "The Third Generation in America" and Werner Sollors' "First Generation, Second Generation, Third Generation...: The Cultural Construction of Descent" in *Beyond Ethnicity: Consent and Descent in American Culture*.

[7] Mangione's Mount Allegro describes Gerlando's trips to the confessional stall, where an Irish-American priest, a harsh disciplinarian with a "booming voice," chastises the children as they disclose their sins (85). Cappello's Night Bloom posits a relationship between prohibited sexuality and Catholicism, equating being Catholic with a coming-out story that tells of "humiliation and abjection... The Catholic coming-out story is equivalent to a sharing of stigmata: 'Here's my wound, my battle scar, my badge'" (147).

References

Baker, Aaron and Juliann Vitullo. 1996. "Mysticism and the Household Saints of Everyday Life." *Voices in Italian Americana* 7.2. 55-68.

Barolini, Helen. 1999. "Another Convent Story." *Chiaroscuro: Essays of Identity*. Madison: University of Wisconsin Press. 10-24.

Barreca, Regina. 1992. *They Used to Call Me Snow White But I Drifted: Women's Strategic Use of Humor*. New York: Viking Penguin.

Bona, Mary Jo. 1999. *Claiming a Tradition: Italian American Women Writers*. Carbondale: Southern Illinois University Press.

Cappello, Mary. 1998. *Night Bloom*. Boston: Beacon P.

De Rosa, Tina. 1980. *Paper Fish*. Chicago: Wine P. 1996. New York: Feminist P.

_____. 1985. "An Interview with Tina De Rosa." By Fred Gardaphé. *Fra Noi*. May: 23.

_____. 1986. "My Father's Lesson." *Fra Noi*. Sept.: 15.

_____. 1987. *Bishop John Scalabrini: Father to the Migrants*. Darien, CT: Insider P.

Donofrio, Beverly. 2000. *Looking for Mary: Or, The Blessed Mother and Me*. New York: Penguin.

Edwards, Jonathan. 1741. "Sinners in the Hands of an Angry God." 1998. *The Norton Anthology of American Literature*. Eds. Nina Baym, et al. 474-85.

Ermelino, Louisa. 2001. *The Black Madonna*. New York: Simon & Schuster.

Foucault, Michel. 1980. *The History of Sexuality: Volume I: An Introduction*. New York: Vantage.

Gambino, Richard. 1974. *Blood of My Blood: The Dilemma of Italian Americans*. 1996. Toronto: Guernica.

Gardaphé, Fred. 1996. *Italian Signs, American Streets: The Evolution of Italian Americana Narrative*. Durham: Duke University Press.

Gordon, Mary. 1991. *Good Boy and Dead Girls: And Other Essays*. New York: Penguin.

Hansen, Marcus Lee. 1952. "The Third Generation in America." *Commentary 14*: 492-500.

Lang, Amy Schrager. 1987. *Prophetic Woman: Anne Hutchinson and the Problem of Dissent in the Literature of New England*. Berkeley: University of California Press.

Leonardi, Susan J. 1994. "Bernie Becomes a Nun." *The Voices We Carry*. Ed. Mary Jo Bona. Montreal: Guernica.

Levi, Carlo. 1947. *Christ Stopped at Eboli*. New York: Farrar, Straus.

Lloyd, Susan Caperna. 1992. *No Pictures in My Grave: A Spiritual Journey in Sicily*. San Francisco: Mercury.

Mangione, Jerre. 1942. *Mount Allegro: A Memoir of Italian American Life*. 1998. New York: Syracuse University Press.

Maniscalco, Nancy. 1979. *Lesser Sins*. New York: Avon.

Meyer, Lisa. 1999. "Breaking the Silence: An Interview with Tina De Rosa." *Adjusting Sites: New Essays in Italian American Studies*. Eds. William Boelhower and Rocco Pallone. Stony Brook, NY: Forum Italicum. 221-43.

Romaine, Elaine. 1985. "you were always irish, god." *The Dream Book: An Anthology of Writings by Italian American Women*. Ed. Helen Barolini. New York: Schocken. 306.

Maso, Carole. 1986. *Ghost Dance*. Hopewell, NJ: Ecco.

Russo, John Paul. "DeLillo: Italian American Catholic Writer." Unpublished paper.

Savoca, Nancy, dir. 1993. *Household Saints*.

Sollors, Werner. 1986. *Beyond Ethnicity: Consent and Descent in American Culture*. New York: Oxford University Press.

Turner, Kay. 1980. "The Virgin of Sorrows Procession: A Brooklyn Inversion." *Folklore Papers* 9: 1-26.

Vecoli, Rudolph. 1969. "Prelates and Peasants: Italian Immigrants and the Catholic Church." *Journal of Social History* 2: 217-68.

Waldo, Octavia. 1961. *A Cup of the Sun*. New York: Harcourt.

Part IV
Modern and Post-Modern Models and Images

Chapter 10

✦

THE PSYCHOANALYTIC INVESTIGATION
OF ITALIAN CATHOLICISM

For a long while the study of popular religion in Italy was dominated by what Gabriele De Rosa (1983: vii) called the "two monads" model. This model suggested that Italy was divided into two camps. On the one side stood the Church, which worked to promote "official" Catholicism, and on the other side were a variety of economically subordinate groups, mainly peasants, who clung to beliefs and practices inherited from dim pagan pasts. Under this view, "popular religion" was what resulted when these economically subordinate groups modified the official Catholicism passed onto them by the Church in light of their own superstitions and beliefs. Under this view, popular religion was seen to consist mainly of pagan practices to which a thin veneer of "real" Catholicism had been applied. Although this "two monads" model still has adherents, it has increasingly been supplanted by a more dialectical model. This newer model sees popular religion as emerging from the mutual interaction between distinct social groups each of which is associated with a different variant of Catholicism.

The basic idea here is not entirely new. Quite some time ago, Antonio Gramsci (1966 [1948]: 120) suggested that:

> Every religion, even Catholicism (in fact especially Catholicism, precisely because of its efforts to maintain a superficial unity...) is really a multiplicity of religions that are distinct and often contradictory: there is a Catholicism of the peasant, a Catholicism of the petty bourgeoisie and urban workers, a Catholicism of women, and a Catholicism of the intellectuals.

Gramsci's original insight here is important because it suggests that the particular form of Catholicism we embrace is shaped by the needs and desires that emerge in us as the result of our being a member of some particular group. Still, what has been added to Gramsci's insight is the possibility that the needs and desires that arise from group membership, and that shape the Catholicism which group embraces, can change over time. In the end, the emphasis in the newer model on group interaction and on adaptation to changing conditions has fostered a view of popular Catholicism that allows far more room for "creativity" and for "change over time" than under the older two monads model.

Some of the best-known examples of the work done in this newer tradition, at least in Italy, have been concerned with particular bishops who worked in the immediate aftermath of the Council of Trent and the ways in which these bishops had to adapt Tridentine reform to the conditions that prevailed in their dioceses. Gabriele De Rosa, for example, has demonstrated the ways in which *Mezzogiorno* bishops had to contend with things like a royal government often at odds with Rome; disputes with other bishops; the penetration of capitalism into the countryside; emigration, etc. De Rosa is especially proud of his discovery that popular Catholicism in southern dioceses was shaped more by the *chiese ricettizie* (a type of local church in which a group of priests, all of whom had to be native to the area, worked land held in common) than anything else.[1] Daniele Montanari (1987) has demonstrated that northern bishops were similarly creative. Montanari's study of Domenico Bollani, Bishop of Brescia and suffragan to Archbishop Carlo Borromeo of Milan, makes it clear, for example, that Bollani also took local conditions into account in implementing Tridentine reform. In some communities, for instance, enforcing the Tridentine rules relating to matrimonial impediments would have deprived too many people of a potential mate and so in these communities Bollani was relatively tolerant of local practice even though in other areas he could be quite rigorous in applying those same rules.

Unfortunately, most of those working in this newer tradition, while clearly acknowledging that Catholicism is shaped by the mutual interaction between different groups, continue to privilege the official Church. In the work of De Rosa and Montanari, for example, the only individuals granted any significant amount of agency and creativity are the particular bishops who are the focus of their studies. What is still not well-developed is a concern with showing that the creativity which results in the emergence of new forms of Catholic devotion can come not simply from Catholic bishops but also from any of the other social groups who embrace Catholicism. Demonstrating this is one of the goals of this article.

So: where does the "psychoanalysis" of my title fit into all this? Very simple. Once we accept that much of what constitutes popular Catholicism in Italy consists of beliefs and practices that have emerged as different groups respond to changing social conditions, we need to explain the "fit" between these new beliefs and practices and the changing social conditions that gave rise to them — and as a purely practical matter, I have found psychoanalysis to be especially useful in doing just this.

Suppose, for example, someone reported a dream in which they felt compelled to drag and lacerate their tongue on the ground in order to please their mother; or a dream in which their mother threatened to kill them if they did not cater to her every wish; or a dream in which they found themselves unable to walk or talk every time they found themselves in their mother's presence? Would we think it inappropriate to turn to psychoanalysis in an effort to make sense of such things by relating it to the life experiences of the individual involved? I doubt it. Yet, at one time or another, each of the behaviors that I have described has been observed at some particular sanctuary dedicated to a particular local madonna in Italy. If psychoanalysis can help us to understand why someone might associate such behaviors with their "mother" in a dream, it can help us understand, I suggest, why such Italian Catholics have some associated these same behaviors with their madonnas.

In the remainder of this article, then, I want to demonstrate the value of merging psychoanalysis with the "dialectical" model now routinely encountered in the study of Italian Catholicism by looking carefully at three examples of popular Catholic practice in Italy — and the first example concerns the liquefying blood relics of Naples.

The Blood Miracles of Naples[2]

For several centuries now, two vials containing the dried blood of San Gennaro, a bishop martyred in 305 A.D., has been exposed to the faithful at Naples on several occasions throughout the year. The vials are shaken, and the dried blood inside does — generally — liquefy. When I first reviewed the scholarly literature on this phenomenon, I was recurrently struck by two things. First, most of the existing literature was concerned with deciding if the liquefaction of San Gennaro's blood was truly miraculous, or rather the result of some purely natural process (and if so, *what* natural process). Second, in virtually all discussions, regardless of what position the commentator took on the "miraculous origins" issue, it was invariably pointed out that Neapolitan Catholics believed that the relic's occasional failure to liquefy was a portent of disaster.

This literature struck me as problematic for two reasons. The first has to do with history. The liquefaction of San Gennaro's blood was first recorded as occurring in 1389, and until the mid-sixteenth century it was the *only* liquefying blood relic in the Naples area. Over the next two centuries, however, more than a dozen other blood relics started liquefying on a regular basis in Naples and the surrounding area and each became the focus of a popular cult at some particular church. (Some of these other relics are still liquefying, including here the relic of San Pantaleone in the Cathedral at Ravello and the relic of Santa Patrizia in the Church of San Gregorio Armeno in Naples.) None of the existing discussions of blood relics made any attempt to explain this historical patterning. Second, notwithstanding the common claim that Neapolitans regarded a failure of San Gennaro's blood to liquefy as a portent of disaster, careful inspection of the statistical data kept by Church officials at the Cathedral in Naples over the last several centuries made it clear that "failure to liquefy" never really happens. Thus, traditionally, the blood relic was exposed to the faithful on eighteen different occasions honoring the saint during the year: nine days in May, eight days in September) and on December 16. While there were certainly particular days when the relic did not liquefy, in any given year it virtually always liquefied on most of the days on which it was exposed. Given this, the recurrent emphasis in the scholarly literature on "failure to liquefy portends disaster" seemed (to me) to be nothing more than an attempt by modern commentators to suggest that Neapolitans were intrinsically superstitious and so by extension, to suggest that *this* is why they found these blood relic cults appealing.

I want to suggest that something more than simple superstition is involved and, in particular, that if blood relic cults proliferated between the mid-sixteenth and mid-eighteenth centuries, this was likely connected to something that was happening in Naples and the surrounding area during this period.

Why "Damage to the Head?"

Most of the blood relics that we know about are associated with five particular saints: San Gennaro, San Pantaleone, St. John the Baptist, St. Stephen and Santa Patrizia. Interestingly, inspection of the legends surrounding these saints reveals that they share a number of common features: all five saints sustained damage to their heads, this damage provoked bleeding and it was this blood from their damaged heads — according to legend — that was collected and which became the focus of a blood relic cult. Thus, three of these saints — San Gennaro, San Pantaleone, and St. John the Baptists — were decapitated and blood which flowed from their decapitated bodies was gathered up. St. Stephen might not seem to fit this pattern since the New Testament tells us simply that he was stoned to death. Nevertheless, from the Middle Ages on, a tradition arose that the stones which had killed Stephen struck his head. Indeed, during the Middle Ages and the Renaissance, it was routine to depict Stephen either with a stone attached to his head and/or with blood dripping off the top of his head.

The case of Santa Patrizia is particularly interesting. According to legend, Patrizia was a niece of the Emperor Constantine who fled an arranged marriage in order to preserve her virginity. She ended up at Naples and died a natural death a few months later. A year after her death, however, a Roman aristocrat, who had been cured of a painful illness after praying to Patrizia and who wanted a personal relic, pried loose a tooth from Patrizia's skull. Blood immediately started pouring from the socket and continued pouring forth until the next day. This blood was gathered into vials by Church authorities and it is this blood that continues to liquefy to this day at the church of San Gregorio Armeno.

To anyone familiar the psychoanalytic literature, the imagery here, involving as it does "damage to the head" and "bleeding," will seem familiar. From Freud on forward, psychoanalytic investigators have argued that our unconscious thoughts and concerns are routinely disguised before being allowed into our conscious mind and that one of the most commonly-encountered of these mechanisms of disguise is what Freud (1900), in his classic study of dreams, called "displacement upwards." What happens, Freud argued, is that imagery and feelings associated with those parts of our body that have a strong sexual connotation, like the genitals, are "displaced upwards" usually onto the head. Freud's own investigation, for example, led him to conclude that both "decapitation" and "extraction of a tooth" when they appeared as elements in a dream were a disguised representation of concerns about castration, or more simply, a disguised representation of a concern with damage to the genitals. I want to suggest that this is precisely the concern that underlies the imagery surrounding the blood relic cults of Naples *and* the concern that explains their popularity.

Although Freud himself only discussed the dynamics of the standard nuclear family, subsequent investigators have explored the psychodynamics of other family types, including what can be called the "father ineffective family." This is a family type in which there is an ideology of male dominance but in which fathers are absent from the home for long periods of time, though they do return home on a regular basis, and in which there is a *de facto* concentration of authority in the hands of the mother. Anne Parsons (1969), a psychoanalytic anthropologist who did fieldwork in the slums of modern Naples, has suggested that this sort of family structure has a number of effects on sons. First, it increases the son's sexual attachment to the mother, by which is meant only that the son finds contact with the mother, even if that means only being in the same room with the mother, to be physically pleasurable. Although this sexual attachment to the mother comes to be repressed, it continues to exist in the son's unconscious and produces a sort of free floating guilt. This guilt in turn produces in sons a desire for self-punishment. Given that it is the son's physical desire for the mother which has given rise to this desire, the most fitting form of punishment would be damage to the one part of his body that the son most associates with physical pleasure, his genitals.

Other investigators have been less concerned with the psychological consequences of the family ineffective family than with the conditions which give rise to this family type. In many under-developed areas, for example, it seems clear that the father ineffective family proliferates when agriculture is organized such that males must tend fields at some distance from their homes. In other areas, the father ineffective family becomes more common when mass unemployment among class males forces them to travel far afield in search of what few employment opportunities exist.

If we now look carefully at the events of Neapolitan history, we find that in the last half of the sixteenth century, the increasingly oppressive nature of Spanish rule (most notably manifest in a dramatic increase in taxation) lead to widespread impoverishment of the Neapolitan lower classes, a massive depopulation of the countryside, and a dramatic increase in the number of people living in urban slums. In other words, what we see in this period is an intensification of precisely those conditions that give rise to the ineffective family. What happened I suggest is that the father ineffective family did indeed become far more common during this period, and that this did lead to an increasing sense of guilt among sons and a consequent desire for self punishment. It was this, I suggest, that predisposed Neapolitan males toward cults — like the blood relic cults we have been discussing — that were pervaded by castration imagery.

I am of course fully aware of the fact that terms like "castration imagery" and "son's sexual attachment to the mother" are typically *not* found in popular discussions of religion. Do I really mean to suggest that these things are important? Yes, I do. Not only does the argument I have just presented provide a way of explaining the content of these cults (why they are organized around blood relics associated with "damage to the head"), but it also provides a way of understanding why these cults enjoyed a surge in popularity during precisely that period of time when Neapolitan Catholics were confronted with conditions that facilitate the rise of the father ineffective family,

More generally, and quite apart from the matter of blood relics, the sort of argument sketched here provides a theoretical foundation for explaining something else that other investigators have avoided confronting: the strongly masochistic emphasis that for so long was a feature of Italian Catholicism. As bizarre as it sounds to us today, in this post Vatican II era, things like tongue-dragging, flagellation, self-mutilation, and so on, were for centuries central to the lived experience of Catholicism for the great mass of Italian Catholics in many areas of Italy, males in particular. Whatever else it does, seeing a "desire for self punishment" as an outcome of the father ineffective family provides a way of identifying the source of this widespread masochistic emphasis.

Finally, this sort of argument provides a way of explaining regional differences. In recent centuries, masochistic practices of the sort that I have mentioned here have most of all been associated with communities in Southern Italy. If we grant that the father-ineffective family does make sons receptive to "self punishment," the increasingly masochistic emphasis that came to pervade religious practice in southern Italy as we move from say, the sixteenth to the early nineteenth century, can be seen as resulting from the increasing impoverishment of the South relative to the North, and the consequent proliferation (in the South) of the father-ineffective family structure and the masochism that this sort of family structure breeds.

My next two examples will of necessity be shorter (and, I promise, will not involve castration imagery).

An Explosion of Saints and Madonnas[3]

Something that emerges quite consistently from statistical studies of the cult of the saints in Western Europe is that Italy tends to "produce" more saints than most other European countries. In Delooz's (1969)study of the cult of the saints in Western Europe, for example, Italy accounts for more saints than France, Germany, and the Iberian peninsula combined. This Italian proclivity for "saint making" has nothing to do with the Vatican's control over the canonization process since it is a pattern that emerges even when considering only those European saints whose cults emerged solely as the result of local tradition.

Further, the Italian predisposition to "create" saints is matched by a similar tendency to "create" madonnas. Thus, while the official Church may maintain that there is only one "Mary," the fact is that Italy has long been a land populated by a plurality of madonnas, each of which is venerated separately, usually in association with a particular image at a particular location, and each of which is seen to be independently powerful. True, popular tradition always recognizes that there is some link between these different madonnas, but this linkage is just as likely to find expression in the view that the madonnas are related — for example, that "madonnas are like sisters" (Provitera 1978: 343) — as in the official view that these madonnas are all versions of a single "Mary." While I know of no statistical study of madonnas, comparable to the statistical study of saints mentioned above, the areas the number of separate cults, each organized around a particular madonna, that exist in particular areas of Italy is enormous.

What accounts for this proliferation of saints and madonnas?

The Birth of Our Gods and Goddesses

In trying to explain the origins of religion, Freud (1927) suggested that as adults we come to perceive that the world around us is pervaded by danger and uncertainty. This, however, is not entirely unfamiliar to us; on the contrary, it inevitably reminds us of the helplessness that we experienced as infants. The difference is that as infants we had the advantage of being able to appeal to our parents, who we perceived, quite justifiably, as beings who were both powerful and transcendent. As adults, then, we come to yearn for beings, similarly powerful and transcendent, who can protect and care for us now. The result is that we create a god or gods modeled on the infantile images of our parents, both father and mother, that lie buried in our unconscious.

What Freud failed to appreciate, however, but what subsequent investigators have established, is that children not only see their parents as powerful but also as potentially threatening, i.e., children come to believe that the immense power of the parent can be used both to help them *or* to harm them. It doesn't matter that this may or may not be true; what matters is only that this is the conclusion that emerges in the infantile mind and as a consequence, children regularly develop defense mechanisms designed to protect themselves (in their own minds) from the threatening power of the parent. Melanie Klein (1975; 1980), in particular, suggested that one of the most important of these defense mechanisms is "splintering." What the young child does, Klein argued, is to "splinter" their image of say, the mother, into a number of separate and distinct images, each of which is then "projected" onto things in the child's environment, which in turn are then brought back into the child's mind as separate objects. Klein saw this process, which she labeled "projective identification," to be especially important since it established a template that functioned to mold the adult personality.

If we now revisit Italian Catholicism with these arguments by Freud and Klein in mind, what do we see? Well, let's start with the question that Freud's argument leads us to ask: when Italian Catholics have wanted a miraculous cure, either for themselves or for someone in their family, or when they have wanted protection from danger, both seen and unseen, to what supernatural being do they appeal? The data here are clear. They do not go to Christ or to the God of the Tridentine Church, they go mainly to madonnas and sometimes to saints.

Italy has hundreds of churches that house (or have housed) a miraculous image, that is, a statue or painting seen to be infused with supernatural power. Sometimes people travel to these churches as pilgrims seeking a cure for an illness; more commonly, as the painted *ex voto* at these shrines indicate, people travel there *after* having confronted some great danger and having been delivered from that danger by appealing to the image resident there. What's interesting is that the vast majority (over 80%) of the miraculous images found in Italy have been images of a madonna and, less often, images of some particular saint. There are few if any miraculous images of Christ.

Sometimes, of course, what you want is not a cure but rather protection from unseen future dangers, and in Italy such protection is provided by supernatural patrons, that is, supernatural beings charged with the protection of that group — and supernatural patrons are overwhelmingly saints, male saints in particular, rather than madonnas or representation of Christ. Virtually every city or village in Italy, for example, had a patron saint, and many had more than one. Cities like Florence, Siena and Padua, for example, each had four patron saints. The record for patron saints, however, must surely go to Naples. In the early 1600s, Naples had twenty-one primary patrons and by the mid-1700s, that number had grown to thirty-two.

In summary, then, the most powerful supernatural beings in Italy fall easily into two *gendered* categories: female madonnas who cure and heal, and saints, overwhelmingly male, who protect. This binary split, involving two categories — one female and one male — seems roughly consistent with Freud's original argument, which is that we model our gods on the images of our parents, mother and father, buried in our unconscious. In turn, the later work by Klein and others helps us to understand why each of these two basic categories, "madonna" and "saint," would be "splintered" into range of separate and independent personalities. In other words, faced with immensely powerful supernatural beings whose power might be used to harm us as well as help us, Italian Catholics make use of the same defense mechanism that we all used in confronting powerful images of parents: they split each category into a range of separate personalities and give each of these personalities an independent existence. Under this argument, the plurality of madonnas in Italy results from the splintering of the infantile image of a powerful mother, just as the proliferation of male saints results from a splintering of the infantile image of a powerful father.

If the argument that I have just sketched is correct, of course, we might expect that Italian Catholics would see the madonnas and saints on whom they rely for cures and for protection as simultaneously as source of danger. In fact, it is relatively easily to establish that this has long been true. As strange as it may seem to modern audiences, the madonnas who appear in popular legend are often quite unlike the gentle, loving Mary of the official Church. Italian madonnas have been known to send cholera epidemics against local communities that have not maintained the proper level of veneration; to maim people to whom they have appeared if the person involved does not move fast enough to build the shrine that the madonna wants; and even to cause the premature death of some local priest who put obstacles in the way of building her church.

Saints too can be a source of danger in Italy. Saints, for example, like madonnas, crave devotion and so are often seen as punishing people who work on the saint's feast day. Sometimes this craving for devotion is sufficiently strong that saints send illness as a way of causing people to turn to them for help. "The sickness of San Donato [which is epilepsy]," says one of the informants in Annabella Rossi's (1969: 25) study of popular religion in Puglia, "our saint sends it, and he takes it away; he wants it that way... He sends it to the poor like us to make us suffer." Another informant in the same study noted more generally, that "Every saint sends his illness" goes on to say that San Paolo sends the taranta, supposedly a type of spider in the Mezzogiorno whose bite causes people to dance uncontrollably, so that those bitten will go to the saint's sanctuary in Galatina; San Donato sends epilepsy so that people will travel to his sanctuary in Montesano (Puglia); that San Rocco sends the plague; and so on.

Faced with such supernatural beings, who are extremely powerful and yet potentially dangerous, it hardly seems surprising that Italian Catholics make use of the same defense mechanism — splintering — used by the infantile mind as a defense against the parental prototypes on whom those beings are modeled.

Whole and Entire

I would like to start my third and final example[4] by considering an event which occurred at the Council of Trent in August, 1562. On that occasion, as Adriano Prosperi (1983: 101) has pointed out, the Archbishop of Palermo rose and made a speech to the assembled delegates which attributed the religious unrest in Germany to the fact that the pope had denied a matrimonial dispensation to Frederick of Saxony and to the revocation of Luther's permission to preach on indulgences. It was a remark that clearly confused Frederick of Saxony with Henry VIII of England and Luther himself with Johann Tetzel, whose preaching of indulgences had been *attacked* by Luther. How could the good archbishop have made these mistakes? After all, by this time the Council of Trent had been in session, off and on, for almost 20 years and Protestantism had always been an important topic of discussion. Moreover, like all delegates, the Archbishop had at his disposal a veritable army of theologians he could have called upon for advice. In fact, as Adriano Prosperi (1983) points out, errors of this sort were commonplace in Italian-language commentaries on the Reformation. Why? Well, partly because it appears because that most Italian prelates were just not interested in theological issues, and so paid little attention to the details surrounding the great theological debates of the time. What *was* important to most prelates, however, and what comes through over and over in their pronouncements was that the great disaster of Lutheranism was that it had threatened the organizational integrity of the Church. This explains, by the way, why Italian prelates consistently used the word "luterani" to describe *all* Protestants, including both Lutherans, Calvinists, etc. After all, though these groups may have differed quite dramatically with regard to doctrine, the one thing that they shared in common — and so the thing that justified a common label — is that they, like Luther, had split apart the Church. Remember Gramsci's insight: the Catholic tradition is quite at ease with diveristy so long as that diversity coexists with a "superficial unity" and historically that superficial unity has been provided by an emphasis on the organizational integrity of the Church, that is, on the willingness of the diverse groups within Catholicism to acknowledge they all belong to the same organization. Italian prelates, in order words, might tolerate a Church characterized by diversity but they could not tolerate a fragmented Church.

Given this concern on the part of Italian bishops with the organizational integrity of the Church it hardly seems surprising that they would develop and promulgate this emphasis in ways that ordinary Catholics would understand. One of the ways they did this — I now want to suggest — is by emphasizing cults organized around the bodies of incorrupt saints.

An "incorrupt" saint is one whose body has been exhumed and found not to have suffered the processes of decay and decomposition typical of most corpses. Although saintly bodies have been found incorrupt in several locations, such bodies have an affinity, it seems, for Italy. In her survey of saintly incorruption, for example, Joan Carroll Cruz (1977) was able to identify 107 cases of saintly incorruption in the Western Church and of these, Italy alone accounted for 53 cases. Italy, in other words, accounted for about as many cases as every other country in Western Europe combined. In my own investigation, I was able to identify another 22 cases of incorruption in Italy, bringing the total number of cases (for Italy) to 75.

In thinking about the cults that developed around these incorrupt bodies, we need to be careful about dates. The problems here are well illustrated by considering the case of St. Cecilia. According to legend, Cecilia was a Roman aristocrat who converted to Christianity and who was subsequently martyred around the beginning of the third century AD. She was laid to rest, we are told, in the catacombs at Rome. In the early ninth century, Pope Paschal I transferred her body to the Church of St. Cecilia at Rome (there is no mention in the historical record, I might add, that her body was incorrupt at this time). Nothing more is heard about Cecilia's body until 1599, when Church officials decided to renovate the Church of St. Cecilia. During the renovations, workmen came upon a sarcophagus of white marble near the main altar. Church officials ordered it unsealed, and found inside the incorrupt body of a young woman — whom they promptly declared to be St. Cecilia. This body was then reburied and became the focus of a popular cult.

If our concern is with the popularity of cults organized around incorrupt bodies, then it doesn't matter if St. Cecilia was a real person or not, just as it doesn't matter if the body discovered in the 9th century and the body discovered in 1599 were in both cases the same body. What matters only is that a cult organized around the incorrupt body of St. Cecilia first appears in the historical record in 1599.

With this in mind, what happens if we look at the 75 cases of saintly incorruption in my Italian sample and determine the date at which a cult organized around those incorrupt bodies first appears. In fact, although a few such cults appeared in the Middle Ages, the vast majority of these cults appeared during the fifteenth, sixteenth and seventeenth centuries. Why?

Here I think it important to pay attention to two things. First, unlike, say, cults organized around miraculous images, which often emerge "from below" (that is, from the experience of ordinary Catholics who first attribute miraculous powers to some particular image), cults organized around incorrupt bodies almost always emerge "from above." After all, do ordinary Catholics really have the chance to stumble across incorrupt bodies? No. On the contrary, these bodies are almost always discovered when Church authorities, for one reason or another, order the exhumation a saint (or suspected saint) and it is these same Church authorities who typically make the first attribution of incorruption. Second, we need to pay attention to the way in which these incorrupt bodies are marketed to the public, which means paying attention to something that I have ignored until now, namely, the precise criteria used for determining incorruption.

Devotional commentaries often suggest that incorrupt bodies look fresh, as if the individual has just died. In fact, that is not usually the case. Visitors to the Church of San Marco in Florence, for example, can view the body of Antonio Pierozzi lying in a crystal case at a side altar. Pierozzi, who died in 1459, was a Dominican who founded and priory of San Marco and who became Archbishop of Florence. Only Pierozzi's hands and face are visible but it seems clear that is his skin is gray and desiccated, not fresh. Nevertheless, his facial features are clearly delineated and his corpse even now conveys an impression of what he must have looked like. This sort of incorruption is common. In other cases, however, the appearance of incorrupt bodies leans more toward the hideous than anything else. The incorrupt body of Santa Rosa di Viterbo, for example, can be found in the church of the same name in Viterbo (Lazio). In this case, the face is severely flattened; her nose is all but gone; the exposed skin looks like dark brown leather: and the skin around her mouth has receded to expose her teeth and create the impression of a grotesque grin. Not all incorrupt saints have survived to this day, of course, and so not all such saints can be examined directly. Still, careful reading of the reports surrounding the original exhumation of incorrupt saints will almost always reveal evidence that their skin was discolored in some way right from the start. So: if it is not the appearance of just having died that defines incorruption, what does?

Reading through the reports of incorruption that have appeared over the centuries, the one term that appears over and over again, and that clearly defined "incorruption" for those involved, is "intiero" ["entire"]. Thus, over and over again, we read that some body was found "tutto intiero" or "intiero e incorrotto." What "intiero" seems to mean is that skin had not split apart to reveal bone and that appendages (arms, legs, heads) had not separated from the torso. In short, what "incorruption" connoted to believing Catholics was simply that a body was characterized by a "lack of separation." I suggest that it was this psychological association with "lack of separation" which predisposed Italy priests and bishops to promote cults organized around incorrupt bodies when they stumbled across such bodies during, say, church renovations.

In other words, given their one great preoccupation, which was their preoccupation with the organizational unity of the church, the clergy would have been predisposed to promote cults and devotions which provided themselves and others with a concrete metaphor for thinking about the organizational unity of the Church that had been lost during the Reformation. Saintly bodies were a metaphor for the Church, and bodies preserved "whole and entire" through the agency of supernatural actions were therefore a metaphor for the organizational unity of the Church that had always been a valued feature of the Catholic tradition.

I grant that none this would explain why there was an upsurge in the discovery of incorrupt bodies in the fifteenth century. My own suspicion is this has a purely prosaic explanation. Incorruption, in the sense of "lack of separation," is probably something that happens quite regularly. Most of the time bodies are not exhumed and so incorruption does not come to light. The fifteenth century, however, was a period in which Italian churches were being renovated on a massive scale and in which exhumations — and so the discovery of incorrupt bodies — was common. What the argument I have developed *does* help explain, however, is why cults organized around incorrupt bodies continued to proliferate in the sixteenth and seventeenth centuries as bishops grappled with the loss of unity that followed upon the Reformation.

Conclusion

One of the things that I have tried to do in this brief presentation is to take seriously the suggestion that popular Catholicism in Italy has been shaped by group experience and is often a response to constantly changing social conditions. On the one hand, this has meant paying attention to historical patterns that are easily established but whose explanation has been ignored by previous commentators. Why did blood relic cults proliferate after the late 1500s and why specifically did they proliferate in Naples and the surrounding area? Why are Italian Catholics predisposed to create so many distinct saints and madonnas?

But in answering these questions I have also wanted to demonstrate the value of psychoanalytic theory. In some cases, this has meant merely pointing out, and then building upon, the similarity between patterns routinely encountered in the psychoanalytic literature and patterns encountered in the study of popular Catholicism in Italy. To anyone familiar with the routine discussion of "displacement upwards" in the psychoanalytic study of dreams, for example, the fact that the saints most often found at the center of blood relic cults all suffered "damage to their heads" is a pattern hard to miss. Similarly, Klein's discussion of a child's "splintering" of parental imagos seems to bear an obvious and clear similarity to the Italian predisposition to create a plethora of distinct and separate madonnas and saints. In other cases, using psychoanalysis has meant only observing the methodological injunction (implicit in all of Freud's own case studies) that "meaning" is best uncovered by uncovering the psychological associations that people have with the thing being investigated. Doing that in connection with the study of cults organized around the incorrupt saints is what led, ultimately, to the suggestion that it was an emphasis on "lack of separation" that constituted the basis of the popularity of these cults. Finally, and perhaps most important, the study of family dynamics and its effects on the mental and emotional life of children has always been central to psychoanalytic theorizing. I have tried to show, most notably in connection with blood relic cults, how this body of theory can be used to establish linkages between changing social conditions and new forms of devotions that emerge concomitantly with these changing conditions.

There is of course no "one way" to explain anything, and so it would be entirely inappropriate to suggest that psychoanalysis should always be privileged in the study of Italian Catholicism. Nevertheless, when all is said and done, I would argue that psychoanalysis is useful, if only to bring to light patterns and possibilities that would otherwise be ignored, and yet is a theoretical perspective that is underutilized in the study of religion.

MICHAEL P. CARROLL

---

[1] For an overview of De Rosa's work, especially as it relates to the *chiese ricettizie*, see Carroll 1996: 6-10, 200-204).

[2] The argument developed in this section is based on a number of earlier discussions; see in particular Carroll (1989: 57-78); Carroll (1992: 115-120, 155-59).

[3] My discussion here of splintering and its relevance to Italian Catholicism is derived from Carroll (1989: 154-63; 1992: 145-155).

[4] This final example is a truncated version of the analysis developed in Carroll (1996: 208-225).

References

Carroll, Michael P. 1989. *Catholic Cults and Devotions*. Montreal: McGill-Queen's University Press.

_____. 1996. *Veiled Threats: The Logic of Popular Catholicism in Italy*. Baltimore: Johns Hopkins University Press.

_____. 1992. *Madonnas That Maim: Popular Catholicism in Italy since the Fifteenth Century*. Baltimore: Johns Hopkins University Press.

Cruz, Joan Carroll. 1977. *The Incorruptibles*. Rocford, Ill.: TAN Books.

Delooz, Pierre. 1969. *Sociologie et Canonisations*. La Haye: Martinus Nijhoff.

De Rosa, Gabriele. 1983. *Vescovi, populo e magia nel Sud*. 2d ed. Naples: Guida Editori.

Freud, Sigmund. 1900. *The Interpretation of Dreams*. In *The Standard Edition of the Complete Psychological Works of Sigmund Freud*. Volumes 4 and 5: 1-751. Edited and translated by James Strachey. London: Hogarth, 1953.

_____. 1927. *The Future of an Illusion*. In *The Standard Edition of the Complete Psychological Works of Sigmund Freud*. Volume 21: 1-56. Edited and translated by James Strachey. London: Hogarth, 1953

Gramsci, Antonio. 1966 (1948). *Il materialismo storico e la filosofia di Benedetto Croce*. Turin: Giulio Einaudi editore.

Klein, Melanie. 1975. *Love, Guilt and Reparation and Other Works, 1921-1945*. New York: Delta.

Klein, Melanie. 1980. *Envy and Gratitude and Other Works, 1946-1963*. London: Hogarth.

Montanari, Daniele. 1987. *Disciplinamento in terra veneta: La diocesi di Brescia nella seconda metà del XVI secolo*. Bologna: Società editrice il Mulino.

Parsons, Anne. 1969. *Belief, Magic and Anomie: Essays in Psychological Anthropology*. New York: Free Press.

Prosperi, Adriano. 1983. "Lutero al Concilio di Trento." In Lutero in Italia, Lorenzo Perrone (ed). Casale Monferrato: Marietti.

Provitera, Gino. 1978. "L'edicola votiva e le sue funzioni." In *Questione meridionale, religione, e classi subalterne*, edited by F. Saija, 337-45. Napoli: Guida.

Rossi, Annabella. 1969. *Le feste dei poveri*. Bari: Editori Laterza.

Chapter 11

✦

SECULAR ITALY AND CATHOLICISM, 1848-1915: LIBERALISM, NATIONALISM, SOCIALISM AND THE ROMANTIC IDEALIST TEMPTATION

In many respects, the history of Italy since the collapse of the Roman Empire has been one of a "Dark Peninsula." True, there are a certain number of bright spots familiar primarily to scholars, while the artistic and dramatic characteristics of ages like the Renaissance are well known even on a popular plane. Nevertheless, much of what happens in Italian history is seen by people, both scholars and laymen alike, through a glass, and very darkly indeed. At the top of the list, in this regard, is the entire, crucially important sphere of religious-secular and Church-State relations, whose study is badly vitiated by ideological and cultural prejudices.

When one is speaking of the *Risorgimento* and immediate post-*Risorgimento* eras, both Italian and non-Italian knowledge of the general intellectual, political, and social environment, not to speak of the more specific problem of religious-secular clash, is still, to a large degree, lost in this black hole. Description of the nineteenth century context of Italian life often descends into caricature; indeed, even into cultivation of the most simplistic "good guy — bad guy" myths. This is particularly unfortunate for the sense of historical perspective of Italian Americans, since the *Risorgimento*, the decades following thereafter, and the difficulties of Church-State relations accompanying the movement for Italian unification provided the framework and much of the explanation for the great migration of the population of the *Mezzogiorno* to the New World in the period 1890-1914.

Clarifying this nineteenth and early twentieth century context involves first of all noting the reality of an anti-clerical and even outrightly anti-Catholic spirit existing all through Italian history. That spirit was the product of many causes, beginning with a heritage of State absolutism nurtured by the ancient Roman bureaucracy, which continued its influence, even in the most obscure years of the Middle Ages, through the important function

195

fulfilled by such officials as notaries. Hostility to the Church was also fed from the eleventh century onwards by two other factors: the ever deeper revival of understanding of Roman legal principles, and the increasing lay anger over papal recourse to interdicts, excommunications, and the calling of political crusades against its internal enemies, the most significant of which were those unleashed in the long-lasting struggle to obtain a friendly regime in the Kingdom of the Two Sicilies.

Such stimuli continued to embitter many influential laymen still later, in the Renaissance and Reformation eras, though their irritation was now aroused even further by the involvement of Spanish, Austrian and French dynasties in Italian affairs, and the association of their quarrels with the international Protestant-Catholic battle. Florentine and Venetian anti-Romanism could be extraordinarily heated, as evidenced by the writings of Niccolò Machiavelli in the early 1500's, and the seventeenth-century exchanges between Paolo Sarpi and the Roman Curia. Jansenist anti-papal reformism, Enlightenment dismissal of religious interference in natural life in general, and, finally, the Napoleonic reordering of much of Italy along the lines of the secularizing measures of the French Revolution, each with the aid of certain segments of the nobility and bourgeoisie, all played their role in feeding these tendencies down to the era of the *Risorgimento*.[1]

The strange coalition of monarchists, liberals, and moderate-nationalists which led the movement for independence and unification, created the Kingdom of Italy in 1861, and then engineered the annexation of Rome in 1870, inherited all of these various strains of anti-clericalism and anti-Catholic feeling, giving them a greater chance than ever before to triumph on a peninsula-wide plane. More radically democratic, anarchist, socialist, and nationalist groups, opposed as these might be to the particular combination of conservative and liberal secularist forces dominating Italy in the late nineteenth century, could still share with them a basic, vigorous, historically-rooted, world-view hostile to some or all of the structures and teachings of the Roman Catholic Church. The combustible anti-clerical material was, therefore, abundant and varied.

Moreover, it was stirred to fire heat in the nineteenth century due to another fact of life which was as vexing as it was unexpected: the revival of Catholicism as a consciously supernatural religion after its own bleak flirtation with naturalism in the years preceding the French Revolution. That peculiar tryst had seen the weakening, and, in some cases, entire disappearance of ideas, groups, and phenomena long intertwined with the explanation and practice of Catholicism, which now, in the course of the 1800's, came back into active life with renewed *élan*. These included scholastic theology and philosophy, the Society of Jesus, and popular devotions such as pilgrimages, novenas and eucharistic adoration, all of which were horrifying to the Jansenist, Enlightenment, and basic secularist mind, and had confidently been presumed to be dead and buried. Most upsetting of all, the restoration movement in Catholicism in the nineteenth century brought with it a revival and centralization of that papal power which secularists associated with danger to the stability and independence of Italy, and, in certain respects, to a degree that was greater than any known even at the height of the Middle Ages. Noticeable in the pontificate of Pius VII (1800-1823), and more apparent still in that of Gregory XVI (1831-1846), the pace and significance of Catholic rebirth was represented, above all else, by the person and reign of Pius IX (1846-1878).[2]

Four points can be made about the character and pontificate of Pius IX which explain why his figure and labor stood out as symbols of the "Catholic Question" to secularists in the Italy shaped by the *Risorgimento*. To begin with, the pope was perceived by many supporters of the unification-independence movement as having betrayed what seemed to be his initial, providentially-guided openness to the success of both its specific goals and their root inspiration. Secondly, Pius IX's charismatic, effusive personality made of him a truly modern public figure, fit for demonization after his initial fall from grace, and even more so after his elevation to the status of confessor by Catholics horrified at the overrunning of the Papal States. Next, the pope's commitment to a clarification of Catholic differences with the *Risorgimento*, which he understood to be necessary once the misinterpretation of his position had led to revolution and war with Austria in the years 1848-1849, not only entailed the direct attack on liberal and nationalist principles found in the *Syllabus of Errors* (1864), but also began the process of official Church investigation of politically-charged social questions. Such concerns were stigmatized by the victorious alliance of moderates as a sign of sympathy for radical revolutionaries and peasant "brigands" operating out of the former Kingdom of the Two Sicilies. A Papacy encouraging discussion of them was a Papacy promoting rebellion and subversion.[3]

Finally, and, perhaps, most importantly, Pius instituted the *non expedit* rule, the papal policy which, while allowing Catholic participation in local political life, prohibited it on the national level in the new "robber" Kingdom. Calls for a boycott of the central, parliamentary establishment began with the politically aware Catholic Press and not through the Vatican. The *non expedit* had its roots in experiences dating back to the 1850's, when Catholic activists like Dom Giacomo Margotti (1823-1887) saw that legally-elected deputies who did not subscribe to liberal nationalist ideals were e xcluded from the Sardinian Parliament. They soon realized that "when we took part in elections and in many places won a victory, we called down upon ourselves all manner of vexations, and our work went up in smoke."[4] Gradually, however, the Papacy came to share with Catholic activists a determination to turn what was perceived as a temporary abstention from a sham participation in the existing system into a serious preparation for a real participation in a future, better disposed Italy.[5]

Out of this ripened what secularists eager to build a strong and uncontested Italian State could consider one of the worst of the "rotten" fruits of the *non expedit* policy. "Preparation in abstention" meant the creation of a kind of shadow national government, through the construction and elaboration of a centrally-organized and nearly comprehensive Italian "Catholic Action" movement. For, although the *Società della gioventù italiana* (1868) always retained a certain autonomy, a multitude of other Catholic organizations and local parish committees were coordinated by the *Opera dei congressi e dei comitati cattolici*, founded in 1874 and given its definitive name in 1881, into a Kingdom-wide tool of serious importance. As that name indicates, the *Opera* met in regular congresses and carried out routine work through five permanent sections established in 1884: Organization and Catholic Action, Christian Social Economy, Instruction and Education, Press, and Christian Art. A generation or more of Catholic lay leaders was trained by the *Opera*, with the second section, headed, towards the end of the century, by Giuseppe Toniolo (1845-1918), Professor of Political Economy at the University of Pisa and founder of the *Unione cattolica per gli studi sociali*, being especially active.

> The Opera defined itself as intransigent, since it accepted the Syllabus and the pontifical directives; lay, since it was founded and presided over by laymen; papal, since it concentrated all Catholic efforts and organi-

zation in the service of the pope; and hierarchical, since its organization replicated the hierarchical Constitution of the Church.[6]

None of this substantially changed in the 1880's and 1890's. Leo XIII (1878-1903) is frequently depicted as the antithesis to Pius IX, and did, indeed, explore various strategies for dealing with the new Kingdom of Italy. Nevertheless, Leo's whole political approach was consonant with the main lines of the movement for Catholic revival as adopted in his predecessor's reign, and clearly manifested a wish to extend the influence of the Church in daily life still further. Nothing substantive came of dreams of official reconciliation. Leo always insisted upon the retention of some kind of temporal power, his diplomatic maneuvering with the new German Empire and other European nations frequently involving speculation regarding its side effects for restoration. His criticism of the errors accompanying the drive for national unity, as reflected in an address of 26 July, 1887, was reminiscent of the anti-*Risorgimento* articles of *La Civiltà Cattolica* from the 1850's and 1860's, and still quite biting. The Pope maintained the *non expedit* policy, and allied the Church more solidly still with concern for the Social Question through his encyclical letter, *Rerum Novarum* (1891), and its justification of the trades unions movement, nuanced though this might have been.[7]

The nature of Catholic political action was, however, transformed considerably under the direction of Pius X (1903-1914). Most remembered for his sanctity, and a doctrinal firmness displayed in his battle against Modernism, this northern-born pope, the first since unification who did not come from the former Papal States, was not as troubled as his two predecessors by the specifics of the Temporal Power question. He was, moreover, a man of pronounced democratic temperament, and impatient with many of the formalities and unwritten traditions guiding the behavior of the aristocratic-minded Leo XIII and Roman Curia, viewing them as obstacles to effective action. All this led to his sympathy for alterations in practical Catholic Action, with major, complicated, and, perhaps, unintended consequences for Church-State relations, as will be catalogued in more detail below.[8]

The Kingdom of Italy confronting these popes was a centralized but constitutional monarchy under the House of Savoy. Although the Court did exercise a certain influence in political affairs, Italy was governed much more by the interplay of forces in its Parliament, its municipalities, and in the Press. Before turning to a discussion of the specific attitudes towards the Church and Catholicism expressed by any of these elements of Italian political and social life, it would be wise to take a glance at the different parties and factions shaping their views in more or less organized fashion: the Right and Left factions of the Liberals, the Anarchist-Socialists, and the Nationalists.[9]

The many difficulties in understanding the Italian *Destra* can be clarified by realizing that it is a completely different beast than a rightist element in most other European countries, such as France. No supporters of the Kingdom of Italy could ever be identified as legitimists or proponents of a traditionalist program in which religion played a central role, given that they had accepted and worked for the creation of what itself was in its essence a moderate revolutionary State. Real rightists either abstained from participation in the government after 1861, or were extremely few in number, including some nostalgists for pre-1848 Piedmont, or followers of the other, fallen peninsular dynasties. Italian "rightists," like the Count Camillo Cavour (1810-1861), Count Stefano Jacini (1816-1891), and Marco Minghetti (1818-1886), were actually men who had adopted liberal and moderate nationalist principles that had proven to be useful to the strengthening of the power of the House of Savoy, and were in no way averse to a vigorous and multiform utilization of the authority of the State.

1876 saw the end of the dominance of this so-called "Historic Right" in the government of Italy, power then falling into the hands of the *Sinistra*, which was, as intimated above, simply another segment of the basically liberal party that had created the new nation. Associated at its origins with such pronounced anti-clericals as Urbano Rattazzi (1808-1873), the Left was now directed by Agostino Depretis (1813-1887), a freemason, like many of the other members of his faction. It was much more wedded to the ideals of the free market and the minimal State than its rightist opponent, and, dimly reflecting its earlier republican tendencies, advocated a modest expansion of the suffrage. Troubled greatly by internal governmental scandals, the mainstream of the *Sinistra* was worried by the increasing disaffection of the *Estrema*. This force included intransigents of republican sympathies, radicals focused on the domestic injustices of the new Kingdom rather than the foreign affairs that seemingly obsessed their more temperate leftist *confrères*, and national heroes such as Giuseppe Garibaldi (1807-1882) who considered themselves to be vaguely socialist in sentiment. In fact, opponents within the *Sinistra* feared that the *Estrema* was a potential parliamentary conduit for the expression of all manner of advanced socialist ideas already stirring popular and press circles.

Socialism, one needs to remember, was an extremely broad and ill-defined term for most of the nineteenth century. It included among its supporters people who simply wanted steady work, and others, the developers of the trades unions movement, who wanted that steady work to be honorable and justly paid. Socialism also was the goal of the small, but active bands of anarchists, intellectually stimulated by the Russian, Michael Bakunin (1814-1876), who had come to the peninsula in 1864 and remained there as a force for the next ten years. One segment of that movement, headed by Carlo Cafiero (1846-1892) and Errico Malatesta (1853-1932), formed the *Federazione italiana dell'Internazionale anarchista* (1872), taught its message through various rather short-lived journals, as well as the latter's book, *l'Anarchia* (1891), and aimed down the path of direct action versus the State and other authorities. Another branch shed Bakunian principles and took up the cudgel of legal action. This group included Andrea Costa (1851-1910), who was elected to the Italian Parliament in 1882, as a result of the Depretis suffrage reform. It ended up working alongside Leonida Bissolati (1857-1920), Anna Kuliscioff (1854-1925), Professor Antonio Labriola (1843-1904), who lectured on Marxism at the University of Rome, Claudio Treves (1869-1933), and Filippo Turati (1857-1932). Its journal of intellect was *Critica Sociale*, founded in 1891, and its political organ the Italian Socialist Party, which emerged in Genova in 1892, three years after the establishment of the Second International in Brussels.

Socialist deputies, upon entering Parliament, did, indeed, sympathize with by the *Estrema* faction of the Left, especially its so-called radical wing, just as the mainstream of the party feared. Severely tested by government repression from the 1880's onwards, Italian Socialism continued to reflect the mixture of influences leading to its birth; i.e., republican, anarchist, Marxist, trade unionist, and pragmatist elements simultaneously. Such a mixture enabled men of widely different temperament and intellectual concerns to express sympathy for it. Not surprisingly, therefore, the party conference of 1900 approved two approaches to achieving the Socialist program, those of Maximalism and Minimalism. Maximalists gathered round Labriola and the idea of a more critical break with the existing order of things; Minimalists around the parliamentarians and the possibilities of the parliamentary system, with both factions seeking to gain control of the party newspaper, *Avanti!*.

Personalities of extremely individualist bent took part in the growth of a new nationalist movement, leading to the creation of the Italian National-ist Association in Florence in December of 1910. These included such figures as Francesco Coppola (1878-1957), Enrico Corradini (1865-1931), Luigi Federzoni (1878-1967), Giovanni Papini (1881-1956), Giuseppe Prezzolini (1882-1982), and Scipio Sieghele (1868-1913), publishing in journals like *Il Regno*, *La Voce*, *Leonardo*, and *L'Idea Nazionale*. Their anger was directed against the "legal" Italy which had abandoned the call to greatness of the *Risorgimento* era expressing what they saw to be the deepest sensibilities of the Italian spirit. Legal Italy had dedicated itself in the positivist, materialist, post-Risorgimento decades to the petty ambitions of the unadventurous bourgeoisie, realized through the soul-killing parlia-mentary machinery identified and attacked by Angelo Camillo de Meis (1817-1891), Gaetano Mosca (1858-1941), Alfredo Oriani (1852-1909), Vilfredo Pareto (1848-1923), and Pasquale Turiello (1836-1902). National-ists claimed to speak for the "real" Italy, the land and population that had not yet realized its manifest destiny, either by completing itself geographi-cally or in answering the needs of its suffering southern peasantry. Armed conflict would be the means by which Italy would escape from its corrupt, legal shell and perfect itself, "redeeming" Italians living within the borders of Austria-Hungary. War would also restore to its rightful Italian owners the old Venetian territories now under Ottoman control, and offer lands to the men of the *Mezzogiorno* in those parts of Africa, such as Libya, still in need of a European colonial master. What was essential to affect this per-fection was true, energetic, charismatic leadership, the kind offered by leftists such as Francesco Crispi (1819-1901), Prime Minister on two occa-sions in the 1880's and 1890's, but cut off in his labors due to the disaster at Adowa in Abyssinia in 1896. Nationalists, like socialists, gained great sympathy in varied strata of Italian society, among former anarchists, ad-mirers of modern technology following Filippo Tommaso Marinetti's (1876-1944) famous Futurist Manifesto of 1909, and, ultimately, Italy's most curious literary mixture of quixotic influences, Gabriele d'Annunzio (1863-1938), the author of the jingoist *La Nave,* and future patriotic icon.[10]

All these politically active Italians of the decades preceding and en-compassing the Great Migration had present before them the picture of a

Church that still maintained and wished to retain a formidable hold upon Italian life; an institution that, nevertheless, experience had shown could contemplate adoption of different strategies in its attempt to survive and prosper. Three positions regarding what to think and do about that Catholic grip and will to power grew up among the members of the various Italian parties, and the Court, Parliament, municipalities, and Press that they utilized. I will label these the approaches of the "rejecters," the "pragmatists," and the "palingenesists." While a clear theoretical distinction among all three attitudes may easily be traced, there was no iron curtain separating movement from one to another. It was always especially possible for representatives of the first two approaches to slip into attitudes characteristic of the third.

By "rejecters," I obviously mean those for whom any true reconciliation with the Church and Catholicism as they actually existed was not a serious consideration. Giosuè Carducci (1835-1907), with his *Hymn to Satan* (1863), might be said to have provided a literary manifesto for the most vehement proponents of this approach, and Pius IX certainly believed the outlook expressed therein to be the logical result of cultivating the secularist doctrines of the "robber Kingdom." Indeed, any overall examination of Italian public life would illustrate that the rejecter's camp still held many cards in its hands in the late nineteenth and early twentieth centuries, and this from the Court down to the level of the man on the street.

To begin with, although King Umberto I (1878-1900) was as friendly as possible in his dealings with the Church under difficult circumstances, and Queen Margherita positively effusive in her demonstrations of religious conviction, Vittorio Emmanuele III (1900-1946) showed an outright disdain for everything Catholic. A man of classically nineteenth century positivist convictions, he avoided religious ceremonies and used only Protestant and Waldensian nurses for his children. Within the confines of his very pronounced sense of constitutional decorum, the King demonstrated sympathy for Italian political figures who shared his basic secularism, such as the leftist, Giuseppe Zanardelli (1826-1903), who was Prime Minister very early in his reign, and the socialist, Leonida Bissolati. Vittorio Emmanuele's public dealings with men of the cloth were limited to the most perfunctory and inescapable level on inevitable state visits throughout the country. It was said that the only religious building that he inaugurated during his reign was the Synagogue of Rome.

Moreover, the king was also close to Masonry, which had opened itself to penetration by a much more determined secularism in the latter part of the nineteenth century, once belief in some kind of Supreme Being had been struck from the requirements for membership. Masonry was a strong force in most elite Italian circles, its significance increased by the fact that politicians in all Latin countries often found masonic lodges to be suitable settings in which the members of the loose party coalitions of the day might privately and quietly come to compromises which would be difficult to arrange in the public eye. Zanardelli, under whose name the penal code of 1889 chastizing priests for "abuses" connected with their ministry was promulgated, was a prominent, masonic anti-clerical, as was Ernesto Nathan (1845-1921), the Mayor of Rome at the time of the troubled commemoration of the fiftieth anniversary of the creation of the Kingdom in 1911.[11]

Vexations emerging from the Council of Ministers, Parliament, and municipal governments, as well as from all political factions, from the Right to the Socialists, were still extremely common after the initial spate of anti-clerical legislation promulgated at the time of the establishment of the Kingdom, and then expanded and extended to the Eternal City after 1870. Thus, Prime Minister Francesco Crispi was responsible both for the removal of the Duke Leopoldo Torlonia (1853-1918), Mayor or Rome in 1887, as punishment for his enthusiastic message of congratulations to Pope Leo on the occasion of the latter's priestly jubilee, as well as for the creation of the electoral machinery that would wrest the city's government from the hands of Catholics and philo-Catholics. Prominent statesmen often wished to prosecute those suggesting that further changes, such as the obtaining of international guarantees for the Papacy, were required for the peace of Church and State in Italy. The penal code of 1889 was followed by a law on pious legacies of July 17, 1890, and the Zanardelli-Cocco Ortu proposal for legalizing divorce of 20 February, 1902. All these were bitterly opposed by the Church, using the parish and *Opera* organization to mobilize petitions against them. In fact, the entire decade preceding the First World War was filled with polemic, favorable and hostile, regarding governmental measures concerning everything from control of primary school education and the place of religious instruction within it, to supervision of seminary educational reforms, the rights of Catholic organizations to be represented in governmental councils, and the criminal pursuit of those contracting a religious marriage before passing through a civil ceremony. The speech of Ernesto Nathan, on September 20, 1910 praising the superiority of that lay civilization which had triumphed in the Eternal City in 1870 and would be celebrated in the Roman exposition of the following year, was typical of much anti-Catholic municipal rhetoric, arousing the protests of Pius X himself, and contributing to the decision to prohibit Catholic mayors from participating in the commemorations in the capital in 1911.

Probably the most important of ministerial interferences in the life of the Church took place in the troubled atmosphere of the last decade of the century. It was at this time that agricultural hardships led to the revolts in 1893 of the Sicilian *fasci,* disturbances among peasants in other parts of the peninsula, and, ultimately, to the 1898 riots in Rome, Florence, and Milan. Disorder was quelled with particular ferocity in the Lombard metropolis in May of 1898. Crispi and his subsequent imitators, the Marchese di Rudinì (1839-1908) and General Luigi Pelloux (1839-1924), struck hard on such occasions at all those groups perceived as being friendly to "Socialism," including the supposedly "red" leaning Catholics. On May 27, 1898, Marchese di Rudinì ordered the closing of practically all of the constituent associations of the *Opera dei congressi,* three thousand in number, the prohibition of numerous Catholic journals, and even the arrest of leaders of the Catholic movement, like Davide Albertario (1838-1902).

Anti-clerical journalists representing all political factions from the Right to the Socialists still plied their wares to a sizeable audience, with stories ranging from the classic uncovering of unnatural clerical lusts to Church persecution of intelligent and courageous dissenters. Stories of this kind were to be found in *Il Secolo, Gazzetta del Popolo, Messaggero, Vita,* and *Avanti!,* as well as in journalistic "histories," such as Benito Mussolini's (1883-1945) biography of Jan Huss. Perhaps most blatant in this regard was *L'Asino,* subtitled *è il popolo, utile, paziente, e bastonato,* run by Guido Podrecca (1865-1923) and Gabriele Galantara (1865-1937), and won over to the socialist camp soon after its creation in 1892. Such journals helped to unleash the kind of anti-Catholic incidents marring the transfer of the remains of Pius IX to San Lorenzo in 1882, the dedication of the statue of Giordano Bruno in the Campo dei Fiori in 1889, the commemoration of the discovery of America by Christopher Columbus in 1892, and the fiftieth anniversary of the foundation of the Kingdom of Italy in 1911, the "anno di lutto," as *La Civiltà Cattolica* called it in protest against its pronounced anti-clerical character. Writers like Jessie White Mario (1832-1906), in *La Riforma*, and articles in *La Nuova Antologia*, spread the notion that the Church, more than any other force, lay behind the revolts of the 1890's. Many others taught the simple positivism and scientism that played an important role in Social Darwinism and Marxism, following their counterparts in the rest of Europe in dividing the world into the camps of those infallibly aiding progress and those hindering it, and taking it for granted that the opening of every new train station indicated a victory over obscurantist religion.[12]

Nevertheless, even contemporary observers noted that under the threat of more pressing problems, open, direct, persistent "excommunication of Catholicism" was very much on the decline as the old century turned into the new. What always remained strong, however, was the psychological obstacle preventing some people who came from a background of hostility to the Roman Church from moving from a passion for her destruction to a serious contemplation of an alliance with her in the face of new dilemmas. This potentially inhibited all those of Jansenist or other anti-papal religious heritage, survivors of the most heated Church-State battles of the *Risorgimento,* industrialists involved in a capitalist development which was indifferent to the problems of those whom it displaced and disgruntled, and even the children of families living in fervently Catholic areas, for whom anti-clericalism became, as Jemolo notes, a non-conformist necessity. It helps to explain the desire of press moguls like Luigi Albertini (1847-1941), with his *Corriere della Sera*, to try as best they could to act as though Catholicism simply did not exist at all.

One example of a political figure of this type is the Baron Sydney Sonnino (1847-1922), half Jewish, raised as a Protestant, rightist in temperament, proponent of a powerful lay state, painfully aware of the weaknesses of the "legal" organs of the new Kingdom in the face of the "real" problems of the South in particular, and yet lacking substantive psychological stimulus to engage in anything other than half-hearted bridging of the secular-religious abyss. Such men were souls in agony. Sonnino was very much on the hunt for a means of building deeper support for an Italy which possessed little in the way of solid historical and emotional roots, but in a manner that could circumvent the need to treat Catholicism and the Church as equal partners in the enterprise.[13]

A late nineteenth and early twentieth century activist eager for a non-Catholic intellectual position that might come to his aid in constructing a stronger, secular-minded Italy generally found the crude positivism of the day unappealing and insufficient to his needs. It was this intransigent and simplistic materialism which was brutally criticized by Benedetto Croce (1866-1952) in "A proposito del positivismo italiano," in his journal, *La Critica,* in 1905. Practically all the existing parties in 1900 paid court to such positivism, the *Estrema* and the Socialists in perhaps the most pronounced fashion, and all, in consequence might intellectually be found desperately wonting.[14]

Serious men could, however, turn to another fountain from which to drink, that which was fed by the broad stream of nineteenth-century thought which may be labeled Romantic Idealism. This outlook reflects a heady combination of concerns for feeling, passion, freedom, will, and identification with the masses, presented in a charismatic, prophetic, and seemingly spiritual framework that nevertheless can also justify the exercise of the most brutal, physical force. It emphasizes the importance of the individual as a means of underlining the superiority and dignity of the human person against mechanist insistence upon inflexible mathematical and scientific laws, and the need to display individual "energy" and "action" in order to confirm the justice of a man's convictions. Discussion of various aspects of Romantic Idealism entails what might appear to be a lengthy digression from a precise historical argument into a hazy realm of philosophical and psychological speculation, but it is one which I believe greatly assists in clarifying the confused international climate of opinion from which the more thoughtful Italian strain of anti-Catholicism gained much of its intellectual inspiration. I will, therefore, take the liberty of steering Italian History into this speculative continental whirlpool, in the hopes of drawing substantive, if mystifying fruit from it by the end of the chapter .

Perhaps the most potent of the variety of sources of Romantic Idealism can be found in the writings of the Swiss thinker, Jean-Jacques Rousseau (1712-1778). Rousseau felt himself to be the prophet of an infallible mission based upon his certain possession of "virtue." For Rousseau, virtue was not something which was attained, even if only in part, by personal action, as a Catholic might think. Rather, it was a "state of being" completely separate from each of man's activities but one: that of sincerely stripping himself of all that was not "natural"; i.e., all that was not spontaneous to him, the non-spontaneous being identified by Rousseau as masquerade, pretension and hypocrisy. This stripping-down action he considered himself to have successfully performed, especially when answering critics of his behavior in his *Confessions* (published posthumously, 1782), where he revealed to the world everything deepest within his soul, without consideration for the effect that such disclosure might have upon his personal fortunes. Having through such an action become "virtuous," Rousseau had no further need to be ashamed of deeds that others thought to be reprehensible; deeds which he himself would have considered to be reprehensible in a "non-virtuous" man, who did not openly proclaim a consistent commitment to spontaneous "nature." Rousseau had come to terms with himself as the consistently passionate, natural man; Rousseau was, therefore, good. He was also perfect, because truth and virtue could not help but allow the completely liberated person to reach complete self-fulfillment.[15]

More than this, however, Rousseau was actually Everyman. Anyone who sincerely stripped himself down to his natural state, and thus became truthful, virtuous and free, as Rousseau had done, would have to be indistinguishable from him. This is why the various lovers in his widely-read *Nouvelle Héloise* (1761) are actually only loving themselves as they see their images in other people, and the teacher in his enormously influential *Emile* (1761) can be said by Rousseau to both liberate the child and make the youth into himself at one and the same time. For "the whole art of the master is hiding this constraint under the veil of pleasure or of diversion in such a way that they think they want everything that one obliges them to do....There is no subjection so perfect as the one which retains the appearance of liberty; thus one captivates the very will itself.[16] Conversely, anyone who is not Rousseau-like, anyone who criticizes him and his actions, anyone who fails to pity him in his trials, is neither free, nor virtuous, nor truthful. In fact, he is not human. Blum describes the situation well in commenting on Rousseau's discussion of himself as the "spectator-animal" contemplating the pointless being — the "suffering animal."

> The Spectator animal was denied pleasurable pity in regarding the suffering animal because the suffering animal was evil and hence unworthy of sympathy. Since Rousseau knew that mankind was, like him, good, he was forced to the awful but inevitable realization that the creatures who treated him so heartlessly were not really people at all, that the key to the mystery was that 'my contemporaries were but mechanical beings

in regard to me who acted only by impulsion and whose actions I could calculate only by the laws of movement.' He was now really alone, the only human being left amid a throng of automatons; the human race existed solely in him.[17]

Rousseau was convinced that the non-virtuous and non-human world around him was basically hostile to the effort to perfect it. The duty of Everyman-Rousseau was to make that world into himself or cause it to disappear before it do him any further damage. The question of an initial flaw undermining the value of this entire argument could not even be imagined; the sincere, virtuous, free, liberated Everyman was free from error. No discussion concerning the ground and justification of this underlying truth was permissible. It was a self-evident given. Doubt regarding his position would in effect mean allowing the sham world of the hypocrite to influence him once more. A critique of his obvious rejection of the doctrine of Original Sin, such as that offered by Archbishop Christophe Beaumont of Paris, had no meaning in the Rousseauian universe whatsoever. It simply proved the fact that the prelate, by belief and profession a slave of a supernatural religion, was not thinking naturally. He was not really human. Logically speaking, he was one of the suffering animals for whom no sympathy could be felt, and who could be eliminated.[18]

A Rousseau-like conviction of the infallibility of the free, non-hypocritical, virtuous, natural man, and the simultaneous reliability of this perfect being as the key to understanding the Will of the People, was so endemic to nineteenth century revolutionary thought as to defy any attempt to exhaust depiction of its incidence. It regularly appears in literature, in the novels of men like Hugo and Stendhal, in appeals to the example of Napoleon, and in the manifestos of the leaders of liberal, democratic, nationalist, anarchist, and utopian or Marxist socialist movements. We are constantly told that proponents of one cause or another are "sincere" (i.e., spontaneous, non-hypocritical, and natural) in their beliefs, and, therefore, virtuous and infallible; that their sincerity is revealed by an energy and consistency of often inconsistent and passionate action that only the Enemy of the People, destined for the rubbish heap of history, could fail to recognize as being good. Again, others cannot be judged by the same standard as the Rousseauian Hero if they are not incorporated into that Hero's Mystical Body — his immediate entourage, or the organization that he has created to carry out his and, by definition, the People's will. Thus, for Hugo, the revolt of "The

People" accepting his message is redemptive; the revolt of the mass of the inhabitants of the Vendée (or the Italian resisters to the French revolutionary invasion in the 1790's) is a vile riot. The massacres perpetrated by the former are redemptive; a tap on the finger by the latter in self-defense is the most wretched of crimes. Elimination of the Enemy of the People is a cleansing by the actively virtuous, perhaps the most noble of spiritual measures in an obscurantist universe that uses a supposedly supernatural spiritual sense as yet another justification for base, hypocritical sham.[19]

Italy was very sensitive and open to all these arguments. In fact, it developed a love affair with romantic, idealist maxims of Rousseauian flavor. The ground for them had long been prepared by the Jansenists, who were unshakeably convinced of the infallibility of their interpretation of Catholicism, and outraged by the suffering that they had endured at the hands of papally-backed hypocrites for remaining true to their (self-evident) virtuous state. Jansenists were themselves an influence in Rousseau's understanding of Christianity in his brief period of flirtation with the Roman Church. It comes as no surprise that Pisa, in that Grand Duchy of Tuscany which was perhaps the most important center for the dissemination of Jansenist ideas in Italy, was already an eighteenth century foyer for the spread of the Swiss radical's teaching as well, and one that influenced Filippo Buonarotti (1761-1837), the first great Italian revolutionary agitator, active in the life of *Carboneria*. The attraction of Napoleon as the charismatic man for all seasons, whose energy and action justified his transforming the world around him, was strong in Italy. Much of the peninsula had been swept up in the general-consul-emperor's whirlwind, and many of its bourgeois inhabitants were given new ambitions through the influence of his revolutionary changes, which were favorable to their interests.[20]

Giuseppe Mazzini (1805-1872), the foremost Italian nationalist thinker and organizer, grounded his certainties about God, man, nationhood, and democracy in a theism whose precise roots have been hotly debated by different historians. Whatever their roots, they were backed by infallible, prophetic utterances that exude the omnipresent Rousseauian motifs, and the conviction that energy and action are the sure signposts guiding men to truth.[21] Italian nationalists of even a moderate spirit had little difficulty proclaiming the indefectibility of their program in similar fashion. Massimo d'Azeglio (1798-1866) and the editors of *Il Cimento*, for example, thought that they could adequately defend themselves against the charge of unwittingly unleashing an amoral nationalist crusade with which the Jesuits of the Roman journal, *La Civiltà Cattolica,* taxed them, by pointing to the sincerity, generosity, and *ipso facto* correctness of the *Risorgimento's* intentions.[22] Even Italian Marxism in its most positivist form spoke frequently not with the voice of science, but with that of the author of the *Confessions* — prophetic, charismatic, natural, and absolutely certain of accurately representing the popular will.[23] To be an educated non-Catholic Italian, as to be an educated, non-Catholic European, was to breathe a climate of opinion permeated by ideas best expressed by Rousseau. To be a Catholic, in the mind of someone raised in this *zeitgeist,* was to be the proponent of an unnatural, hypocritical religion disguising base motivations under the cover of the supernatural, and to appeal to a counter source of infallibility that could only make him an enemy of spirit, freedom, and the people at one and the same time. And even Catholics themselves were affected by this atmosphere, through the writings and example of the Abbé Félicité de Lamennais (1782-1854), whose *Paroles d'un Croyant* "directly inspired Mazzini's *Faith and the Future* of 1835, which he considered his best work." [24]

But, here, someone might object that Rousseauian influences in Italy were far overshadowed by those coming from Georg Wilhelm Friedrich Hegel (1770-1831). Indeed, no one can deny that Hegel had an enormous and much more demonstrable vogue in official circles of both the *Risorgimento* and the new Kingdom of Italy. Hegel's books were smuggled into prisons in pre-unification days for the inspiration and encouragement of righteous suffering nationalists, who needed to be shown that history would vindicate them. His ideas were promoted through the work of Francesco de Sanctis (1817-1883), Minister of Education in the 1870's, and the academic and literary circles in Naples surrounding Bertrando Spaventa (1817-1883) and his brother, Silvio (1822-1893). Their influence was central to the development of the greatest of Italy's early twentieth century intellectuals, Benedetto Croce and Giovanni Gentile (1875-1944). Michael Bakunin, the seminal anarchist teacher, sang paeans to his Hegelian heritage, while no self-conscious Marxist, like Labriola, a student of the Spaventas, could do anything but confirm the German's significance as well.[25]

Be this as it may, Hegel's argument, in practice, is, nevertheless, a variation on the theme dear to Rousseau. It presumes that nature is shaped by a spiritual principle of freedom, incarnating and working itself out in history through the clash of energetic manifestations and counter epiphanies, as these are charismatically revealed and commented upon by the Prussian professor. Hegel proclaimed the nineteenth century standard bearer for spiritually propelled action-for-freedom to be the coercive authority of the modern State. His teachings were, therefore, immensely useful in defending the righteousness of the Italian *Risorgimento* doctrine of the necessity of building, through violence, a unified, independent State. They were also handy in support of the measures of Italian rightist lay authorities who valued the "transcendent" qualities of that institution in its battles with a Church, a pope, and a *Syllabus of Errors* which they saw as hopelessly out of touch with the world of the truly spiritual and undeniably infallible. The combined *Risorgimento*-rightist appeal to spirit and freedom on the one hand, and encouragement of secularization and police repression on the other, proved to be a potent tool for confusing the more logical, Aristotelian, Catholic mind seeking cogent arguments to oppose it. Anyone interested in the complications of the intellectual battle thus unleashed should consult the exchanges between Luigi Taparelli d'Azeglio (1793-1862) of *La Civiltà Cattolica* and Bertrando Spaventa.[26]

Practical consequences are especially noticeable when these theories were applied to the issue of education. Hegelian adulation of the use of the State to achieve freedom in a way that was offensive to Catholics was omnipresent in journals and parliamentary reports from the 1850's through the 1870's, and underlay much of what was done in this realm thereafter. "Unregenerated" people were said not to be allowed the opportunity to succumb to the temptation of entrusting their children to the care of monks and nuns. Antonio Gallenga (1810-1895), writing in *Il Cimento* in June of 1855, well demonstrated the kind of approach that Catholics loathed, when he claimed for the State the total right to educate, religious being kept from this task until the people had been given "the discernment of good and evil." Up till the moment that "national regeneration" was completed, he insisted, the State had the duty to exercise, "let us say it frankly, the *tyranny* of educating." A proper State required "the unity in one person of the attributes of highest magistrate and supreme pontiff," in order to root out the long centuries of servitude with which the Church was associated. It was useless to cite the example of England and America as models of freedom, he concluded, since in these countries, nothing positive was demanded from the individual:

But here among us the citizen is the property of the State: the law of conscription binds them to the soil of the fatherland during the most florid period of their life. The State has therefore the right and the duty of exercising over him an almost paternal tutelage. It would scarcely be able to consider him responsible to the laws of the land if it neglected or permitted others to pervert its moral and political education. That State that did not claim for itself the sole right for educating would only half understand the duty of legislator.[27]

And Hegel, after all, was himself but a product of the other wing of the anti-mechanist tradition, that which was shaped by one of the greatest of Rousseau's contemporary admirers, Immanuel Kant (1724-1804). Kant, awakened by David Hume (1711-1776) from his "dogmatic slumbers," was as concerned as the Genevan prophet for rebuilding order upon nature, and a nature that could not be expressed by conforming oneself to the faulty information coming from outside the human person. He, too, was ultimately forced back upon an assertion of the infallible, sincere, non-hypocritical will that universalizes and proves itself in energetic action. However traditional the kind of order that Kant might have thought would emerge from this process, it seems to me that its practical, historical effect was to give *carte blanche* to different thinkers to assert varied and conflicting universalizing wills as infallible guides to order in a universe lacking objective scientific and logical laws: the natural, passionate, irrational individual; the conspiratorial, nationalist organization guided by the charismatic, spontaneous prophet; or the State shaped by the liberated, the strong-willed, and the consistently energetic men of action.

All the concepts discussed above responded to something embedded in the Italian anti-Catholic's mindset. Hence, Italian admirers of the lay State trained in Hegel's thought were always potentially open to Kantian or Rousseauian influences, while the same is true the other way around. Even the appearance of certain more traditional rightist treatises on the library shelves of turn of the century non-Catholic Italians has generally to be understood in the context of an attempt to utilize them to romantic idealist purpose. This is why Italian thinkers eager to escape the barren silliness of mechanist, positivism were susceptible to the arguments of other Europeans groping in the same direction, but finding it impossible to do so without falling into the Rousseauian-Kantian anti-mechanist camp; men of "energy" and "action" like Henri Bergson (1859-1941), with his concept of *élan vital*, or Georges Sorel (1847-1922), in his *Reflections on Violence* (1908).[28] This is why one can find in the statements of all the varied inheritors of Romantic Idealism a frequently unconscious mixture of what would appear to be conflicting themes: on the one hand, devotion to an anarchic freedom based on a non-hypocritical energetic action which defines consistency as firm commitment to willfulness and the changeability of personal whim or a spiritualized "history"; on the other, an appeal to the use of State organs of physical repression to destroy that liberty in those deemed incorrigeably tied to a slave mentality. And, finally, this is also why none of them ever really raises himself to an appreciation of a truly spiritual Catholic position. For Romantic Idealism, of both Rousseauian and Kantian-Hegelian origins, is itself a by-product of the same naturalist, anti-Catholic, Enlightenment outlook that gave birth to mechanist positivism; a *weltanschauung* which was held together, at its outset, only by a seemingly transcendental Deism begging to be logically refuted and brought down entirely to earth.

Several factors were thus coming together in the late nineteenth and early twentieth centuries to indicate the possible creation of new, lay movements that could conceivably bring together the many heads of Romantic Idealism, and, in particular, the combination of irrational will with private violence or coercive state power. Anarchism of the anti-parliamentary Errico Malatesta variety, as well as that promoted by part of the syndicalist movement, intent on escaping Marxist mechanism and finding some immediate tool by means of which to wipe out the corruption of authority, were very much susceptible to such developments. So was the more impatient strain of Socialism, represented, among others, by Benito Mussolini. Both of these sources provided recruits for the force that would most effectively profit from the latest appeal to energy and action, the new Italian nationalist movement. And this, in turn, would pave the way for the emasculation of Parliament in 1915, and the victory of fascist *squadrismo* after the First World War. D'Annunzio, the Futurists, and other nationalist devotees all worshipped at the shrine of a fascinating and explicit anarchic cult of violence and coercion which sometimes reached absolutely grotesque proportions, the profoundly anti-Catholic character of which was not lost on them:

> ...the Nationalists could hardly preach of the mysterious powers of war and violence without rejecting the Christian view of peace and humility. A D'Annunzian, Nietzschean pose of neopaganism was very much in evidence among the imperialists. Papini in the *Leonardo* indulged in

this, scoffing at what he called *pecorismo nazareno*. Corradini spoke of Christianity as 'pathological and economic,' whereas life consisted of conquest and struggle. In this, Papini and Corradini, as well as D'Annunzio, were capitalizing on Carducci's 'Romanism' as expressed in the *Odi Barbare* long after Carducci himself had abandoned the idea as artistically sterile.[29]

Nevertheless, the winds of the times seemed more to be blowing away from the "rejecters" and their efforts to circumvent Catholicism, towards the second position, that of the "pragmatists," those who wished to reach some open reconciliation with the Church and the Catholic position in society. Although this tendency claimed much of its constituency from men who were either indifferent to religion or merely disgruntled by the more exaggerated manifestations of clerical power, it also included individuals of firm anti-Catholic belief who were convinced that the deeper importance of other questions required a practical change of policy on their part. One could find supporters of the pragmatic viewpoint among practically all of the forces active in Italian political life.[30]

By 1876, rightists had become greatly concerned about the potential demagogic effects stemming from the Left's proposal of an expansion of the suffrage, modest though this actually was, and the weakening of State power accompanying its more pronounced espousal of free market principles. They turned, in consequence, to urgent appeals to Catholics, to abandon the *non expedit* policy, and join their "natural allies" in a campaign against a resurgent "Jacobinism." Problems arose from what the orthodox judged to be the dubious professions of Catholicism coming from some men of the Right, as well as from the impossibility of accepting their infallible, spiritualized Hegelian State as the final arbiter of what constituted abuses in governmental *contretemps* with Church. It was difficult for Catholics to forget the mass of anticlerical legislation passed between 1861 and 1876 under rightist auspices, and the fact that Minghetti had himself boasted, when faced with criticism on this score from the Left of the high level of royal interference with the free action of the Papacy and Catholic organizations during his tenure as Prime Minister. Additionally, insofar as Catholics were active politically, on the local level, they found themselves more in tune with the leftists, and sharing with them a desire for an extension of the suffrage.

In fact, that historically anticlerical Left proved itself to be open and pragmatic in national matters also, as its awareness of the weaknesses of the new Kingdom, which they desperately wanted to play a major international role, became more vivid. The Left did nothing to harm the *modus vivendi* for cohabitation with the Papacy which followed Pius IX's rejection of the Law of Guarantees of 1871. Even Zanardelli, whose anticlerical convictions we have already noted, refused to entertain the suggestion of more rabid opponents of the Church, who wanted those bringing up new plans for reconciliation prosecuted. As the social crisis matured, leftists hoped that Catholic concern for private property and order would be a tool for bringing the Church into unified action with them versus Anarchism and Socialism.

In this, they were joined by a former member of the *Estrema*, Francesco Crispi. Crispi, a freemason of deist convictions who had opposed the Law of Guarantees, had warned Bismarck and Gambetta of the international danger of the Papacy in 1876, and had sacked Torlonia as late as 1887, gradually emerged as the leader of the effort to form an alliance with Catholics in defense of the established order. He refused to take part in the ceremonies inaugurating the creation of an organization honoring the atheist, Giordano Bruno in 1889, and claimed, now, to be happy with the system of practical cohabitation with the Church which maintained the existing equilibrium. Other anticlerical heirs of Angelo Brofferio (1804-1866), Giuseppe Ferrari (1811-1876), Francesco Domenico Guerrazzi (1804-1873), and Ferdinando Petruccelli della Gattina (1815-1890) in the *Estrema* faction could still be venomous, however. And this fact, along with serious moral concerns for the social responsibility of property, doubts regarding the commitment to free-enterprise of a party that had began by confiscating Church property, conviction that materialist socialism was merely a development of the same naturalism that had produced materialist capitalism, and fears of merely being used by a basically unaltered Crispi, led Catholic activists and prelates to back away from the Prime Minister's advances. His sharp reaction to their recalcitrance, as the crises of the 1890's increased, seemed to justify their reluctance.

Anarchism was closed to offers of cooperation with an authoritarian Church, but the still vague socialist movement allowed for some pragmatic proposals to emerge from its undefined ranks. The young Francesco Saverio Nitti (1868-1953), the future radical statesman, invited the Church, in *Il socialismo cattolico* (1891), to consider the way in which friendship for the socialist cause would benefit it. Despite the violent rants of *l'Asino*, disdain for cooperation with outrightly religious organizations, and the often bitter ridicule by socialist workers of their Catholic comrades, the party was never officially hostile to individual believers. Bissolati, Turati, and Treves all made it abundantly clear that such issues troubling Catholics as the divorce proposal of 1902 were purely upper class bourgeois concerns. Still, the reality of the role of materialist Marxist principles in the socialist movement, and the potential competition for control of the masses that its growing organization threatened, made activists like Toniolo and Romolo Murri (1870-1944) more concerned to draw stimulus and lessons from it for the purpose of better opposing it.

Liberals of both rightist and leftist complexion made new efforts to obtain pragmatic cooperation from Catholics in the period of social calming following the assassination of Umberto I at the hands of the anarchist, Gaetano Bresci, in 1900. These were the years most associated with the dominance of that dispassionate, peace-loving, Piedmontese supporter of the organs of legal Italy, Giovanni Giolitti (1842-1928). Giolitti was open to compromises with Catholics, but saw no need for any dramatic reconciliation as a prelude to joint action. *La Nuova Antologia*, which had published articles in favor of repression, turned away from an approach that seemed to menace all freedom of association. Rightists, by now, generally lacked the old *Risorgimento* interest in Church matters, perhaps due to an indifference to all theological and philosophical issues that would have seemed impossible to a Cavour, a Bettino Ricasoli (1809-1880) or a Ruggiero Bonghi (1826-1895). Indifference was demonstrated by their general lack of concern for the Modernist Controversy, which they considered to be an internal Church issue, and one that interested the State only insofar as it required defense of the civil rights of those who were excommunicated by the pope. The American example of Church-State relations, already appreciated by Crispi in the 1890's, was held up to ever further praise. A group of senators, disturbed by radicalism and socialism, began to pursue Church assistance versus enemies of individual freedom and property, while putting Catholics on warning regarding such "utopians" in their ranks as Toniolo and Murri. More modest proposals for divorce were presented by leftists in Parliament, while old rightists like Sonnino rejected the divorce project entirely. Alessandro Fortis (1841-1909), a Prime Minister from the ranks of the *Estrema*, was even willing to appoint the Marchese Nerio Malvezzi, "a

Catholic... who dared to say that the Law of Guarantees had not closed the Roman Question," into his cabinet in 1905, although he paid for it with opposition bringing his ministry down.[31]

While Liberals appealed to Catholics to join them against madcap colonial ventures, proponents of precisely such enterprises also made their voices heard by the second decade of the new century. Not all nationalists labeled the Church a divisive, effeminate institution, or were averse to exploring ties to it. Pragmatist proponents of conciliation could appeal to a joint nationalist-Catholic disdain for a legal Italy shutting out the legitimate demands of the real nation. Some of those favoring North African colonization also argued that this would solve the social problem of the *Mezzogiorno*, which troubled Catholic reformers as well. Nationalists were pleased by the reality of support from prelates, priests, and laymen, especially children of old Catholic families, and hoped to build upon the interpretation of the war with the Ottomans over Libya that broke out in October, 1911 as a Crusade against Islam.[32]

The chief practical method by means of which Italian governments accomplished the work of accommodation was that of *trasformismo*. This term, utilized by Agostino Depretis in 1876 to describe efforts to work with, and "transform" the existing rightist government into one which reflected the will of its new, but divided, leftist masters, had a history in Italy extending back to the days of the Count Cavour. Giolitti was one of its master practitioners in the period before the First World War. *Trasformismo* sought to avoid radical divisions by, in effect, co-opting the representatives of all important political movements, well-established or embryonic, and winning them over to actions acceptable to the governmental majority. If the most energetic "enemy" figures could be won over to support the ruling coalition, the troops that they commanded would be left without leadership. While functioning nicely to maintain stability, successful *trasformismo* prevented the voice of serious opposition from being effectively heard in the organs of "legal" Italy even more than limited suffrage did. Pragmatists might appreciate it, but it was detested by men of strong conviction in the *Estrema*, as well as the more uncompromising Socialists, Nationalists, and Catholics, all of whom often viewed it as *the* most cynical tool of a thoroughly cynical parliamentary system.[33]

More thoughtful pragmatists could easily find themselves becoming supporters of the third position, that of the "palingenesists." Palingenesis, formed from the Greek words "again" and "birth," was the idea that a new and much better society was "emerging" in the nineteenth century, one that was destined to develop out of the earlier forces dominating Western life, and one that would eliminate past divisions in a higher unity releasing energy for ever happier and humane projects. Such a vision rejected the pessimism of the apocalyptic thought popular in some religious-minded circles of the day, and reflected much more the hope for a "third age of humanity," "the arrival of the age of gold under the sign of universal fraternity through social justice," redolent of the prophecies of Joachim of Fiore.[34] Palingenesis could be appealing to any defender of "modern" ideas who still possessed a spiritual sense and did not want to jettison his personal Christian baggage and that of European civilization as a whole.

One group of people extremely interested in palingenesis and the uncovering of its mysteries was the Saint-Simonians: Claude Henri de Saint Simon (1760-1825), Barthélemy Enfantin (1796-1864), Saint-Amand Bazard (1791-1832), Auguste Comte (1798-1857), and their many fellow-travelers, like Charles Fourrier (1772-1837). Horrified by the violence and rejection of social order represented by the Revolution, the Saint-Simonians set out to illustrate the laws underlying community life, and the principles by means of which it had grown and developed, organically, through the ages. In this manner, they could show how one historical era had been the prelude for the next, and how the teachings of Jesus, the cult of the Virgin, a hierarchical priesthood, a liturgy, and many other elements of Western civilization still had a role, transfigured though it might now be, in modern life. Liberals, democrats, and nationalists all around Europe could, and did, all tap into palingenesist visions, along with fervent supporters of more particular causes dear to one or another nineteenth-century group.[35]

Catholics also responded to palingenesist arguments. To understand why is to have to turn once again to the figure Lamennais and his Rousseauian respect for the sparks unleashed by the non-hypocritical, natural order of things. Lammenais' connection to the energy syndrome was "traditionalism," that philosophical-political school which disdained the role of reason in grasping the truth, and, for that matter, in passing on the Faith as well. For Lammenais and traditionalists as a whole, the truth could not be understood, and the faith could not be taught, rationally; rather, they were handed down by the power of the very nature of things themselves, by means of a solid society's culture, institutions, and people. For traditionalists, the Revolution's basic untruthfulness was evident in that, in order to accomplish its program for the salvation of France and her people, it had tried, against the opposition of the majority of those people, to destroy the culture and traditional institutions whose energy had created and defined France and Europe in the first place.

Problems for Lamennais came from the fact that one part of the tradition — the Catholic part — was not as active as it had to be. It itself lacked energy. If it continued to lack energy, then it would run the risk of not shaping the environment and not passing down the truth in the only way that truth effectively could be transmitted: namely, through society and social action. Hence, the need to shake institutions and people out of their torpor and "indifference," and restore their desire to fight energetically for what was true. *The Essay on Indifference*, his first important work, was really not an apology for the Catholic Faith; it was much more a call to energetic action.

Torpor could scarcely be hounded out of the Catholic soul if the Church as an institution were unable to function as she must. Thus, the Papacy, so cruelly harassed by the Enlightenment and the Revolution, had to have its just and full powers appreciated and restored. In addition, the Church in each nation had to be unchained for action under papal guidance. Unleash the Papacy and the national Church together, and indifference would retreat. With the retreat of indifference would come the willingness to commit energy to struggle, and with more energetic struggle would come advancement in truth.

Unfortunately, both the traditional monarchy and the national episcopacy fettered by it themselves lacked commitment to truth and the requisite courageous energy to free the Church. Casting about for an alternative force to lead the defense of traditional culture and Catholicism within the nation, Lamennais was led, ironically, to an untraditional conclusion. The Catholic People, which as a whole had also been unjustly and absurdly enchained by a legitimate monarchy playing the Enlightenment game, became, almost by default, *the* pillar on which to energize the tradition, politically and socially, inside the nation.

But here yet another unexpected problem intervened. The Papacy failed to see the truth. The very institution whose energies Lamennais was most seeking to release, and which would have most benefited by the liberation of the Church from the legitimist State through the work of the Catholic People, rejected his logic. It had lacked energy and erred. This left the Catholic People as a whole on their own as the sole defenders of the Truth within a given nation, a "silent majority" destined now to do its work even in opposition to those who were thought but yesterday to be its leaders. Palingenesis gave Lamennais the means to explain this phenomenon. In *Paroles d'un Croyant*, he argued that "the republicans of our days would have been the most ardent disciples of Christ eighteen centuries ago," thus, in effect, teaching that contemporary Catholics should see in them the best guides to the meaning of Christianity in the nineteenth century.[36]

Two themes appearing more clearly than any others in the nineteenth century palingenesist vision were adopted by Lamennais. One insisted that the age that was "emerging" from the European past was one that would overcome the confessional differences of what Lamennais called a mere "diplomatic Christianity." The third epoch of Humanity would see society enjoying the communion of a universal religion transcending an historical Faith which had outlived its usefulness, whose standard-bearers, again, as Lamennais noted, would be the People:

> How far we still are from that religion of devotion, of self-forgetfulness for the good of all; in sum, of that fraternity of which one speaks so much! I only find it in the People; the People surround the cradle of the future, just as the shepherds at Bethlehem surrounded that of the God about to be born. Blessings on the little ones, the simple of heart. It is those who will save the world.[37]

Moreover, the age emerging out of the Christian past would be "socialist" in character. We have seen that the precise definition of "Socialism" in the nineteenth century was a tricky question at the very best, given the wide variety of understandings and aspirations then attached to it, but the idea that it involved a concern for economic inequality and social injustice was universal. Secularists, spiritualists, Protestants and Catholics alike were involved in Socialism's birth and evolution. For our purposes, one ought to note that Catholics like Philippe Ballanche (1776-1847) and Philippe Buchez (1796-1865) were fervent proponents of the concept of the emergence of Christianity into Socialism, the former seeing the essence of the Christian mission as the abolition of inequality in his day, and the latter elevating the Revolution, Robespierre, and the People into the instruments of a constant battle against tyranny until the great day of the final liberation of all should arrive. Similar themes were elaborated in the Catholic Parisian journal, *L'Ere nouvelle*, of 1848. And Lamennais himself also evoked some of the language of the Socialist palingenesists in his commentaries on the Gospel of Mark, which he read as a kind of allegory of his own historical fate.[38]

Mazzini and Garibaldi were sympathetic to the palingenesist approach, the former having been deeply influenced by Lamennais, and both men having traveled in Saint-Simonian circles. The founder of Young Italy sought to turn the former priest from violent revolutionary writing to violent revolutionary action. Mazzini reproached him for his inactivity by letter from London, urging him, at the very least, to lead a regenerated priesthood basing itself purely on God's love, to guide a "Church of Precursors which I should like to see you found while waiting for the People to rise," one that would embrace the heavens and the earth:

> Why do you only write books? Humanity awaits something more from you...Do not deceive yourself, Lamennais, we need action. The thought of God is action; it is only by action that it is incarnated in us....So long as you will be alone, you will only be a philosopher and a moralist in the eyes of the masses; it is as a priest that you must appear before it, a priest of the future, of the epoch which is beginning, of that new religious manifestation of which you have a presentiment, and which must inevitably end in that new heaven and new earth which Luther glimpsed three centuries ago without being able to attain it, since the time had not yet come.[39]

Lamennais' influence in Catholic Action, even after his excommunication, was immense. Liberal Catholicism was founded by men who came from his camp and expressed openness to the "energetic" reform movements of their day, while claiming to reject the theoretical principles that had brought him into ill repute with the Papacy. His ideas, the ideas of a palingenesis connected with expressions of energy and action, and t he Romantic Idealism lying behind them, revived, again, in Catholic circles, in Italy as elsewhere, in the 1890's and 1900's, through the example of movements like the *Sillon*, the development of Modernism, and the influence of writings like Maurice Blondel's (1861-1949) treatise, *Action*. They would continue, despite papal condemnation, to gain further strength after the First World War.[40] Mazzinians, Risorgimento nationalists in general, and men of the Right in particular, expressed a mix of Liberal Catholic, Hegelian, and Saint-Simonian sentiments that easily fit into a palingenesist perspective. Even Croce, an a-religious defender of lay culture with no place for true transcendence in his vision, gave a palingenesist timbre to some of his comments in *La filosofia della pratica,* by speaking of the religious man as the philosopher's little brother. His one time collaborator on *La Critica,* Giovanni Gentile, dabbled in efforts at reconciliation with Catholic thought, claimed appreciation for the logic of men like the *Civiltà* editor, Luigi Taparelli d'Azeglio, but happily jettisoned earlier Church doctrine which had been superceded by the demands of modern "energies."[41] In short, there was always some support in Italy for the idea of the "emergence" of a "higher" religious viewpoint, and this from both Catholic and secularist starting points.

Whether for pragmatic or palingenesist reasons, many Italian Catholics were eager for a change in the Church's prohibition of participation in national political life from the 1860's onwards. Aside from the movement of "patriotic priests" of the *Risorgimento* era, one can note the yearnings of such prominent writers as Fr. Carlo Curci (1810-1892), a former editor of the determinedly anti-unification *La Civiltà Cattolica,* who paid for his change of heart about reconciliation and alliance with the more conservative elements within the government by expulsion from the Society of Jesus. Other enthusiasts included the contributors to the Florentine journal, *La Rassegna Nazionale,* founded by Count Stefano Jacini, a rightist statesman, in 1879, which sought to bring both Catholic and moderate non-Catholic thinkers together, and the novelist, Antonio Fogazzaro (1842-1911). Hopes were particularly high among these men in 1886-1887, as can be seen in the exchange of ideas on this issue between Monsignor Geremia Bonomelli (1831-1914), the Bishop of Cremona, and Leo XIII, and their expectations following the pope's conciliatory letter of May, 1887, *Episcoporum ordinem.* Church participation in the public sorrow for the colonial losses at Dogali in Abyssinia (1887), Father Luigi Tosti's (1811-1897) pamphlet, *La Conciliazione,* with its exuberant prophecies for reconciliation in 1888, and the joy of many Catholics over the friendly expressions of King Umberto I (1878-1900) on the occasion of Leo's priestly jubilee, all attest to a similar hope for change. We have seen how these dreams came to naught in the tense social atmosphere of the late 1880's and 1890's.

By the late 1890's, however, many more activists within the *Opera dei congressi* were seriously divided over their future attitude towards participation in national politics. One group insisted upon continuing business as usual, neither compromising with the existing authorities, nor opposing them in politics directly, lest the Socialists, whom it considered to be simply the more radical child of an erring liberal parent, pick up the pieces in a bitter public conflict with the government. A second force, many-headed in character, thought that business-as-usual was no longer opportune. One of its constituent elements, from 1899 onwards, wished boldly to declare liberal economic policies to be erroneous and immoral, and longed for the creation of a distinctly popular political party. Although priests like Don Romolo Murri were prominent in its ranks, it was nevertheless convinced that its social concerns would give it a broad appeal beyond the immediate camp of believers that would require operation outside the constraints of the ecclesiastical hierarchy.

Another faction, which came to be known as the clerico-moderates, wished, by 1908, to take advantage of the signs of a weakening of liberal opposition to the Church to see if a broader "conservative party" might be created. Catholic abstention from national politics would end, and leaders who had been prepared during that abstention could move forward to exercise direct influence over Italian political life. This group included the old conciliarists around *La Rassegna Nazionale,* younger proponents of cooperation such as Filippo Meda (1869-1939) and, eventually, even the editors of *La Civiltà Cattolica*, which began to speak sympathetically of the work of leftist Prime Ministers like Luigi Luzzatti (1841-1927) and the need for devotion to the Italian Army.[42]

Papal involvement with these three approaches became more intense after the turn of the century. To begin with, Leo XIII, on January 18th, 1901, published the encyclical letter *Graves de communi,* in which he made it clear that he was not in favor of the creation of a distinctly Catholic mass party in Italy. This may well have been because of his recognition of a recrudescence of Mennaisian tendencies, in France as well as in Italy. If the words "Christian Democracy" were employed at all, Leo insisted, they could only legitimately be employed to indicate "a beneficent Christian action in favor of the people"; not as a statement committing the Church *qua* Church to a party involved in democratic politics. Moreover, as the first of its two names emphasized, "Christian Democracy" could only exist with reference to a grounding in the Christian Faith; a direct appeal for the votes of non-Catholics was thereby excluded. Even what today would be called a "preferential option for the poor" was dismissed as unacceptable by the pope, since a true concept of "the People" had to include all social classes, coordinated into one harmonious whole. Leo did, however, praise those who were engaged in a democratic action that did all it could to lessen the sufferings of the ordinary man.

But papal action did not end here. It was further stimulated by an intensification of the debate within the *Opera*. In September of 1902, Giovanni Grosoli, a man who was rather favorable to the more social-minded democratic elements of the movement who were looking forward to the eventual creation of a political party, became President of that organization. At the XIX Congress in Bologna, from November 10-13, 1903, it became clear that the supporters of Grosoli and the much more committed Christian Democrat, Romolo Murri, had gained the edge over the faction which was eager to continue abstention from politics and maintain a joint anti-liberal and anti-socialist approach. An imprudent circular from Grosoli indicating that "old questions," presumably such as those surrounding the Temporal Power, no longer mattered that much to contemporary Catholics, stirred a second intervention, this time by the new pope, Pius X. While personally content to let the Temporal Power issue die, he was disturbed by what he considered to be the *Opera's* lay-clerical insubordination to higher ecclesiastical authority. It was dissolved, by order of the Secretary of State, Merry del Val (1865-1930), on July 28, 1904. Section II, dealing with Social Economy, was alone maintained to emphasize the fact that "beneficent action in favor of the people" was not being punished by this severe measure.

Pius X, like Leo XIII, clearly disliked the idea of a creating a distinct, Italian Christian Democratic Party appealing for non-Catholic support. He, too, felt threatened by Mennaisian concepts. This can be seen not only by his condemnation of the French *Sillon*, but also by his reproach of Bonomelli of Cremona, who had begun to praise the separation of Church and State, and his chastisement of Romolo Murri, the journal, *Cultura sociale*, the *Lega Democratica Italiana*, and every other initiative that envisaged a democratic cooperation of officially constituted Catholic organizations with non-Catholics who were inspired by the concept of 'social justice' rather than by religion.

Still, Catholics, by the time of the general election of 1904, had already shown little respect for the *non expedit*. Pius X's restructuring of the Catholic Movement on June 11, 1905, with the publication of an encyclical letter, *Il fermo proposito,* took account of this. Section II of the *Opera* became the *Unione Economico-Sociale dei Cattolici Italiani.* An *Unione Popolare tra i Cattolici d'Italia* was established on the model of the German *Volksverein.* Most importantly, however, the *Unione Elettorale Cattolica Italiana,* designed to prepare Catholics for participation in political life on the national level, also now made an appearance. With the "business as usual" position abandoned and the hopes for a Catholic Party squelched, Rome ended by opting for the clerico-moderate line, pushing the *Unione Elettorale* towards the kind of contractual agreement with more conservative-minded liberals already utilized in other countries. While the *non expedit* remained on the books, *Il fermo proposito* empowered bishops to dispense from its strictures, and it was quite clear to everyone that its days were numbered, officially, as well as on the practical level.[43]

The great chance to put the plan into operation came with the introduction of universal male suffrage, which Giolitti felt could no longer be resisted, in 1913. This increased the value and impact of the Catholic vote for "conservatives," resulting in the famous and highly successful "Pact" of 1913 of the President of the *Unione,* Vincenzo Ottorino Gentiloni (1865-1916) with the Giolittan Liberals; an "alliance" that some moderate Nationalists would also not have been averse to joining. Seven "commandments" laid out by the *Unione* were subscribed to by a large number of individual Liberal candidates for office, all of them assuring support for Catholic policies in exchange for Catholic votes. The *non expedit* was lifted entirely to allow election of the men in question, guaranteeing victory for over two hundred deputies.

Revealed by a variety of sources, ranging from the *Corriere della Sera* to *La Civiltà Cattolica,* the Gentiloni Pact aroused the last anticlerical storm of the period with which this article is concerned. A mass of Rightists, Leftists, *Estrema* supporters of both republican and radical hues, and Nationalists took up the cudgel in defense of secularist principles. These included Sonnino, Albertini, Luzzatti, two future Prime Ministers, Antonio Salandra (1853-1931) and Vittorio Emmanuele Orlando (1860-1952), most of the Press, and numerous student organizations. The State was endangered by its ancient enemy once more. Much was made of the address of Antonio

Rossi, the Archbishop of Udine, at the VIII *Settimana Sociale* at Milan, in November-December, 1913, as proof of the evils that cooperation could engender for the civil authority. Rossi had merely argued that the Temporal Power issue could be adequately resolved through adding international backing to the Law of Guarantees. This reopening of what secularists considered to be the closed Roman Question, was said to show a rebelliousness and even lack of patriotism which was a portent of the further clericalist demands that Catholics emboldened by the Pact were sure to make. Men like Meda, who had been elected to Parliament in 1909, protested that Catholic patriotism was in no way in question; the problem was simply that honest Italian citizens had decided that they need not view themselves merely as bulwarks against Socialism, and wanted a reconciliation and political participation that would fully protect their religious rights. Still, the anticlerical outcry was so great that Giolitti tried to repair the damage to his government by reviving the call for enforcement of civil before religious marriage. In doing so, he alienated the Catholics alongside his other enemies. With the radicals of the *Estrema* especially angry, the fourth Giolittan Ministry thus came to an end.[44]

Nevertheless, the Church's option for the clerico-moderate position held firm, and, with it, a desire to influence an Italy governed by conservative minded Liberals of basically capitalist mentality. It was this encouragement of candidates of the industrialist interests that aroused *Estrema* complaints regarding the Gentiloni Pact as much as any anticlerical sentiment. In fact, the same sous-text alarmed men of Christian democratic sympathies like Murri and Dom Luigi Sturzo (1871-1959) as well.[45] As time went on, the answer of the call for cooperation with Italians of moderate liberal background would lead to the "emergence" of a clerico-moderate, Liberal Catholicism. This, like Lamennais, would come to insist upon a pragmatic, separation of Church and State that rendered the Catholic doctrine on that issue theoretical and harmless. It would also subscribe to a palingenesist "higher" religion of all men of good will reducing the anti-modernist Catholicism of Pius X to the world of the catechism; a world that could conveniently be forgotten when dealing with the "real world" of practical politics. This "rebirth" of Catholicism could overcome quarrels of Church and State by effectively bending the former to the demands of the "higher religion" of the latter, whose value was demonstrated by its "energetic" practical "action" for peace and prosperity. But it would also encourage those whose understanding of palingenesis required a new Catholicism which was much more democratic or socialist in character, one allying itself with liberation theologies of varied types. All this, however, is the stuff of another chapter in another book.

What concerns us here is the Italian environment out of which the migrants to the United States from the *Mezzogiorno* arrived. These were mostly peasants lacking formal education, religious as well as secular, that would have introduced them to the complications of the problems discussed above. Such problems did, indeed, affect them, secularization and liberal economic theories playing a serious role in the rural disruptions in the former Kingdom of the Two Sicilies. Interestingly enough, the American experience to which the uprooted southern peasantry would be exposed, the experience praised by men like Crispi, was one that was guided by a vision of pragmatic pluralism whose attitude towards religion and Church-State relations had much in common with ideas and experiments that were important in Italy. Their New World Order, when transported as a model to the Republic of Italy (and the globe) after the Second World War, was to find a soil in which it could easily take root and grow, whether for good or for ill, time alone will tell.

JOHN C. RAO

---

[1] For an introduction, see Waley; Bouwsma; also, Anderson, in Scott, pp. 37-55.

[2] Mayeur, Vol. 11, pp. 349-366, also, *passim*; Rao, pp. 6-38.

[3] Mayeur, Vol. 11, pp. 272-278, 611-636.

[4] Invernizii, *Il movimento cattolico*, p. 22.

[5] Kalyvas, pp. 179-183.

[6] *Ibid.*, p. 217

[7] Jemolo, pp. 47-79.

[8] *Ibid.*, pp. 80-160; Romanato, pp. 223-291.

[9] Mack Smith, pp. 95-108, 157-173.

[10] Thayer, pp. 86-143, pp. 192-233.

[11] For Risorgimento anticlericalism, see Pellicciari; Jemolo, pp. 101-109; Mack Smith, 202-203.

[12] See Jemolo's whole discussion of the problem, pp. 47-160.

[13] Thayer, 59, 69-70, 79, 89, 124-126, 174-175, 210; Jemolo, pp. 76, 96, 106, 132, 148, 149.

[14] Jemolo, pp. 87-94.

[15] See Blum's whole discussion of Rousseau's thought, pp. 27-132.

[16] Rousseau, in Blum, p. 67.

[17] *Ibid.*, p. 99.

[18] *Ibid.*

[19] Billington, pp. 155-157, 206-226, 234-242.

[20] *Ibid.*, pp. 88, 98, 149-150, 206-226; 248.

[21] See, for example, Mazzini, p. 3.

[22] See *Il Cimento*, vi, ii (1855), 110-111; Taparelli, in Pirri, pp. 182-185.

[23] Thayer, pp. 89-91.

[24] Billington, p. 161.

[25] Thayer, pp. 49, 53, 128-138, 188, 199; Mack Smith, 236, 240.

[26] S paventa, in Ge ntile, p p. 278- 300; T aparelli d' Azeglio, in *L a Civiltà Cattolica*, ii, viii (1854), *passim.*

[27] Gallenga, Il Cimento, v, xii (1855), 1080, for extended quotation; otherwise, 1079-1081.

[28] Thayer, pp. 13, 106, 133-141, 195-198, 201, 258. 388-389 ; Billington, pp. 425-427.

[29] Thayer, p. 202.

[30] Jemolo, pp. 47-160; Thayer, pp. 124-133; Mack Smith, pp. 78-232.

[31] Jemolo, pp. 131-132.

[32] *Ibid.*, pp. 153-154; Thayer, pp. 201-230.

[33] Mack Smith, pp. 103-107, 123-128; Thayer, pp. 44-45, 48, 66-67, 82-83, 112, 130, 139, 141, 149, 372.

[34] Mayeur, X, p. 864.

[35] *Ibid.*, pp. 837-904; Billington, pp. 217-224.

[36] Mayeur, p. 848; also, pp. 837-904. See the complete discussion in Bowman as well.

[37] Mayeur, p. 866.

[38] *Ibid.*, p. 892.

[39] *Ibid.*, p. 893, Billington, pp. 161, 217-224.

[40] Petit, pp. 15, 135, 192-197, 200, 211; See, also, Dubarle, Meinvieille, and Sarasella.

[41] Jemolo, pp. 91-95; Gentile, pp. ix-xiii; Minghetti, Vol. iii, 17-18; *passim.*

[42] Jemolo, pp. 47-160; Invernizzi, *Il movimento cattolico*, pp. 15-58.

[43] Kalyvas, p. 89 ; Invernizzi, *L'Unione Elettorale*, pp. 11-20; Jemolo, pp. 108, 122; Mack Smith, 251-252; Invernizzi, *Il movimento cattolico*, pp. 15-58; Agócs, pp. 165-199; Launay, pp. 106-111, 183-191.

[44] Jemolo, pp. 132-139; Invernizzi, *L'Unione*, pp. 24-38; Agócs, pp. 165-199.

[45] Invernizzi, *L'Unione*, pp. 33-36; Agócs, pp. 197-198.

References

Agócs, Sándor, 1988 *The Troubled Origins of the Italian Catholic Labour Movement* (Detroit: Wayne State).

Billington, James H., 1980, *Fire in the Minds of Men* (New York: Basic).

Blum, Carol, 1986, *Rousseau and the Republic of Virtue* (Ithaca: Cornell).

Bouwsma, William J., 1968, *Venice and the Defense of Republican Liberty* (Berkeley: University of California).

Bowman, Frank Paul, 1987, *Le Christ des barricades* (Paris: Cerf). *Il Cimento* (Torino, 1852-1855).

Dubarle, D., ed., 1980, *Le Modernisme* (Paris:Beauchesne).

Gentile, Giovanni, ed., 1911. *La politica dei gesuiti* (Milan).

Invernizzi, Marco, 1995, *Il movimento cattolico in Italia* (Milan: Mimep-Docete).

Invernizzi, Marco, 1993, *L'Unione Elettorale Cattolica Italiana* (Piacenza, Cristianità).

Jemolo, Arturo Carlo, 1977, *Chiesa e stato in Italia* (Torino: Einaudi).

Jedin, H., and Dolan, J., eds., 1981, *History of the Church* (Vols. VII, VIII, IX, Crossroad).

Kalyvas, Stathis N., 1996 *The Rise of Christian Democracy in Europe* (Ithaca and London: Cornell).

Launey, Marcel, 1997, *La papauté à l'aube du xx siècle* (Paris: Cerf).

Mack Smith, Denis, 1997, *Modern Italy* (Ann Arbor: University of Michigan).

Mayeur, J.M., ed., 1995/1997, *Histoire du christianisme* (Vols. X, XI, Paris: Desclée).

Mazzini, Giuseppe, 1865, *Address to Pope Pius IX on His Encyclical Letter* (London).

Meinvieille, J., 1949. *De Lamennais a Maritain* (Paris: La Cité Catholique).

Minghetti, Marco, 1888-1892). *Miei ricordi* (Three Volumes, Turin).

Pellicciari, Angela, 2000, *L'Altro Risorgimento* (Casale Monferrato: Piemme).

Pellicciari, Angela, 1998, *Risorgimento da Rescrivere* (Milan: Ares).

Pirri, P., ed., 1932. *Carteggi* (Milan: Biblioteca della storia italiana recente, xiv).

Rao, John C., 1999, *Removing the Blindfold* (St. Paul, The Remnant Press).

Romanto, Gianpaolo, 1992, *Pio X* (Milan: Rusconi).

Sarasella, Daniela, 1995, *Modernismo* (Milan: Editrice bibliografica).

Scott, H.M., ed., 1990, *Enlightened Absolutism* (Ann Arbor: University of Michigan).

Thayer, John A., 1964, *Italy and the Great War* (Madison and Milwaukee: University of Wisconsin).

Waley, Daniel, 1988. *The Italian City-Republics* (London & New York: Longman).

Chapter 12

✦

EMERGING ITALIAN AMERICAN CATHOLICISM: A PERSPECTIVE FROM THE SOCIAL CONSTRUCTION OF REALITY

The magnitude of the impact made by Peter Berger and Thomas Luckmann's *The Social Construction of Reality*, published in 1966, can hardly be overstated. Integrating as it did the classical sociological tradition (represented quintessentially by Emile Durkheim, Georg Simmel, and Max Weber), American pragmatist sociology (George Herbert Mead being the seminal figure in this movement) and an emerging phenomenological perspective on society (as developed by Alfred Schutz and others), the work was acclaimed by many commentators as a *tour de force*.

In the first section, the social constructionist theory of Berger and Luckmann will be placed within the context of the intellectual streams that preceded them; namely the seminal contributions of Durkheim, Weber, Simmel, Mead, and Schutz. The thought of Berger and Luckmann, it will be shown, is part of a conversational chain that extends from at least the late 19th century (and one could certainly argue well before then). In the second section, the way in which the general theory was "historicized" in the form of an analysis of modernity is discussed. The extent to which the theory of modernity presented by Berger and Luckmann builds on and departs from the theories of modernity put forward by Durkheim, Weber, Simmel and others is assessed. The implications of the theory of modernity, as laid out by Berger and Luckmann, for individual identity (particularly religious identity) in general, and for Italian American identity in particular are addressed, and in the course of this discussion areas of convergence with and divergence from recent scholarship are identified.

An Overview of Berger and Luckmann's "Social Construction of Reality" Perspective

Arguably at the heart of Berger and Luckmann's general theory of society is their model of the three moments. The first phase or moment in the evolution of society is *externalization*, in which social and cultural forms are created by human beings. Berger and Luckmann are clearly indebted to the Hegelian ideas of "alienation" and "negation" in explaining the meaning of "externalization." This innovative side of the human condition was developed most insightfully in the classical sociological tradition by Georg Simmel, who presented an elaborate theory of social forms. Human beings, according to Simmel (1971), create new aesthetic, moral, scientific, linguistic, and technological forms on an ongoing basis. But why does this happen? Simmel's answer in essence is that the creation of new forms represents a response as it were to previous forms; that a critical aspect of the human spirit is to resist its imprisonment by social forms. This resistance thus manifests itself in a rejection of or distancing from existing social forms. However, Simmel argues that human existence cannot remain formless for any extended period of time; thus resistance to a preexisting form coalesces around a newly emerging form. For Durkheim (1965), human creativity reflects a deep-seated sacred impulse in human beings. "Externalization," from a Durkheimian perspective, can be seen as a carving out of sacred space. It is the effervescence and galvanizing of this human desire to sacralize the world that explains the creation of new forms. As with Simmel, Durkheim's approach to innovation revolves around the idea of "resistance," but with one significant difference. From Durkheim's perspective, innovation can be understood in terms of the resistance to the "profane," that is, to all that is ordinary, mundane, and deadening. Human beings thus construct worlds that catapult them to a higher realm of existence, to their self-transcendence. Durkheim does not see human creativity in terms of a resistance to form as such, but only to a form that has become profanized. Simmel's view, on the other hand, is that the emergence of new forms constitutes a distancing and disengagement from the idea of form itself. Let Weber enter at this point. Clearly for Weber (1946), human creativity is linked to the charismatic process. Human creativity occurs against the backdrop of the forces of tradition and rationalization, which stifle creativity and spontaneity. Its "carrier" is a leader defined as "extraordinary" and to whom is attributed the power to bring followers to the Promised Land. Through the exercise of charismatic leadership, human beings are able to transcend the suffocating effects of tradition and bureaucracy.

The second "moment," according to Berger and Luckmann, is that of *objectivation*. This involves the institutionalization of the newly created

forms. What is the difference between the creation and the institutionalization of forms? Berger and Luckmann draw on Durkheim's analysis of collective representations, which for the latter is defined in terms of their exte-
riority and coercive presence in everyday life. They tap into Durkheim's
conception that social institutions possess a reality *sui generis* by virtue of
their facticity. This quality is attributed to them by virtue of their enforce-
ability, accomplished through mechanisms of social control, both formal
and informal. Put simply, a form is institutionalized if it cannot be "wished
away" or "thought away," language being a prototypical example of this.
For Berger and Luckmann, institutionalization involves, indeed requires,
"legitimation," that is, a system of symbols that gives the form in question
"plausibility."[1] According to Simmel, forms become part of the realm of
"objective culture," acknowledged as part of a moral, scientific, technological and aesthetic landscape that is "out there" and taken-for-granted.[2] From
the perspective of Weber, objectivation (in the sense in which it is presented
by Berger and Luckmann) is understood in terms of the "routinization of
charisma," which can assume either a traditional or a legal-rational form of
domination. Sacrificed in the process are human spontaneity and freedom.

Internalization represents the third and final moment in Berger and
Luckmann's model. Individuals "internalize" an objective form to the degree that it is subjectively appropriated by them. What does it mean to
appropriate any given social institution, to take it on as one's own? Berger
and Luckmann are certainly correct in arguing that significant others "represent" the objective culture in particular, to some degree, even idiosyn-
cratic ways. People with the same "objective" social location (whether
based on gender, ethnicity, religion, or class) thus are socialized very differ-
ently. Significant others inevitably pass along their "world" to those whom
they are responsible for socializing. While there is an objective culture in
a general sense, it is transmitted in a highly selective manner. In addition,
individuals are capable of engaging in a process of self-reflection, in which
these selective "interpretations" of the form are either assented to, dissented
from, or modified in some fashion. In this process, the objective culture
becomes meaningful to individuals. It is no longer "out there," but a reality
that is worthy of commitment. Objective plausibility is transformed into
subjective plausibility. The forms are now seen as self-legislated scripts that
guide choices and conduct, and these scripts have an ultimate grounding in
beliefs that individuals adopt as part of a personal life philosophy. Individu-

als do not create their own reality from scratch; rather they draw on preexisting cultural frameworks and forms in the process of constructing their
identity. From the perspective of Mead (1964), individuals engage in an
ongoing internal conversation between the "me," consisting of the expectations of objective culture (as refracted by significant others) and the "I"
comprising the individual's authentic or core self. According to Mead (1962:
273-281), individuals tend to seek a "fusion" of the two, to construct through
active reflection (by bringing taken-for-granted reality to the discursive level)
social spaces in which objective scripts can be harmonized with their authentic selves. There are two basic strategies for achieving this: the first is to
identify elements of objective culture that are consistent with the authentic
self. The second is to change one's self-image so that it corresponds more
closely with the expectations of the objective culture.

Berger and Luckmann argue that the three moments are part of a continuous and dialectical process. That is, the way in which cultural forms are
appropriated has implications for the subsequent nature of the cultural form
itself, for the ways in which it is re-created, modified, or replaced. It is
instructive to juxtapose Berger and Luckmann's theory with the more recent
formulation of Bourdieu (1992), according to whose theory human beings
draw on their life philosophy (which Bourdieu refers to as *habitus*) to enable them to navigate their way through and around social and cultural
fields. To make the translation, objective culture can be seen as the fields
within which individuals operate and to which they actively respond; and
the demands of these fields are filtered as it were through the "habitus"[3] of
individuals. Innovation and creativity (that is, "externalization") occur when
existing fields are not responsive to the needs, values, and dispositions as
structured by the habitus. From the perspective of Schutz (1967), a major
influence on Berger and Luckmann, actors actively pursue "projects" that
they carve out for themselves. In a similar vein, Mead (1964:3-18) argued
that the evolution of cultural forms can be best understood in terms of the
attempts of human beings to solve life problems. For Bourdieu (1992), in
striving to bring projects oriented to problem-solving to fruition, the fields
within which individuals operate come to be defined either as an enabling
resource or a constraining fact. From this perspective, cultural appropriation
is followed inevitably by strategies that are in the service of life projects.[4]
Forms that provide enabling resources are sustained; forms that constrain
the successful completion of projects are modified or replaced with new

cultural forms. Thus, internalization leads to actions that either re-create existing forms (by affirming their legitimacy and plausibility), or transform them (by challenging their legitimacy and plausibility). In either event, externalization has occurred; and the dialectical process continues unceasingly.

Berger and Luckmann's Theory of Modern Consciousness

Berger and Luckmann in their collaborative and individual writings have developed a comprehensive and penetrating diagnosis of the modern condition, arguing convincingly that it is historically unique, both in terms of its institutional structure and its social psychology. However, the issue of the precise sense in which modernity represents a historically specific playing out of externalization, objectivation and internalization processes remains largely unexplored. I would argue that such continuity can be established. While the three moments are universal features of the human condition, they are lived out in distinctive historical contexts, settings that give them color, texture, and shape. That is, each historical setting provides unique opportunities and challenges to those who live out their lives in it. The question then is: What do externalization, objectivation, and internalization mean in a modern (or even postmodern) context? Put another way, how does the modern situation give shape and form to the three moments and how they relate to each other? If these questions have not been sufficiently fleshed out in the writings of Berger and Luckmann, certain inferences can be drawn which could clarify how their theory of modernity draws on, illustrates, and concretizes their general model.[5]

Of central importance to Berger's social-psychology of the modern life is the *"pluralization of life-worlds."* An individual's life-world, for Berger and Luckmann, consists of the "totality of meanings" shared with others (what phenomenologists refer to as "inter-subjective reality"). These meanings are "reality definitions" that (1) have both cognitive and normative knowledge components, (2) tend to be pre-theoretical in character, and (3) "can't be thought away." For Berger, a modern institutional order (in which technological production and bureaucracy predominate) has specific implications for the ways in which everyday knowledge is organized and for the "cognitive style" exhibited by individuals who participate in it. Berger argues that social change can best be understood as a change in "plausibility

structures," and people functioning in a modern social order adopt a "horizon" that enables them to plausibly navigate their way through it.

But, what does it mean to say that life-worlds become pluralized under
modern conditions of life? For Berger, life in modern society is segmented
into "public" and "private" spheres. In the public sphere, knowledge is
mechanistic and componential in nature, with its reproducibility, measura-
bility, and predictability being given primacy. Social relations are anonymous, abstract, and emotionally controlled. While there is considerable
overlap in terms of the "packages" of consciousness "carried" by the realms
of technological production and bureaucratic organization, there are important differences. One has to do with the issue of the separability of means
and ends: technological production requiring such a separation and bureaucracy often conflating the two. The other has to do with the bureaucratic
emphasis on justice and due process. The private sphere, on the other hand,
is the arena in which personalized relations are pursued, emotions become
unrepressed, and there is an openness to mystery, uncertainty, and spontaneity. While "carryover" may exist (e.g. the private sphere becoming bureauc-
ratized, and the public sphere personalized), the tendency is for the two
spheres to remain segregated. Thus, modern society lacks the cognitive and
normative unity that traditional societies experienced, characterized as they
were by an overarching "symbolic universe" that was grounded in a religious world-view, one that served as the "background" that gave meaning
to life activities and conferred plausibility and legitimacy on those activities. The burden then, for Berger, falls on individuals to carve out a meaningful existence in the private sphere, t o c onstruct, to use Christopher
Lasch's (1977) term, "a haven in a heartless world." However, the very
absence of an overarching symbolic universe, of an ultimate grounding
makes such a project "hazardous" and "precarious." Even if they are able
somehow to "pull it off" in their private spheres, they would remain alienated from their public sphere roles. But their capacity to construct a meaningful private life is compromised by the lack of an overarching symbolic
universe, the existence of which is made unlikely by the relativization of
consciousness that is fostered by contemporary urban life and the mass
communication media.

This dilemma leads Berger to characterize modern identity as fundamentally, perhaps hopelessly, "homeless." The "self-anonymization" of
modern identity has been pushed to an extreme. The self itself becomes

componentialized, with the roles that we play, both public and private, being defined as parts of a potential "do-it-yourself universe." There is an emphasis on the long-range planning of identity, in light of a much heightened level of self-conscious awareness. Identity is open-ended, fluid, unstable, unreliable and conversion-prone. People in a modern context need to fabricate a meaningful self-identity against the backdrop of an objectively given "map of society." Planning such a life trajectory requires "multi-relational syncretization," a juggling of different role options (work, family, leisure, community-based, etc.) at any given point in time and into the future. For Berger, the "accent of reality" shifts from an objective social order to the realm of subjectivity. "Honor," defined in terms of conformity to social roles is replaced by "dignity" and "authenticity," defined in terms of self-actualization and the disengagement of the individual from all social roles (both public and private, but especially the former). It is important to point out that the idea of human dignity has both a public and private dimension. The public dimension revolves around the expectation of justice noted previously. While it is true that a bureaucratic form of social organization treats people "like a number," thus stripping them of their dignity and humanity, it also celebrates the dignity of the person by guaranteeing basic rights of citizenship without regard to the ascriptive criteria of gender, race, age, or family background. In the private sphere, the idea of human dignity involves according individuals the right to design their identity. The sentiment, "It's his or her life," has an almost deafening ring of plausibility in the modern era.

Which strategies are available to individuals on whom the burden of identity-construction falls? According to Berger, two obstacles need to be overcome. The first involves the "de-institutionalization" of the objective order (that is, its loss of plausibility and legitimacy), a condition that, given the essential human need for cognitive and normative order, leads to the creation of new institutions, which Berger labels as "secondary institutions" (e.g. health food stores, hiking clubs, electronic communities). As stated previously, these institutions are highly unstable and precarious. The other obstacle has to do with how individuals can address effectively what Berger calls "collisions of consciousness." For Berger, an "expansion of the individual's social horizon" fostered by the urbanization of life and the widespread diffusion by the mass media of the full range of life plan options makes the individual acutely aware of how his or her life-world dovetails or clashes with other life-worlds. Individuals in a modern context must to come to terms with the relativization of consciousness brought about by the expansion of their social horizons. In this connection, Berger focuses on the "collision" between traditional and modern forms of consciousness: for example, the traditional emphasis on the past vs. the modern emphasis on the "makeability" of the future. Those with a traditional sensibility have essentially three options in encountering a modern world-view. The first is *separatist*, in which case the individual rejects wholesale modern forms of consciousness and retreats to or solidifies a traditionalist plausibility structure. ("Fundamentalism" can be analyzed in these terms.) At the opposite pole is the *modernist* option, which entails a rejection of traditionalist modes of thought and a complete embracement of a modern consciousness. An intermediate position involves a process that Berger labels "cognitive bargaining," in which case the individual adopts selectively elements of a modern frame of reference while holding on to a traditionalist horizon, in the process harmonizing in some way modern and traditional reality definitions. But, there is also the situation in which individuals actively confront the "discontents"—anomie, instability, unreliability—of modernity. Seeking an escape from their "homelessness," they adopt what Berger calls a "de-modernist" attitude. There are various expressions of such a radical transformation of consciousness: joining established traditional social forms is one possibility (e.g. the stockbroker who enters the seminary) or latching on to newly created social movements and secondary institutions (e.g. electronic communities, communes). The common thread here is immersion of oneself in a world that is anticipated to yield an integrated identity and interpersonal solidarity.

In terms of the three moment model, technological production and bureaucracy represent "externalized" forms that replaced feudal social forms. These forms (or megastructures) became "objectivated," in the sense that intrinsic to them is a "package" of reality definitions and meanings necessary to the performance of functional (work and client) roles. These rationalized social forms, however, are "internalized" by individuals only in the sense that recognition of their facticity is required for what Lemert (1977) calls "sociological competence." Individuals do not for the most part identify themselves with the functional roles that they play. Public sphere roles are legitimated but do not form a basis for personal identity. At best, they are evaluated in neutral terms, at worst in negative terms, and rarely in

positive terms. But, the objectivated order consists as well of a differenti-
ated private sphere that cannot be thought away. The private sphere is itself
a creation, an externalized product of human will. The privatization of
family, religious, and ethnic roles is itself constitutive of an objectivated
social order. It too cannot be thought away. However, while the existence
of a privatized life-world takes on aura of legitimacy and plausibility, it
becomes a form without content due to the paucity of reliable and stable
structures.[6] Thus, the idea of life planning and constructing a do-it-yourself
universe is thoroughly internalized by individuals. The form is internalized,
but there is little or no content to appropriate, this a function of the under-
institutionalization of the private sphere. This condition of "homelessness"
leads to a new iteration of the three-moment process. Structures (that is
secondary institutions) are externalized and then become part of the
objectivated social map, but the degree to which they are meaningfully and
durably appropriated by individuals is highly precarious. This prompts yet
another iteration. This is consistent with Simmel's position that formless-
ness (with respect to private life) becomes the quintessential form in a modern context.[7] The iterative process will not cease until a meaningful and
subjectively plausible existence is "achieved." For Berger, this is unlikely
to ever be accomplished successfully.

Implications for Religious and Ethnic Consciousness

For Berger, the implications of the modern situation for religion and
religiosity are clear.[8] Permit me to quote him at length on this issue:

> The pluralization of social life-worlds has a very important effect in the
> area of religion. Through most of human history, religion has played a
> vital role in providing the overarching canopy of symbols for the mean-
> ingful integration of society. The various meanings, values and beliefs
> operative in a society were ultimately "held together" in a comprehen-
> sive interpretation of reality that related human life to the cosmos as a
> whole. Indeed, from a sociological and social-psychological point of
> view, religion can be defined as a cognitive and normative structure that
> makes it possible for man to feel "at home" in the universe. This age-
> old function of religion is seriously threatened by pluralization. Differ-
> ent sectors of social life now come to be governed by widely discrepant
> meanings and meaning systems. Not only does it become increasingly
> difficult for religious traditions, and for the institutions that embody
> these, to integrate this plurality of social life-worlds in one overarching

and comprehensive world view, but even more basically, the plausibility of religious definitions of reality is threatened from within, that is, within the subjective consciousness of the individual...*Pluralization has a secularizing effect.* That is, pluralization weakens the hold of religion on society and on the individual...Institutionally, the most visible consequence of this has been the *privatization of religion.* The dichotomization of social life into public and private spheres has offered a "solution" to the religious problem of modern society. While religion has had to "evacuate" one area after another in the public sphere, it has successfully maintained itself as an expression of private meaning (1973:79-80).

Berger argues that modern religiosity, like all aspects of the private sphere, is "open-ended," subject to life planning. A variety of religious options is available to individuals. He states:

In the absence of consistent and general social confirmation, religious definitions of reality have lost their quality of certainty and instead, have become matters of choice. Faith is no longer socially given, but must be individually achieved. (1973:81)

It is obvious to me that the same line of analysis could be applied to ethnicity. It would be consistent with Berger's overall theory of modern consciousness to see ethnic identity as privatized, open-ended, a matter of choice, as something to be achieved.

Mapping and Assessing Berger and Luckmann's Theory of Modern Consciousness

In this section, Berger and Luckmann's theory of modern consciousness is placed within the context of the literature on the social-psychology of modernity. The standing of their theory in relation to the contributions of their predecessors, contemporaries, and successors is sketched. The strengths and deficiencies of the theory are in the process identified.

As stated above, for Berger and Luckmann, modernity is a double-edged sword. On the one hand, it produces "homelessness," a psychically untenable state. On the other, it provides an opportunity for individuals to construct their own identity. It celebrates the dignity and worth of the individual, while denying the individual the cultural resources necessary to carve out a meaningful self-identity.

On whose shoulders did Berger and Luckmann stand in constructing their theory of modern consciousness? Weber's (1946) theory of "rationalization" is clearly reflected in Berger and Luckmann's approach to modernity. The idea that means-ends rationality (functional rather than substantive in nature) pervades the economic and political order is a critical aspect of Berger and Luckmann's theory of the public sphere. What Berger perhaps gives insufficient attention to is the degree to which the private sphere has been rationalized as well. Recall, that Berger allows for the possibility of "carryover" from the public to the private sphere, but argues that this is contained by the individual's compulsive search for a "haven in a heartless world." Individuals seek out mystery and awe, and revel in ecstatic experiences that punctuate the dreariness, deadening quality, and lifelessness of their public roles. There are those in the Weberian tradition who emphasize the "colonization" of the life-world (that is the realm of meaning and identity) by the forces of rationalization.[9] For Weber, the individual needs to mount a "heroic" effort to impose meaning in a world of pervasive and all-encompassing rationalization.[10] Weber argued that the modern condition is one of value-plurality, with each sphere (economic, political, aesthetic, and erotic) operating in accordance with its own logic. The burden falls to the individual to construct an integrated and overarching life philosophy against the backdrop of value-pluralism. This can only be achieved through a personal assertion of will (Nietzsche's influence here is obvious). An "ethic of brotherhood" is put forward by Weber as a possible integrative and overarching value system in the modern world, although Weber recognized that it is in considerable tension with economic, political, aesthetic, and erotic logics.[11] One key difference between Weber and Berger is that for the former, the individual's search for value-integration includes what Berger would call public realms of existence. For Berger, by contrast, any effort to integrate public and private value systems is subjectively implausible and futile. Both share a profound skepticism as to the capacity of the individual to "pull it off" and each argues that in the very effort to construct a meaningful identity lies the dignity of the person, whether or not the project is in some ultimate sense "successful."

Berger along with Durkheim[12] emphasizes how the dignity of the person is celebrated and elevated in modern society; both see it in a positive light, while recognizing the great potential for "homelessness" (or anomie in the language of Durkheim) intrinsic to this value to some degree. Durkheim takes the position that it is under conditions of organic solidarity, characterized by a pronounced division of labor and interdependence based on differences rather than likenesses, that the "cult of the individual" flourishes. He takes great pains to distance the idea of the "cult of the individual" from the utilitarian conception of individualism of British social contract theory (Spencer's in particular). For Durkheim, the "cult of the individual" is the collective conscience in modern societies, representing as it does an overarching symbolic universe. On this point, Durkheim and Berger clearly part company. Berger as we have seen vehemently rejects the notion that any overarching value system exists in modern society. Parsons (1991) picks up the Durkheimian theme of value integration, although he casts individualism more in terms of the value of "worldly asceticism" than in terms of what Weber referred to as an "ethic of brotherhood" (which is closer to Durkheim's notion of the "cult of the individual). While Berger differs with Parsons on the question of the existence of an overarching value system in modern society, both emphasize the centrality of "differentiation" in the modern social order. The bone of contention is that while for Parsons (1977) the various life spheres (both public and private) are functionally coordinated, for Berger there is nothing holding them together.

The connection between Simmel and Berger is also instructive. Simmel asserts the fragmentation of modern life and the difficulty of authenticity in it. Berger, like Simmel, is pessimistic about the capacity of individuals to construct a meaningful existence in an ever more abstract and expanding social context.[13] Urbanism (later globalization) expands our social contacts, "disembedding"[14] us from "local knowledges" (to use a postmodern concept). Group expansion for Simmel results in a blasé attitude toward human relationships. This would seem to work against an "ethic of brotherhood." Modern individuals assert their resistance to this abstraction (which is epitomized by the money economy and the commodification of life that is intrinsic to it) by countering "form." "Life" thus asserts itself against "form," and in the process, their uniqueness manifests itself. The "tragedy"[15] lies in the fact that human beings are incapable of existing in a formless state. So, the resistance to form is expressed in new forms (what Berger calls "secondary institutions"). Simmel also discusses the obvious tension that exists between "objective" and "subjective" culture. Objective culture (art, science, technology) is subjectively appropriated by individuals in a fragmentary way. Gergen (1991) refers to this dynamic in terms of "the saturated self," a consequence of individuals being bombarded with discrepant cultural elements. Berger's theory follows markedly on Simmel's insights.

How, then, should Berger's theory of modern consciousness be assessed? Let me identify a few problematic features. First, Berger dismisses prematurely the potential meaningfulness of public (especially occupational) roles. For many people in modern society, there is a strong desire to "enchant" or "re-enchant" the public realm. Berger is correct in arguing that to the degree that the work process is rationally organized the likelihood of imbuing one's work with meaning is seriously compromised. However, for many people in modern society, self-identity is tied in with their work, and they seek out work contexts that express their authentic selves. Second, there are many instances in our society in which religion and ethnicity have served as an organizing principle for public involvement. In his later writing, Berger acknowledged that the privatization/secularization thesis put forward overstated the case, that it is not intrinsic to the modern situation for religion and ethnicity to be forever relegated to a private status. Third, on the one hand, Berger argues that the dignity of the person has both a public and a private face, but then rejects wholesale the possibility that an overarching value system might exist in modern society. Fourth, Berger underestimates the shaping role of commodification in modern life and in the construction of personal identity. Fifth, in arguing as he does that the relativization of consciousness results in an unstable and incoherent identity, Berger dismisses the possibility that it might conceivably provide an opportunity for a strengthened and more solidly grounded identity. In defense of this counter-argument, relativization "forces" people to defend and justify their cognitive style and their life-style. The end result of bringing collisions of consciousness to the discursive level, of entering into a reflective process of "cognitive bargaining" may very well be an identity with more cognitive and normative weight, one which is not based on habit and tradition, but rather one based on affirmative commitment. Bateson's (1994) view that personal maturity requires ongoing encounters with the "strange" can be cited in this connection. Finally, perhaps the public/private dichotomy needs to be revisited. A convincing model offered by Boulding (1990) divides social life into three sectors: the threat system, the exchange system and the integrative system, represented in modern society by the state, the market, and civil society, respectively. Boulding argues that a sphere is legitimated to the degree that it is defined in integrative terms, that is, based on love, respect, trust, and fairness. This is consistent with Habermas' (1987) emphasis on "non-distorted communication" as a basis for an identity that is genuinely autonomous. It becomes an empirical question as to whether the political system is capable of transcending "threat" and the market is capable of transcending "exchange" in favor of integration and solidarity. So it is possible for a political relationship to be based on trust, and a personal relationship to be grounded in threat. Such a recasting of the public/private dichotomy opens up new avenues of analysis. These issues notwithstanding, Berger's theory of modern consciousness has made a seminal contribution to our understanding of the social psychology of modernity.

On the Current State of Italian American Catholicism

What does all of this mean for an understanding of the state of contemporary Italian American Catholicism? How does Berger's theory help us to analyze this phenomenon? How should the reservations raised at the conclusion of the last section be incorporated into the analysis of Italian American Catholicism?

Italian American Catholicism has involved "cognitive bargaining" in a double sense. In the first place, Italian Americans have had to "make their peace" so to speak with Irish-dominated Catholicism. The establishment of national Italian parishes in areas where settlement was concentrated (mainly the major urban areas) facilitated this process. Italian Americans practiced their faith and administered their parishes under the thumb of the Irish hierarchy, at the same time expressing their Catholicism within the context of a distinctively Italian plausibility structure. Probably the main "collision of consciousness" that took place involved the distinctively ethnic folk elements that became fused with Roman Catholicism. From the perspective of the Irish hierarchy, this was a serious enough breach to warrant the label, "The Italian problem." What to Italian Americans were legitimate cultural expressions of Catholicism was defined by the Irish hierarchy as superstitious, magical, and paganistic, as indicative of heresy (e.g. the evil eye). Another collision of consciousness had to do with the "amoral familism"[16] of Italian Catholicism as contrasted with the "ethic of brotherhood" espoused by the Universal Church. From the latter's perspective an untempered devotion to one's blood kin attenuates significantly commitments to the common good and to human solidarity in the broadest sense. Finally, there is the gender issue, involving widespread male indifference within the Italian American community to regularized participation in the Sacramental life of the Church.

These "collisions" were not between traditionalists and modernists, but between traditionalist variants within Catholicism itself and one can argue that these cognitive tensions have at this point in time been more or less resolved to the satisfaction of each "camp." More germane to the concern of the present essay has to do with the extent to which Italian American Catholics have bought into a modern world-view. I think that it is safe to say that a counter-modernization posture can be found in a shrinking number of Italian American enclaves. The social and cultural insulation and isolation required to sustain a counter-modernist plausibility structure has been undercut significantly by the "dis-embedding" processes noted earlier in the essay. A cosmopolitan attitude fostered by expanded social and cultural contacts, fueled by the media of mass communication, reduces if not shatters the plausibility of an enclave mentality. So, the key question boils down to the degree to which Italian American Catholics have accommodated themselves to modern ways of thinking. In what sense have they adopted wholesale a modernist frame of reference? Or, have they engaged in and worked through a process of cognitive bargaining with modernity and the cognitive style intrinsic to it.

Obvious tensions exist, at least on the surface. These include the following:

- The Italian-Catholic emphasis on personal relations is in tension with the modernist emphasis on abstraction.

- The Italian-Catholic emphasis on "enchantment" runs counter to the "dis-enchanting" tendencies of a functionally rational society.

- The Italian-Catholic emphasis on "family" conflicts with the modern emphasis on occupational status.

- The Italian-Catholic emphasis on tradition is in tension with the modernist emphasis on makeability.

- The Italian-Catholic emphasis on honor and doing one's duty (particularly in relation to one's family) is in conflict with the modernist emphasis on authenticity

On the other hand, there are intriguing points of intersection. For example:

- The Italian-Catholic belief that meaning can only be derived in family, community, and parish life is in line with modernist indifference if not hostility toward public roles and the importance that modernism accords to the private sphere as the locus of identity and meaning.

- The Italian-Catholic insistence that life events are mystical and mystifying, that they defy rational explanation is consistent with the modern "heroic" effort to introduce a sense of awe and wonder into life experience, this in reaction to the "iron cage" constructed by the forces of technological production and bureaucracy.

What scenarios are possible as Italian-Catholics continue to negotiate their way through modernism? Some are positioned on the modernist (or at least de-traditionalist) side of the continuum, others on the traditionalist side, and the remainder in a zone that harmonizes the two.

To argue that an Italian American Catholic has gone through a process of "de-traditionalization"[17] is to identify a life-world with the following characteristics:

- The colonization of the private sphere o f l ife by the forces of functional rationality. More specifically, life events are largely explained with reference to empirical data and the fruits of scientific research. Explanations that draw on the role of Divine Intervention, saintly intercessions, even fate, no longer have the ring of plausibility about them. This dynamic can also take the form of a coarse commodification of private life, with ostentatious patterns of consumption.

- No effort is made to "carryover" an Italian American Catholic sensibility into the individual's work and other public roles. More specifically, the work place remains "dis-enchanted" in a fundamental sense, and the individual refrains from any attempt to "re-enchant" it . B ut, what form might this take? One example would be to see the absurdity in public sphere involvements, to not take these activities all that seriously. Another would be to attempt to infuse public sphere activities with a personalist dimension, by cultivating meaningful work relationships.

- The individual seeks to carve out a meaningful existence in a heartless, deadening world, but does not draw on an explicitly Italian American Catholic frame of reference in this effort. In this case, the individual does not see the Italian American world-view as being capable of providing a meaningful, stable and reliable self-identity. Other frames of reference (e.g. New Age movements, or Eastern religion) have greater plausibility as sources and candidates for self-identification. The difference between the New Age and Eastern religion as candidates for self-identity is fundamental. While the former represents in essence a "secondary institution," a religion or quasi-religion created in response to the "discontents of modernity," the latter constitutes a preexisting cultural tradition (that is predating the onset of modernity). Thus, to argue that Italian American Catholics have become "de-traditionalized" is not to assert that the individual necessarily abandons a traditional frame of reference, only the Italian American Catholic frame of reference.

It is self-evident that Italian American Catholics adopt a cognitive style closer to the traditionalist end of the spectrum to the degree that they:

- Resist the colonization of the private sphere by the forces of functional rationality;

- Carry over an Italian American Catholic sensibility into public sphere involvements; and

- Seek to carve out a meaningful identity by drawing explicitly on an Italian American Catholic frame of reference.

Under what circumstances could it be said that the individual's life-world represents a hybridization, a co-existence, however uneasy, tenuous and precarious, of Italian American Catholic traditionalism and modernism. Two possibilities can be noted:

- Individuals compartmentalize their Italian American Catholic identity, relegating it to the private sphere: for example, politicians who refuse to let their views on abortion to affect their public stance on the issue;

- Individuals adopt a functional rational approach to the world and to their life experience, while retaining an openness to mystical explanations. In this case, a mystical symbolic universe complements a rational one, a juxtaposition that is plausible to the degree that rational frameworks are found wanting. Such hybridization represents "additive" rather than "substitutive" assimilation.[18] That is, a modern world-view is grafted on to a traditionalist one (or perhaps the other way around): one is not substituted for the other. They are both available to the individual in the cognitive and normative smorgasbord that is the modern condition. Both ways of defining reality are available to individuals as they embark upon the project of designing their identity.

This section concludes by considering an issue that has not been addressed adequately to this point. Recall that one critique of Berger's position had to do with it not entertaining the possibility that an identity that is the result of individual choices can be as substantive, as subjectively plausible as one that is transmitted by "guardians" of a particular cultural tradition.[19] Can a self-identity that is constructed in a "horizontal" society be as reliable as one passed on in a "vertical" society?[20]

To argue in the affirmative is to take and defend the position that it is only out of an internal conversation among different "generalized others" (to draw on George Herbert Mead) that a mature self-identity can emerge, particularly in a pluralist setting. Mary Bateson, as previously noted, argues in a similar vein, emphasizing the indispensable role of encounters with the strange in the development of a viable self-identity in the contemporary world. Developmental psychology has also weighed in on this issue; central to, for example, Kegan's (1994) stage model of the self is the ongoing assimilation of perspectives, with the individual moving from the particular to the universal, from the concrete to the abstract, from the exclusive to the inclusive.

On the other side of the issue are scholars like Robert Bellah, James Hunter and Joseph Varacalli (the latter two being students of Berger), who argue that the development of a reliable self-identity requires immersion into a particular cultural tradition.[21] This is the Durkheimian side of Berger in evidence here. According to Hunter, necessary for the development of character is being steeped in the stories, the collective memory of a tradition. Varacalli argues that the weakening of a Catholic plausibility structure is the responsibility of the tradition's "guardians" and "interpreters"[22] who either were too accommodationist in their encounter with modernity and secularism or began to inculcate a con-tentless and thin Catholicism in the next generation, one devoid of stories and particularistic content. In short, a precondition for an expanded consciousness is the internalization of a "thick"[23] local knowledge.

The gulf between these two positions is not as unbridgeable as might appear at first glance. Bateson and developmental psychologists like Kegan make the point that one can encounter "strangeness" only by being firmly grounded in the familiar. Such a grounding provides individuals with "ontological security,"[24] the pursuit of which is an ineluctable feature of the human condition. Many who emphasize the importance of thick local knowledges as a starting point for character development would agree that an autonomous self (one that self-consciously affirms its social and value commitments) is more solidly grounded than one that "blindly" subordinates itself to the authority of established cultural "guardians" and "interpreters." Durkheim (1961) in his work on moral education argued that a fully developed personality required attachment to groups, discipline, and autonomy.

What are the implications of this disagreement for the present and future state of Italian American Catholicism in the 21st Century? The Hunter-Varacalli thesis is correct in pointing out that an inability or unwillingness on the part of "guardians" and "interpreters" to give the succeeding generation a crystallized "world," an existential "home," a collective memory in all of its richness and depth, does have a self-liquidating effect, for there is little or nothing to pass along after that, and there are no "guardians" and "interpreters" with the authority and expertise to accept and carry out this mandate. Many would argue that the collective memory train has already "left the station," and that we are unlikely to see a resurrection of its plausibility any time soon. To the degree that Italian American Catholics participate in the cultural and sacramental life of their community at all, these activities tend not to be grounded in the symbolic universe of the tradition. The link between the activity and its meaning has in many cases been severed, perhaps permanently so. So while there are numerous instances in which Italian American Catholics get their children baptized, opt for Church weddings, nominally raise their children in this tradition, attend San Gennaro. Mt. Carmel and St. Anthony festivals, have icons of one kind of another (medals, statues) in their homes and on their persons, and have Christmas Eve fish dinners does not mean that the full meaning of those activities is known or understood. These elements of the cultural tradition take on the status of "relics," part of a "living museum."[25] They may engage in such activities out of habit, but their meaning within the context of the tradition completely largely escapes them. Rituals are observed, but for no apparent reason. From the perspective of Varacalli, this becomes a slippery slope toward the rituals not being observed at all. From the Bateson-Kegan perspective, overemphasizing the transmission of the Italian-Amercian Catholic tradition promotes a narrowness of thinking, a ghetto mentality, a less than genuine openness to the "other." But, Varacalli, Hunter, and Berger would respond that one cannot be open to the strange in the abstract. An encounter with the unfamiliar can only take place from the vantage point of a particular "world." Without that world being implanted in the individual at an early age, serving as an internal "gyroscope,"[26] "openness" to the other becomes a meaningless and futile exercise. To focus on openness without recognizing the necessity of first transmitting a richly textured "world" in the primary socialization process is to put the cart before the horse. Bateson and Kegan would concede this point, but would emphasize to a much greater extent the essential value of encountering the "other." What Varacalli, Hunter, and Berger would see as deracination and "homelessness," Bateson and Kegan would define as "maturity."

Many people in our society (Italian American Catholics clearly being among them) are "discontented" with modernity. Yet, their de-modernization project rarely it would seem taps into the rich cultural tradition in which they were raised. Is that because, as Bateson would argue, they have been exposed to the other and have incorporated it in constructing or reconstructing their identity (e.g. adopting Eastern religion or Scientology), or as Varacalli and Hunter would argue, because they have only been thinly exposed to the Italian American Catholic cultural tradition, if at all?

The purpose of this essay has been to lay out the social constructionist perspective of Berger and Luckmann, and to show how it could be fruitfully applied to the current and future state of Italian American Catholicism. While the main emphasis has been on the profound insights into the modern condition and by implication contemporary Italian American Catholicism that are offered by the framework, the limitations of the theory have not gone unnoted.

ANTHONY L. HAYNOR

---

[1] cf. C. Wright Mills' (1940) notion of "vocabularies of motive."

[2] See the essay, "Subjective Culture," in Simmel, 1971.

[3] cf. Bredemeier's (1998) notion of "internal controls," Parsons' (1951) notion of the "need-dispositions" of the personality system, and Haynor's, 2003, analysis of the self.

[4] One could argue for an affinity with the rational choice theories of Coleman (1990) and Bredemeier (1998), although Berger and Luckmann would challenge such a characterization.

[5] The following discussion draws for the most part on Berger's book, *The Homeless Mind* (1973), particularly those elements of the argument most germane to the subject of this essay, namely how a modern social order shapes forms of religious and ethnic consciousness.

[6] The distinction between "form" and "content" is crucial to Simmel's sociology.

[7] See the essay, "The Conflict in Modern Culture," in Simmel, 1971.

[8] See also Luckmann, 1967, on this general theme.

[9] See Habermas, 1987 on this point. Both Berger and Habermas cast considerable doubt on the likelihood of a carryover in the reverse direction.

[10] The term, "heroic," is used by Shilling and Mellor (2001) in their interpretation of Weber's perspective on modernity.

[11] See Weber's essay, "Religious Rejections of the World and Their Directions," in Weber, 1946.

[12] See "Conclusion" of Durkheim, 1964.

[13] See the essay, "Group Expansion and the Development of Individuality," in Simmel, 1971.

[14] See Giddens, 1994 on this point.

[15] Shilling and Mellor, 2001, use the term, "tragic," to describe Simmel's sociology.

[16] See Banfield, 1958.

[17] See Giddens, 1994.

[18] See Yinger, 1994.

[19] Giddens, *op. cit.*

[20] See Friedman, 1999.

[21] See Varacalli, 2000; Hunter, 2000, and Bellah et al., 1985, on this point, their disagreements notwithstanding.

[22] See Giddens', *op. cit.*, discussion of the decline role of "guardians" and "interpreters" in late modernity.

[23] I am drawing here on the term introduced in Geertz, 1973. While he used the term in connection with the ethnographer's responsibility to capture the culture studied in all of its richness and detail, I am using it to describe the symmetry between objective and subjective culture, the degree to which objective culture has been appropriated by the individual.

[24] See Giddens, 1991 on this point.

[25] See Giddens, 1994 on this point.

[26] See Riesman, 1961, on the "inner-directed" character.

References

Banfield, Edward. 1958. *The Moral Basis of a Backward Society* (New York: The Free Press).

Bateson, Mary Catherine. 1994. *Peripheral Visions* (New York: HarperCollins).

Bellah, Robert,Richard Madsen, William Sullivan, Ann Swidler, and Steven Tipton, *Habits of the Heart* (Berkeley: University of California Press).

Berger, Peter L. and Thomas Luckmann. 1966. *The Social Construction of Reality* (Garden City: Doubleday).

Berger, Peter, Brigitte Berger, and Hansfried Kellner. 1973. *The Homeless Mind: Modernization and Consciousness* (New York: Vintage).

Boulding, Kenneth. 1990. *Three Faces of Power* (Thousand Oaks, CA: Sage).

Bourdieu, Pierre. 1992. *An Invitation to Reflexive Sociology* (Chicago: University of Chicago Press).

Bredemeier, Harry C. 1998. *Experience Vs. Understanding* (New Brunswick: Transaction).

Coleman, James. 1990. *Foundations of Social Theory* (Cambridge: Harvard University Press).

Durkheim, Emile. 1961. *Moral Education* (New York: The Free Press).

_____. 1964. *The Division of Labor in Society* (New York: The Free Press).

_____. 1965. *The Elementary Forms of the Religious Life* (New York: The Free Press).

_____. 1973. *On Morality and Society.* Edited by Robert Bellah (Chicago: University of Chicago Press).

Friedman, Lawrence. 1999. *The Horizontal Society* (New Haven: Yale University Press).

Geertz, Clifford. 1973. *The Interpretation of Culture.* (New York: Basic Books)
Gergen, Kenneth. 1991. *The Saturated Self: Dilemmas of Identity in Contemporary Life.* (New York: Basic Books)
Giddens, Anthony. 1991. *Modernity and Self-Identity* (Stanford: Stanford University Press).
_____. 1994. "Living in a Post-Traditional Society," Pp. 56-109 in Beck,Ulrich, Anthony Giddens and Scott Lash, *Reflexive Modernization* (Stanford: Stanford University Press).
Habermas, Jurgen. 1987. *The Theory of Communicative Action,* Volume Two (Boston: Beacon Press).
Haynor, Anthony L. 2003. *Social Practice: Philosophy and Method* (Dubuque, IA: Kendall/Hunt Publishers).
Hunter, James Davison. 2000. *The Death of Character: Moral Education in an Age Without Good or Evil* (New York: Basic Books).
Kegan, Robert. 1994. *In Over Our Heads: The Mental Demands of Modern Life* (Cambridge: Harvard University Press).
Lasch, Christopher. 1977. *Haven in a Heartless World* (New York: Basic Books).
Lemert, Charles. 1997. *Social Things: An Introduction to the Sociological Life* (Lanham: Rowman and Littlefield).
Luckmann, Thomas. 1967. *The Invisible Religion* (New York: Macmillan).
Mead, George Herbert. 1962. *Mind, Self and Society* (Chicago: University of Chicago Press).
_____. 1964. *On Social Psychology,* edited by Anselm Strauss (Chicago: University of Chicago Press).
Mills, C. Wright. 1940. "Situated Actions and Vocabularies of Motive." *American Sociological Review.* Volume 5, No. 6 (December)
Parsons, Talcott. 1951. *The Social System* (New York: The Free Press).
_____. 1977. *The Evolution of Societies,* edited by Jackson Toby (Englewood Cliffs: Prentice-Hall).
_____. 1991. "A Tentative Outline of American Values." In *Talcott Parsons: Theorist of Modernity,* edited by Roland Robertson and Bryan Turner (London: Sage).
Riesman, David. 1961. *The Lonely Crowd* (New Haven: Yale University Press).
Schutz, Alfred. 1967. *The Phenomenology of the Social World* (Evanston, IL: Northwestern University Press).
Shilling, Chris and Philip A. Mellor. 2001. *The Sociological Ambition* (Thousand Oaks, CA: Sage).
Simmel, Georg. 1971. *On Individuality and Social Forms,* edited by Donald Levine (Chicago: University of Chicago Press).
Varacalli, Joseph A. 2000. *Bright Promise, Failed Community* (Lanham: Lexington Books).

Weber, Max. 1946. *From Max Weber: Essays in Sociology*, edited by H. H. Gerth and C. Wright Mills (New York: Oxford University Press).

Yinger, J. Milton. 1994. *Ethnicity: Source of Strength? Source of Conflict?* (Albany: State University of New York Press).

Chapter 13

◆

CATHOLICISM AND SOME ITALIAN AMERICAN WRITERS

In an essay that appeared in the fall 2003 issue of *MELUS* dedicated to Italian American literature (edited by Mary Jo Bona), John Paul Russo identifies seven features of Italian American Catholicism: Immanence, Imagery, Festa, Cultism, The Church and Clergy, The Religion of the Home, and Tame Death. Of the seven he uses to read the portrayal of Catholicism in the work of Don DeLillo, Russo sees DeLillo's use of immanence and analogical image as the two areas that reveal "the highest level of his artistic achievement and probably will mark his main contribution to American literature" (2003: 14). Russo is onto something important in this essay, and he does a great service to the readers of contemporary American literature by exploring an area that has for too long been ignored, not only in DeLillo's work, but also in the scope of American literature.

However, long before Catholicism was ignored in the work of Don DeLillo, it was ignored in the writing of John Fante, Pietro di Donato, Jerre Mangione, and nearly all of the American writers of Italian descent. One reason Russo points to is the idea, promulgated primarily by left-wing critics, that Italian American Catholicism is pagan based and centered on anti-clericalism. While Russo does not deny these elements, he does believe that they have the tendency to dominate critical inquiries into the role Catholicism plays in Italian American culture.

It is quite true that the work of Italian American writers is heavily imbued with an Italian Marianist Catholicism that is in many ways distinctly different from an American Catholicism institutionalized and controlled by Irish Catholics. Discussion of their work thus requires the establishment of new interpretative frameworks that identify and include such contexts. James T. Fisher's *The Catholic Counterculture in America, 1933-1962*, is a good example of a study that examines the role American Catholicism played in the culture of the period. However, Fisher's study does not examine the many Catholic writers of Italian descent. Such frameworks are only now being constructed. In a paper presented in a special session at the

Modern Language Association's 1991 meeting, Thomas J. Ferraro proposed re-examining the period by focusing on the relationship of Mediterranean, Marianist Catholicism to American culture. Such an approach would make it impossible to ignore the writings of Italian Americans, both proponents and opponents of American Catholicism, who documented what the Catholic Church referred to as "The Italian Problem" (Malpezzi and Clements, 1992: 108). In most of the works of John Fante, Pietro di Donato, and Jerre Mangione, a variation of Catholicism is presented and represented that is often anti-institutional and often referred to as being "un-American." The Italian/Catholic backgrounds these three writers, which strongly roots their works, is an area of American literary history that has yet to be adequately examined.

At an early age John Fante was encouraged by Catholic nuns to write. He attended Regis College and the University of Colorado at Denver. Fante's stories represent unparalleled insights into American Catholicism, which are often portrayed through cultural conflicts between an Irish clergy and Italian American parishioners. "Altar Boy," his first publication, presents a young Italian boy's crisis of faith that erupts into childhood pranks played in church and acts of petty thievery in the community. He returned to Catholicism in later stories such as "My Father's God," in which an old Italian succumbs to his wife's pleas and decides, after many years, to reunite with the Church. When told that he must confess his sins, he asks if he can do it in writing, a request the young, Italian American priest allows. When presented with a long confession written in Italian, the young priest, who cannot read the language of his ancestry, can only laugh when he realizes he's been tricked and shamed at the same time.

Fante's four-book saga of Arturo Bandini, of which *Wait Until Spring* is the first, follows a young Italian Catholic who lights out for California with the intent of escaping his family and its ethnicity by becoming a writer. One impediment which continually keeps Fante's protagonists from identifying themselves completely with mainstream American culture is their strong connections to Italian Catholicism. More than half the stories of his collection *Dago Red* (1940) deal with this subject. In *1933* the protagonist writes an essay on the mystical body of Christ (1985a: 14) and believes he has been visited in the night by the Virgin Mary (1985a: 37). Though Fante strays from this strong identification with Marianist Catholicism in *The Road to Los Angeles*, a novel in which Bandini constantly mocks the Christianity of his mother and sister in his attempt to separate himself from his background, he nevertheless does so through a character who is more comical than serious. John Fante, a loyal follower of H.L. Mencken, stayed out of politics and described his attitude toward party-line politics and Marxist aesthetics in a letter to his literary mentor:

> I haven't sucked out on Communism and I can't find much in Fascism. As I near twenty-six, I find myself moving toward marriage and a return to Catholicism. Augustine and Thomas More knew the answers a long time ago. Aristotle would have spat in Mussolini's face and sneered at Marx. The early fathers would have laughed themselves sick over the New Deal. (Moreau, 1989: 103)

Fante's "return to Catholicism" and his choice not to align himself with left-wing ideology would prove to hinder his reception and consideration by cultural critics adhering to Marxist aesthetics. In spite of the fact that Fante shares some of the concerns of those traditionally identified with the modernist movement, his ethnic and religious orientation combine to create philosophical obstacles which prevent critics and historians from including him in the Marxist and New Critical studies that have shaped the definition and thus our awareness of the modernist American literary tradition.

Perhaps the novel that has had, even to this day, the greatest impact on American culture, second only of course to the late Mario Puzo's *The Godfather*, is Pietro di Donato's *Christ in Concrete*. The strong Catholic imagery of the novel was never captured in *Give Us This Day*, the 1949 film by Edward Dymytrk, but will be the major focus of a new film version that just been contracted by a young Italian American producer, Joseph di Pasquale. To help us understand the lack of attention to di Donato's work, which beat out John Steinbeck's *Grapes of Wrath* for the 1939 September selection for Book of the Month Club, we need to take a good look at the history.

Unlike many of the proletarian writers of the period in which he begins publishing, Pietro di Donato continually inserts Italian Catholicism as a force that controls the immigrants' reactions to the injustices of the capitalist system that exploits as it maims and kills the Italian immigrant. di Donato's Catholicism has its roots in pre-Christian, matriarchal worship. Annunziata, the mother in *Christ in Concrete*, controls her son's reaction to the work site "murders" of his father and godfather by calling on him to put his trust in Jesus, the son of Mary. By the end of the novel, Paul's faith is nearly destroyed as evidenced by his crushing of a crucifix offered to him by his mother (1993: 296-7). However, the final image of the novel suggests that the matriarchal powers still reign. The image we are left with is an inversion of the *Pietà* in which son is holding a mother who is crooning a lullaby depicting her son as a new Christ, one that her children should follow (303). Thus, while di Donato's novel depicts the injustices faced by the immigrants, there is no revolutionary solution offered to the reader. The absence of such a solution, combined with the novel's Roman Catholic philosophical underpinnings, keeps *Christ in Concrete* out of the historical purview of proletarian literature, and thus outside those literary histories devoted to such literature.

Only recently have critics begun to realize the true revolutionary figure that di Donato was. In a resurrection of di Donato's contribution to the Third American Writers' Congress, Art Casciato helps us to understand why writers like di Donato have been ignored by the established critics and scholars of the period. As Casciato points out, di Donato, in his brief speech at the 1939 American Writers' Congress convention in New York, (which Malcolm Cowley asked to be rewritten so that it would conform to Cowley's expectations), refused to adopt "the prescribed literary posture of the day in which the writer would efface his or her own class or ethnic identity in order to speak in the sonorous voice of 'the people'" (1991: 70). As Casciato explains, di Donato's style resisted the modern, and "thus supposedly proper ways of building his various structures." The result is that he is "less the bricklayer, than a bricoleur who works not according to plans but with materials at hand" (1991: 75-76). di Donato was the only one of these three writers to join the Communist Party, which he did at the age of sixteen on the night that Sacco and Vanzetti were executed. The following excerpt from his contribution to the Third American Writers' Congress is an example of di Donato's attitude which Cowley found troublesome.

> I am not interested in writing for class conscious people. I consider that a class-conscious person is something of a genius — I would say that he is sane, whereas the person who is not class conscious is insane...In writing *Christ in Concrete* I was trying to use this idea of Christianity, to get an 'in' there, using the idea of Christ. ("1991: 69)

Needless to say, di Donato's paradoxical use of "comrade-worker Christ" (1993: 173) as a metaphor for the working-class man would prove to be quite problematic when viewed from a Marxist perspective. While Fante was busy portraying characters who struggled to fit into American life, Pietro di Donato was rejecting the American dream by documenting the disintegration of the Italian family caused by American capitalism through a rewriting of the story of Christ. Prior to his death in January of 1992, he had completed a novel entitled *The American Gospels*, in which Christ, in the form of a black woman, comes to Earth at the end of the world to cast judgment on key historical figures of contemporary America. The theme of Christ as a woman can be found in much of what di Donato has written. It is a theme that is usually left in the shadow of the more obvious interpretation of the great Italian/American myth created in his first novel. di Donato's Catholicism has its roots in pre-Christian, matriarchal worship. As he admitted, "I'm a sensualist, and I respond to the sensuality of the Holy Roman Catholic Church, its art, its music, its fragrances, its colors, its architecture, and so forth — which is truly Italian. We Italians are really essentially pagans and realists" (1987: 36). Annunziata, the mother in *Christ in Concrete*, becomes the key figure in di Donato's rewriting of the Christian myth. She controls her son's reaction to the work site "murders" of his father and godfather by calling on him to put his trust in Jesus, the son of Mary. This is a trust that has led immigrants to accept poverty as their fate and passivity as their means of survival in a world bent on using and then disposing them. This trust is a myth that di Donato, through his protagonist Paul, refuses to accept.

As a myth it presents an heroic figure, Paul, who searches for God, in the form of Christ, whom he believes, as he was trained to believe, can save his family from the terrible injustices brought upon them by a heartless society. The novel is divided into five parts: "Geremio," "Job," "Tenement," "Fiesta," and "Annunziata," each focusing on the key figures in the myth. In "Geremio" and "Job," di Donato presents JOB (one's work) as the antagonist which controls the Italian workers lead by his father Geremio, through the human forces of Mr. Murdin, the heartless foreman, the State Bureaucracy that sides with the construction company during a hearing into Geremio's death, and the Catholic Church through an Irish priest who refuses to do more than offer the family a few table scraps from his rich dinner. Paul's mother, Annunziata, pregnant at the opening of the novel, serves as the figure of the Madonna and represents the immigrants' faith in God whom she invokes through prayers such as, "God of my fathers, God of my girlhood, God of my mating, God of my innocent children, upon your bosom I lay my voice: To this widow alone black-enshrouded, lend of your strength that she may live only to raise her children" (1993: 63). In the chapter devoted to her "widowhood," Annunziata attempts to raise her children according to the Christian myth, but in the process her son Paul loses the faith she hopes to pass on to him through recollections of her husband's show of faith.

Toward the end of the novel, Paul has a dream in which he is about to die just as his father had. His godfather attempts to save him and is tossed off a scaffold by the foreman who was threatening Paul. Paul fails to save his godfather but remembers who can save him: "it is our Lord Christ who will do it; he made us, he loves us and will not deny us; he is our friend and will help us in need! Bear, oh godfather, bear until I find Him" (1993: 285). Paul then takes off in search of Christ and runs into his father who is on his way to work. The work site becomes a shrine for the workers who have now become saints; Mr. Murdin, the foreman, appears as a magician who "each time he revolves and shouts at Geremio and Paul he has on a suit and mask of a general, a mayor, a principal, a policeman" (1993: 288). Job falls apart and Paul is the only one who tries to save himself. Paul sees himself in his father's crucified form. He is then carried off to "the Cripple," the hag who earlier in the story had conducted a seance for Paul and his mother. Paul sees his father hovering over "the Cripple" and as they embrace Geremio sighs, "Ahhh, not even the Death can free us, for we are...Christ in concrete" (1993: 290). Paul's dream quest ends here with the failure of Christ to save him and his family. Paul realizes that only he can save himself. This realization is dramatized in the last scene of the novel.

By the novel's end, Paul's faith is nearly destroyed as evidenced by his crushing of a crucifix offered to him by his mother (1993: 296-7). However, the final image of the novel suggests that the matriarchal powers still reign. The image we are left with is an inversion of the *Pietà*; the son is holding a mother who is crooning a deathsong/lullaby that hails her son as a new Christ, one that her children should follow (1993: 303). But this haunting image can also suggest that the mother has become the new

Christ, who in witnessing what America has done to her son, dies and through her death frees her son from the burden of his Catholic past. This death is quite different from the death of his father which leads Paul in search of Christ. His rejection of Christ as the means to survive in this world contributes to his mother's collapse. She then becomes the basis for a new faith in himself toward which his mother urges her children as she cries out "love...love ...love...love ever our Paul" (1993: 303). For di Donato, this figure of the dying *Mater Dolorosa* replaces Christ as the figure through which man can redeem himself. There is no redemption through the father; if Paul stays in the system, if he continues to interact with Job, he will share the same destiny as father, Geremio. di Donato's revision of Christ points to the failure of American Catholicism to support the immigrants' struggle. He reveals Catholicism as a force that controls and subdues the immigrants' reactions to the injustices of the capitalist system that exploits as it maims and kills the Italian immigrant. His deconstruction and remaking of the Christian myth forces us to reread his novel as more revolutionary than it has been portrayed by past critics.

Di Donato's rewriting of the Christ myth leads him away from organized Roman Catholicism and toward paganism. This return becomes explicit in nearly all his subsequent work. Di Donato's ability to see through the repression created by a Christianity which aligns itself with a capitalist power structure leads him toward socialism. Cultural critic Louis Fraina foretold di Donato's dilemma in his 1911 essay "Socialism and the Catholic Church:

> the economic suffering of the peoples makes them turn to religion, and the dominant church being allied with the exploiting-properties elements, the toiling masses, too ignorant as yet to embrace Socialism, turn for "relief" to a religion in opposition to the established church. The desire for happiness is the *conditio sine qua non* of religious faith. (1911: 5)

For di Donato, this "religion" would come not in the form of an organized church, but rather through a spiritual quest for truth that would lead him back to a pre-christian pagan (quasi-Sadistic) sensualism that he would record in his next novel.

Much of di Donato's later work continues to portray this conflict of the sacred and the profane. In *The Penitent* (1962) he recounts the story of Alessandro Serenelli, the man who killed the virgin Maria Goretti who was later sainted by the Catholic Church. di Donato attempts to understand the murder as a crime of nature-driven passion committed by a man who as a fisherman lived the pagan life of the sea (1962: 205). "Was purity more important than the denial of nature, the agony and loss of one life, the ruin of another and the sorrow of two families?" (1962: 203)

After destroying the traditional myth of a Christianity corrupted by temporal powers, di Donato builds a new myth for his readers. This new myth portrays man as surviving best within the naturally spiritual institution of family which is constantly threatened by a world corrupted by the artificial institutions a material hungry capitalism creates. Di Donato's final novel is an attempt to resolve the sacred/profane dilemma presented in much of his earlier work. The redemption of the victims of capitalism through the final judgment of a female Christ becomes the matter of his *The American Gospels*, which at this point remains unpublished. Through this novel, which should be read as his primal scream out of the world just as *Christ in Concrete* was his cry into the world, di Donato takes his "revenge on society" by revealing "all the nonsense of authority and of Church" through what he calls a "conscious evaluation of myself" (von Huene-Greenberg, 1987: 33-4). To di Donato, salvation for the world lies in man's ability to become his own god, to take responsibility and control of the world he's created and to act for the good of all. di Donato's fascination with if not devotion to the Catholic faith prompted him to write two religious biographies: *Immigrant Saint: The Life of Mother Cabrini*, (1960) and *The Penitent*, (1962), which tells the story of St. Maria Goretti through the point-of-view of the man who killed her.

Thus, as Russo's notion of immanence suggests, di Donato's Catholicism is indicated by his location of God through the quotidian life of the worker. The concept of God is a product of the worker's labor as well as through the life led by the worker that leads to the creation of a community of workers who help each other. Identifying and understanding this notion of immanence is important in reading not just di Donato, but many other American writers of Italian descent.

One of di Donato's contemporaries was Jerre Mangione, who wrote very little directly about Catholicism. More interested in issues of social justice, Mangione was sympathetic to the Communist Party in America, but never formally joined it because he recognized in it a constraining dogmatism that reminded him of Catholicism. His memoirs *Mount Allegro* (1943) and *An Ethnic at Large* (1978), along with *The Dream and the Deal* (1972) — a study of the Federal Writers Project — barely mention the practice of Roman Catholicism, leading us to believe that Mangione might have been an agnostic at best when it came to religion. However, when we look to his fiction, we can see some very interesting takes on Catholicism. In *The Ship and the Flame* (1948) he creates an allegory for the sorry state of political affairs in Europe prior to America's entry into the Second World War. Aware of the dilemma of the liberal and the fate of the revolutionary in the world, Mangione created a microcosm of the larger world of his time, suggesting that the struggle against fascism could be won through heroic action that would not compromise one's Catholic beliefs and that would require breaking away from a traditional mindset formed for many generations. Stiano Argento, the protagonist of Mangione's strong anti-fascist novel learns that his Catholicism has room for radical beliefs and behavior.

Catholicism figures strongly in the writing of younger Italian Americans who seem to have had an easier time reconciling their political needs and religious backgrounds. While I don't have the time or the space to do a thorough reading of the use of Catholicism by even one of the many American writers of Italian descent, I would like to end this essay by pointing to a few of the uses by some of the contemporary writers.

Tony Ardizzone's first novel, *In the Name of the Father*, takes it name from the Roman Catholic, "Sign of the Cross," the invocation that accompanies the ritual beginning of prayer. Tonto Schwartz, the protagonist of this novel collects holy cards and baseball cards and these two worlds inside and outside the church, collide in the young boy's mind: "God the Father, first person, first base; God the Son, second person, second base; God the Holy Ghost, third person, third base; Ernie Banks, I don't care if he's chocolate, shortstop....Our Lady of Perpetual Help, pitcher, because of her good arms" (1978:14). Catholicism plays a major role in much of Ardizzone's work. His short story, "The Eyes of Children," shows how a Catholic education fashions a child's imagination so that a variety of realities are made possible. When children report a man bleeding in the church, young Gino thinks it must have been Jesus. But what he comes to realize is that when doubt pierces the cloud of his Catholic training life can become even more terrifying and complex than the simple division between good and evil and heaven and hell. In the short story "World Without End," Ardizzone shows how the Church and family create frames for expected and acceptable behavior. A mother's criticism of her son's life style leads into an argument that is solved only when the son tells his mother what she wants to hear. "Holy Cards" is a marvelous series of vignettes that present key teachings of the Catholic church through the eyes of a young boy. From mortal sin in milk bottles, to martyrs and saving the world's pagan babies, Ardizzone's humor is full of respect and reverence for the way it used to be. Ardizzone's second novel, *The Heart of the Order*, presents protagonist Danilo Bacigalupo, son of Italian immigrants, brother of many, growing up on Chicago's northside. In a game of alley baseball he hits a line drive that changes his life. The shot results in Mickey Meenan's death and the adoption of Mickey's spirit by Danny. Ardizzone juxtaposes the mystical world of the Catholic Church to the wonderful world of sports, he comes up with a humorous tale driven by tragedy.

Tina DeRosa's sense of *Italianità* is intertwined with her sense of being a Roman Catholic. In 1987 she completed her second book *Father to the Migrants*, the life story of Bishop Giovanni Battista Scalabrini, founder of the Missionary Fathers of St. Charles, whose sole mission in the 19th century was to follow Italian migrants to the Americas to tend to their Roman Catholic faith. When asked to define *Italianità*, De Rosa replied:

> On any level, *Italianità* is religious. When we look at our Italian American writers we get a different sense of religion from all of them. The work of di Donato, is religious in the deep sensual or pagan sense. The religion he presents is that of family, of father, and of food; religion of the body. I mean, think of the image of his father frozen in concrete. The work of [Jerre] Mangione represents a more refined, educated, aristocratic and intellectual sense of religion. Someone like Bishop Scalabrini represents something closer to a Christ-like version of religion; his work brings the body of di Donato together with the intellect of Mangione. And it is this sense of *Italianità* that I've tried to capture in the biography. (Gardaphe, 1985: 23)

The recovery of her ethnic heritage, both through the research she conducted for the non-fiction book and through the recreation of the immigrant grandmother in the novel, has strengthened DeRosa's sense of *Italianità* and has contributed to the development of her self-identity as a woman and as an American writer of Italian descent. Her contribution to the anthology, *From the Margin: Writings in Italian Americana* (1989) was a poem entitled "Psalm of the Eucharist."

Josephine Gattuso Hendin's novel *The Right Thing to Do*, vividly presents the drama of a young women's attempt to make her own way outside of her family and her father's expectations. Nino Giardello, a Sicilian immigrant shadows his daughter Gina whenever she leaves and when she hugs her American boyfriend, he becomes furious at her defiance of his authority. As Gina rebels, her father weakens in health. As he moves closer to death, she comes closer to understanding his life and the legacy she will inherit. She also comes to understand the fallibility of the Catholic church through the story that Nino weaves to teach her a lesson about straying too far from home.

What I hope to have pointed to is in this brief look at Catholicism in a few American writers of Italian descent is that there exists a variety of takes on Catholicism executed by American writers of Italian descent. Much more work needs to be done in the areas of cultural history and criticism to situate these writers along with their counterparts from other ethnic and racial groups in the development of an awareness of the rich tradition of Catholic writing in the United States.

FRED GARDAPHÈ

References

Ardizzone, Tony. 1986. *Heart of the Order*. (New York: Henry Holt).
_____. 1978. *In the Name of the Father*. (New York: Doubleday).
_____. 1986. *The Evening News*. (Athens, GA: University of Georiga Press).
Casciato, Art. "The Bricklayer as Bricoleur: Pietro di Donato and the Cultural Politics of the Popular Front." *Voices in Italian Americana*. 2.2 (fall 1991): 67-76.
di Donato, Pietro. 1993. *Christ in Concrete*, 1939. (New York: Signet).
_____. 1991. *Immigrant Saint: The Life of Mother Cabrini*. 1960. (New York: St. Martin's Press).

_____. 1962. *The Penitent*. (New York: Prentice Hall).

Fante, John. 1985. *1933 Was a Bad Year*. (Santa Barbara, CA: Black Sparrow).

_____. 1980. *Ask the Dust*. 1939. (Santa Barbara, CA: Black Sparrow).

_____. 1990. *Prologue to Ask The Dust*. (Santa Barbara, CA: Black Sparrow).

_____. 1985. *The Road to Los Angeles*. (Santa Barbara, CA: Black Sparrow).

_____. 1983. *Wait Until Spring, Bandini*. (Santa Barbara, CA: Black Sparrow).

_____. 1985. *The Wine of Youth*. (Santa Barbara, CA: Black Sparrow).

Ferraro, Thomas J. 1995. "Catholic Writers." *The Oxford Companion to Women's Writing in the United States*. Eds. Cathy N. Davidson and Linda Wagner Martin. (New York: Oxford University Press): 155-157.

Ferraro, Thomas J., ed. 1997. *Catholic Lives, Contemporary America*. (Durham, NC: Duke University Press).

Fisher, James T. 1989. *The Catholic Counterculture in America*, 1933-1962. (Chapel Hill: University of North Carolina Press)

Fraina, Luigi. (Pseud. Lewis Corey). "Socialism and the Catholic Church." *Daily People*. 12. 128 (November 5, 1911): 5.

Gardaphè, Fred. "An Interview with Tina DeRosa." *Fra Noi*. 24 (May 1985): 23.

Malpezzi, Frances M. and William M. Clements. 1992. *Italian American Folklore*. (Little Rock, AR: August House).

Mangione, Jerre. 1983. *An Ethnic at Large: A Memoir of America in the Thirties and Forties*. 1978. (Philadelphia: University of Philadelphia Press).

_____. 1972. *Mount Allegro*. 1943. (New York : Columbia University Press).

_____. 1965. *Night Search*. (New York: Crown Publishers).

_____. 1948. *The Ship and the Flame*. (New York: A. A. Wyn).

Moreau, Michael, ed. 1989. *Fante / Mencken: A Personal Correspondence, 1930-1952*. (Santa Rosa, CA: Black Sparrow).

Orsi, Robert Anthony. 1985. *The Madonna of 115th Street: Faith and Community in Italian Harlem, 1880-1950*. (New Haven: Yale University Press).

_____. "The Religious Boundaries of an Inbetween People: Street *Feste* and the Problem of the Dark-skinned Other in Italian Harlem, 1920-1990." *American Quarterly*. 44.3 (September 1992): 313-347.

Russo, John Paul. "DeLillo: Italian American Catholic Writer." *MELUS*. Mary Jo Bona, ed. (Fall 2003).

Stephanile, Felix. 1992. "The Dance at St. Gabriels." *From the Margin: Writing in Italian Americana*. Anthony Tamburri, Paolo Giordano, and Fred L. Gardaphè, eds. (West Lafayette, IN: Purdue University Press). 158.

Tamburri, Anthony Julian, Paolo A. Giordano and Fred L. Gardaphe, eds. 1991. *From the Margin: Writings in Italian Americana*. (West Lafayette, IN: Purdue University Press).

Tusiani, Joseph. *Ethnicity: Selected Poems*. (Boca Raton, FL: Bordighera Press, 2000).

von Huene-Greenberg, Dorothee. "A *MELUS* Interview: Pietro Di Donato." *MELUS*. 14.3-4 (Fall-Winter 1987): 33-52.

Chapter 14

✦

The Church as It Seems to Me: A Second Generation Look at Post-Modernism

Giacomo and Lapa di Benincasa already had twenty-two children when twins Caterina and Giovanna were born, on the feast of the Annunciation. Giovanna died in infancy, as did their twenty-fifth child, a second Giovanna. Fourteenth century Siena did not have the benefits of modern medicine.

They nicknamed little Caterina, "Euphrosyne," Greek for Joy. She was quite a different little girl. As she grew, Caterina fell in love with Christ and with God's people. At seven she promised herself to Christ; at sixteen she asked to become a Dominican Tiertiery. She visited the sick, she helped the hungry, she buried the dead. She prayed in her tiny room day and night. As she moved through her twenties, Caterina began an even larger life within the Church. In 1376 she traveled to Avignon on her own authority and convinced Pope Gregory XI to return to Rome. She also had the reform of the clergy on her agenda.

Caterina, whom we now know better as Catherine of Siena (1347-1380), had had it with the foolish clerics of her day. Clerical celibacy and simplicity of life were a mockery in much of Catherine's Italy. Priests lived in open concubinage, sometimes in great splendor. These priests often did not seem to do much work. The charge of both the corporal and the spiritual works of mercy was often left to mendicant friars, Franciscans and Dominicans, and to Third Order women like Catherine.[1] The great preachers of morality were not the heads of dioceses; they were the religious and lay people who felt they had no more to lose. Crises of belief among the faithful mirrored the laxness of the clergy. Meanwhile, the Pope had repaired to the south of France. Throughout Christendom, subjectivism was overtaking objective understandings of reality. Individualism reigned supreme.

Has anything changed in the last six hundred years?

The Church in the United States, if not worldwide, is imploding of its own volition. What Catherine saw we see again: subjectivism in collision with objective reality, and corruptness of the clergy. What is worse, no one in authority seems able or even willing to do anything about it. The Church in America is at risk of being seen as (if not actually becoming) another Enron: a morally corrupt profit-driven corporation with its eye only on the bottom line.

Meanwhile, the world is in a desperate moral crisis, evidenced by wars both large and small. Warfare among nations destroys lives, property and wealth; the subtle wars against honest people and the poor erode hope and crush spirits. Except in degree, there is really little difference between the destruction of a city with munitions and the destruction of one life by moral turpitude.

While moral pronouncements flow freely, some bishops seem not to "get it." For example, the Bishop of Manchester, New Hampshire complained a few years ago that one of his pederast priests received a long jail sentence, inappropriate he thought because other convicted felons "are serving much shorter sentences for very serious crimes."[2] Such comments have been relatively commonplace.

The problem of subjectivism and subjective morality, combined with radical individualism and a denial of objective reality, both by the world and by segments of the Church, is complicated and caused by ineffectual Church leadership. The result of such subjectivism and subjective morality (moral relativism and deconstructionism), combined with radical individualism (an autonomous individualism with the locus of authority in the individual) and a denial of objective reality, is a communal Catholicism that both bends to pluralism and merges with older philosophies — some would say heresies — to make the Church appear more "modern."

But more modern to whom? Once a religious philosophy accepts relativism, of either morals or belief, it becomes irrelevant to traditional believers and useless to anyone else.

The combined processes of post-modernism arrest the religious inculturation of Italian Americans, and of other traditionally Catholic ethnic groups. As the children and grandchildren of first generation Italian Americans increasingly leave aside traditionally Italian means of celebrating religious heritage, so do they leave aside older and more traditional understandings of ecclesia and of faith. Too often the tradition — whether cultural or theological — meets with opposition from within. And too often the structural and societal morass in which these Italian Americans search for truth is overwhelming.

Subjectivism and Subjective Morality

Not long ago I heard a visiting priest preach on the gospel passage where Jesus is challenged by the Pharisees over his disciples picking ears of grain to eat on the Sabbath. Jesus responds that even David ate the bread of the Presence when he was hungry. "I desire mercy, not sacrifice," Jesus told the Pharisees (Matt. 12, 1-8).

The priest, a chaplain at a well-known hospital, said that this was a very important passage. It helped the Church learn that there is nothing absolutely right and nothing absolutely wrong. In fact, he said, the Church even allows abortion under certain circumstances.

I challenged this priest after mass. Perhaps I had misheard him. I thought he said in certain circumstances the Church allows for abortion, and teaches that there is nothing plainly right or wrong. The priest asked if I had ever heard of double effect. I told him that the principle of double effect allows for the termination of an ectopic pregnancy or removal of cancer of the womb, but none other, and in either case the procedure is not termed "abortion." That is, neither procedure intends or results in direct abortion; the death of the baby is a "double effect" in an untenable situation. Well, he responded, what he said was that "the Church teaches there is nothing absolutely right and nothing absolutely wrong." He turned on his heel and left for his full-time (four days a week) job.

The end of this man's belief is a denial of objective reality, and I will return to this later. On his specific point of moral teaching, he himself is absolutely wrong. The Church has always taught that things can be objectively wrong, while recognizing that sin — the commission or omission of something objectively wrong — may be subjectively reduced in gravity. The Church has never taught, nor will it ever teach "there is nothing absolutely right and nothing absolutely wrong." And then there is the rest of the story.

When the pastor challenged the priest on his statement that "the Church teaches that there is nothing absolutely right and nothing absolutely wrong," the priest refused to budge. He claimed that he was right. But the pastor did not fire him. Nor did he remove the priest's preaching faculties. To do so would place more burdens on the pastor, who hires this man for two or three days each week to help him out. The facts of Catholic teaching appear to be irrelevant. The pastor wants a few days off.

Some might argue that compassion must come into play. The visiting priest needs the extra money. The pastor needs the extra help. But true compassion would recognize that the people of God, who are paying for all this, deserve the gospel seen through a Catholic lens, not a water colored half rendition of the truth.

In fact, the creeping subjectivism in the Church in America is very often rooted in compassion gone wrong. It is an easy leap from an act being objectively gravely wrong, yet able to be forgiven, to the act itself losing its gravity. This is nowhere so obvious as in reports of the national priest-ped-erast cover-up, where a reported 1,400 Catholic priests in the United States have been accused, and many of them arrested, convicted and/or laicized for sexual predatory acts on young males and females. In some cases they have impregnated girls, and arranged for abortions. In some cases they have molested scores of boys, and been protected for decades by their bishops. In some cases they are bishops themselves. At least one man's crimes were so egregious he has received a life sentence.[3]

Given the national media attention to this ongoing problem, one can assume that the Diocese of Rockville Centre, Long Island, New York is no better and no worse than any other. The diocese comprises suburban Nassau and Suffolk Counties. In a 171-page document, a Special Grand Jury of the Suffolk County (New York) Supreme Court found the following:

> The response of priests in the Diocesan hierarchy to allegations of crim-inal sexual abuse was not pastoral. In fact, although there was a written policy that set a pastoral tone, it was a sham. The Diocese failed to fol-low the policy from its inception even at its most rudimentary level. Abusive priests were transferred from parish to parish and between Dio-ceses. Abusive priests were protected under the guise of confidentiality; their histories mired in secrecy. Professional treatment recommenda-tions were ignored and dangerous priests allowed to minister to chil-dren. Diocesan policy was to expend as little financial capital as possi-ble to assist victims but to be well prepared for the possibility of enor-mous financial and legal liability. Aggressive legal strategies were em-ployed to defeat and discourage lawsuits even though Diocesan officials knew they were meritorious. Victims were deceived; priests who were civil attorneys portrayed themselves as interested in the concerns of victims and pretended to be acting for their benefit while they acted only to protect the Diocese. These officials boldly bragged about their success and arrogantly outlined in writing mechanisms devised to shield

them from discovery. These themes framed a system that left thousands
of children in the Diocese exposed to predatory, serial, child molesters
working as priests.[4]

The horrific details in this Grand Jury document are echoed in news reports
and in other Grand Jury documents and prosecutors' statements across the
country. An ostrich-like response, whether by clergy or laity, ignores reality.

These are not fabrications of the media. The bald facts of the matter are
staggering: apparently nearly 1,400 of approximately 46,000 priests in the
United States have been accused of sexual predatory acts.[5] That number
represents 3% of the given population (far higher than the claimed "fewer
than 1%"). And these are they who have been publicly accused or arrested
or convicted and, in some cases, jailed. It is inconceivable that bishops and
others did not know earlier that 3 of every 100 priests in the United States
had a serious psychosexual problem. And every priest, bishop, abbot or
other religious superior who did nothing is complicit in each objectively
gravely morally wrong act committed against an innocent person.[6] The
numbers, therefore, grow quite quickly. How many ordinaries and auxiliary
bishops of the 177 Latin dioceses and 15 Eastern eparchies in the United
States knew? How many provincials, abbots or guardians? How many pastors and co-pastors? How many associates, deans, rectors, spiritual direc-
tors, committee chairs, Church spokesmen and personnel officers allowed
things to become this bad? How many thousands of churchmen created and
then ignored this travesty of justice? I have often asked if every high school
girl on my bus knew which priest to stay away from, how is it the bishop
did not know about him?

Radical Individualism

Whether chicken or egg, subjectivism and subjective morality are inextricably tied to radical individualism. The post-modern Church in the
United States suffers from a tremendous lack of structural accountability,
that is, real accountability outside the clerisy. The diocesan clerical structure
has effectively removed non-clerical eyes from internal workings and,
independent of the trusteeship problems of a century or so ago, this has
proved to be a great mistake. No longer is there internal clerical accountability in the parish; more and more parishes are staffed by just one priest,
with as much part-time help as he can buy.

The rapid evolution of the Boston-based lay group, Voice of the Faithful, bespeaks the outright anger among the people whose money supports the parish and diocesan structure. This lay organization directly emanates from the horrific scandals and cover-ups in the Archdiocese of Boston that led to the resignation of Bernard Cardinal Law in late 2002. The Voice of the Faithful motto, "Keep the faith, change the Church" calls for a maintenance of faith and a reformation of structure. This is not unlike the suggestions of the Protestant Reformation.

The expressed goals of Voice of the Faithful are: to support those who have been abused; to support priests of integrity; and to shape structural change within Church.[7] The Voice of the Faithful complaints about clericalism and its attendant failings are no newer than clerical corruption, but its suggested solutions could be seen as naive. A Voice of the Faithful hope is that new or reinvigorated parish pastoral and finance councils would somehow oversee parish matters in a manner more suited to a community of Christians. Parish pastoral councils comprise the parish staff and selected other members of the parish. Parish finance councils comprise the pastor and at least two others, selected according to norms established by the bishop.

Parish pastoral councils are not new, but in many cases they are moribund or non-existent. Where they exist, parishioners other than clergy and lay parish employees are either elected or (more usually) appointed by the pastor. However, whether a parish pastoral council exists or not its power is the same; that is, parish pastoral councils have no power. They may be established, and may be required, if the bishop allows or requires them (Canon 536), but parish pastoral councils serve a merely consultative (not collaborative) role in advising the pastor on pastoral planning. Juridically, the pastor remains the sole determinant of policy. Additionally, it appears that the manner by which parish pastoral or finance councils come into being or add members does not contradict their essential powerlessness. The pastor typically sets the means of adding members, and retains the right to reject or remove members, in addition to his canonical right to ignore their recommendations.

The parish finance council, however, is required by universal church law (Canon 537) to "assist the pastor in the administration of the goods of the parish." It is the parish finance council that watches — or is supposed to watch — where the money goes. Even so, it also serves a consultative role.[8]

In the collision between lay activism and clerical privilege, both appear to lose. Lay groups in and of themselves exist in opposition to the authority structure of the Church where they are not either first established or later acknowledged by the diocesan ordinary. The clerical structure, by Canon Law, only consults or collaborates with lay structures of its own erection. Hence Voice of the Faithful is *de jure* irrelevant. But in fact, Voice of the Faithful has created a huge discussion in the Church.

In some respects, the Voice of the Faithful argument, when focused upon the scandal of corrupt clergy, creates a new Donatism. Unlike the first Donatists in fourth to sixth century North Africa, who argued that sacraments administered by sinful clerics were invalid, the Voice of the Faithful prefers to support those clerics whom they deem worthy. By implication they do not "support" those whom they deem unworthy. The revolting concept of a pederast celebrating Eucharist or other sacraments notwithstanding, few would doubt the efficacy of such sacraments. Efficacy, however, does not eliminate an abhorrence of what has transpired, especially as it has become clear that the local ordinary allowed celebration of sacraments by individuals whom he knew were morally corrupt and criminally liable. The situation is especially disconcerting insofar as the explicit approval of the bishop can be viewed as his concurrence with, if not approval of, criminal acts of pederasty. In this respect, the actions of the local ordinary bespeak a radical individualism that truly injures the Church. That is, no matter any discussion about "collaborative decision-making," bishops can do whatever they wish. In case after case individual bishops in the United States, disconnected from the norms of society, unilaterally determined to send pederasts back to public ministry. Hence every Eucharist, every reconciliation, every baptism, every anointing by a criminal known to the local ordinary and permitted by him while efficacious is also injurious. How many thousands upon thousands of persons suffer from such episcopal negligence? How many Italian Americans and others have given up their cultural and religious attachment to Catholicism out of sheer nausea?

The attitude (or belief) that allows a bishop to foist a criminal cleric on the faithful is clericalism at its worst. And clericalism at its worst is radical individualism. The law resides in the mind of the cleric, and not in the mind of the Church. I may not be the first to suggest that radical individualism is at the root of clericalism, and in fact it may not be. But autonomous individualism, where the locus of authority is within the individual, is rampant in the world and waits in every rectory in the land to infect its inhabitants. How can it not? The pastor has sole territorial authority. The pastor is the sole representative of the parish as a juridic person. The pastor can consult with others on pastoral matters, and must consult with at least two others on finances, but in either case the consultation is just that, consultative. The others (presumably laity) do not share his authority; they can merely cooperate with it. The pastor has the final authority, and it rests within himself. While the genuine pastor is mindful of his flock, and he hears their requests and acts to the whole parish's benefit, in too many dioceses he has as example the imperiousness of the local ordinary. Hence, the short hop to true individualism is an easy one. It is always easier to autonomously set policy than to listen to the suggestions and complaints of the people.

Even so, suggestions and complaints rarely reach the heart of the matter. Financial irresponsibility and recklessness are not new to any human endeavor, but the more lay members of the Church delve into financial matters, the more they recognize that the money is not going to the poor. Again, the specter of radical individualism: the local ordinary or the parish pastor is so disconnected from the people that the people have no say in how their donated funds are distributed. No longer as in the ancient Church are goods and gold brought to the deacons, who distributed to the needy and the poor. In too many places a skewed concept of "Church" means all collections go to serve the parish plant and priests. Hence the rectory excesses, the cooks and cleaners, the secretaries, the extra clerical help, and the lack of active concern for the poor. The general parish taxation in too many dioceses goes to pay for diocesan administration; it is the special "bishop's appeal" that supports the services to the poor. On the parish level, where special needs arise and money will go directly to charity, a special collection springs from the back of the church.

Denial of Objective Reality

There appears to be a denial of objective reality in the Church on two levels: first, a denial of just how bad things really are on the part of Church authorities and, second, a denial of some basic teachings by clergy and laity alike.

One can wonder if the pope and bishops have any true concept of just how horrific the damage to the people of God in current and non-so-current scandals really is. In the face of major press coverage of public matters, the stonewalling and cover-ups continue, especially on the part of the public relations officers of the dioceses with the most to hide.[9] From the victims' point of view, the intense media coverage has helped. A report of the Rome visit of three Boston victims seems to indicate that the Pope, at least, is beginning to understand. The three unsuccessfully attempted to see Pope John Paul II, and eventually received a visit from a high Vatican official, who said the Pope asked for messages from them in return. The direct messages were: (1) "The Holy Father needs to make sure that this never, ever, ever happens to another child," (2) "The Holy Father needs to heal the church, not just the survivors but the church itself. He needs to realize how the church in the United States is hurting," and, (3) "The Holy Father needs to put a face with the problem, meaning he needs to meet with us.... Only then will he understand the depth of the wound."[10] What is so stunning is that their messages apparently reached the Pope.

Even if the reality of the situation is sinking in, there still exists the denial of objective reality as it is defined and taught by the Church. This is different from, but related to and often the result of, subjectivism and individualism. The Church teaches that there are certain undeniable truths and realities. The Church teaches that there exists an objective reality of truth. Yet whether willfully or through invincible ignorance too many members of the Church are simply ignorant of the teachings of the Church.

Which brings us back to the priest and the local pastor, and even farther back to Protagoras. It was Protagoras who, in ancient Greece, said "of all things the measure is man, of the things that are, that [or how] they are, and of things that are not, that [or how] they are not." Recall that the priest said: "the Church teaches that there is nothing absolutely right and nothing absolutely wrong." For him, of course, this is how things are. That is, he holds that there is nothing absolutely right and nothing absolutely wrong. Such is absolute truth for him, although for him alone. If such moral relativism allows (for him) abortion or any other objectively wrong act, there is no way to convince him otherwise. He alone stands as the measure of the gravity of an act, and therefore of its objective right or wrongness.

The Italian writer and Nobelist, Luigi Pirandello, explains things similarly in his short play, *Così è, se vi pare!*

Never mind your husband, madam! Now, you have touched me, have you not? And you see me? And you are absolutely sure about me, are you not? Well now, madam, I beg of you; do not tell your husband, nor my sister, nor my niece, nor Signora Cini here, what you think of me; because, if you would all tell you that you are completely wrong. But, you see, you are really right; because I am really what you take me to be; though, my dear madam, that does not prevent me from also being really what your husband, my sister, my niece, and Signora Cini take me to be — because they also were absolutely right![11]

For these people, just as for the priest, truth lies within their own perceptions, and consequently within their own minds. We know this as a subjective truth or subjective reality, and for the individual this subjective truth has a certain objectivity. The hilarity of Pirandello's play comes from their insistence that their personal, subjective truths are in fact absolute truth or objective reality. Yet this is what they have been taught.

One can subjectively know objective reality, but there are limits to untested assertions. If one person feels cold in a room, and another feels warm, each is experiencing a certain truth and reality. But if one feels warm, and the other warns the pipes are about to freeze, clearly one or the other is disconnected from what we would agree is an objective reality. That is, either the pipes can or cannot freeze; we can obtain an objective measurement of the temperature.

The priest above is teaching subjectivism as a truth, in opposition to the objective reality taught by the Church. He argues that truth is only perspective, and hence not normative; recall, he said: "nothing is absolutely right and nothing is absolutely wrong." Therefore, the end of his logic is that no truth taught by the Church is normative to him.

Hence the people of God will continue to be fed his view. The pastor will not fire him, however, because his own autonomous individualism allows him to override the truth: the objective fact that the visiting priest is teaching heresy. Here the pastor bends to some skewed view of pluralism. He does not wish to challenge the visiting priest again. Perhaps he wishes to make the Church, or at least his parish, appear more "modern." More likely, thought, he merely wants the help.

For Italian Americans, indeed for anyone whose cherished Church is on the rocks, these questions are extremely serious. They are as serious as the questions addressed by Catherine of Siena. Belief in Church and in Church teachings, so culturally conditioned among Italian Americans a generation or two ago, is fading with each issue of the daily paper.

It is possible that the world may be entering a post-Catholic era. Subjectivism and subjective morality, radical individualism, and a denial of objective reality as complicated by corrupt (autonomous) clergy are combining to create this post-modern Church. The people can only follow where they have been led. Attacks from within are far worse than the past attacks on future generations by Margaret Sanger, who founded Planned Parenthood, it is said, to rid Brooklyn of the scourge of "Italian vermin." Her project has met with success. It is not news that fewer Italian American Catholics are being born; it truly is news that Catholicism is simply melting away. Of course there are ethnic parishes here and there that preserve both cultural heritage and the faith. But they are fewer and fewer, and they are not necessarily Italian.

While throughout history the Church has suffered significant attacks from without, the most significant changes in this country in this century have come from a far more insidious erosion of the principles of belief from within. The media-enhanced crossfire between and among dioceses, victims' advocacy groups, and Voice of the Faithful surely wounds some innocent priests and bishops, along with hundreds of thousands of innocent faithful whose belief is shaken by the spectacle.

These faithful, whose hard-earned money supports each little parish in the land, deserve better. They deserve to know more honest priests and bishops for whom the gospel speaks the truth, and who in turn will preach it. They deserve to know where their money is going, especially since it seems that so little of it goes to help the poor. They deserve an objective truth, unvarnished by subjectivism and untrammeled by individualism, both spoken and lived by their leaders. They deserve to be taught, and argued away from moral relativism.

They are not getting what they deserve.

PHYLLIS ZAGANO

---

[1] There were about one hundred women Dominican Tiertiaries in Siena in Catherine's day, typically widows or older, unmarried women. Called Mantellates because of the black mantles they wore, they served the poor and sick. At sixteen

Catherine challenged tradition by not seeking entrance to a cloister, but rather choosing an essentially active apostolic life.

[2] "Church, States and Children", Editorial, *The New York Times*, March 6, 2003, p. A30. Other examples abound. An obvious indication that the Church has lost touch with reality was the spectacle of Bernard Cardinal Law, before he resigned as Archbishop of Boston, formulating the USCCB policy statement on Iraq. The statement included the sentence: These are…moral [choices] because they involve matters of life and death. Traditional Christian teaching offers ethical principles and moral criteria…." See Michael Paulson, "Law Leads US Bishops' Discussion on Iraq" *The Boston Globe*, November 13, 2002.

[3] See http://www.survivorsfirst.org/, which is regularly updated. Postings on the website indicate the ire of the people involved — and not involved — in the situation. For example, "Please do not validate the Church by turning to her for answers. She knows one thing: $$$Money$$$. That is her god. When the people and the State hit her right, smack dab in the wallet, then you will see how suddenly virtuous and righteous she can be. In the meantime, she is thumbing her nose at her own followers. The very best, most effective protest you personally can make is to withhold every cent of your money until every last record and document is turned over to the secular authorities. Then encourage ALL your friends, workmates, and family to do exactly the same thing. That is the ONLY thing that will make the Church respond. She cares nothing for your problems which are petty dribblings in her eyes."

[4] Suffolk County Supreme Court Special Grand Jury, May 6, 2002 Term ID, Grand Jury Report CPL §190.85(1)(C), January 17, 2003, p. 106.

[5] Of the roughly 1,400 names listed as offenders at: http://www.survivorsfirst.org/, only 53 obviously Italian names appear.

[6] *Catechism of the Catholic Church*, No. 1868. "Sin is a personal act. Moreover, we have a responsibility for the sins of others *when we cooperate in them*… — by not disclosing or not hindering them when we have an obligation to do so."

[7] See http://www.votf.org/

[8] Canon Law is quite clear that both, where they exist, are consultative in nature and require the presence of the pastor. (See John Beal et al. *New Commentary on the Code of Canon Law* (New York: Paulist Press, 2000) 710. In Canon 129, 2, the term "cooperate" (*cooperari*), not "share" (*partem habere*), restricts lay jurisdiction. While some commentators argue that jurisdiction comes with office, and a lay person might thereby exercise jurisdiction by legitimate appointment to those ecclesiastical offices open to laics (finance officer, finance council, parish administrator, administrator of goods, judge, auditor, promoter of justice, defender of the bond), only the ordained may hold offices that require orders or ecclesiastical governance (Canon 274,1).

[9] In some cases they as laypersons are undercut by their clerical bosses: "It seems that the information highway in this organization only runs one way. It is pathetic that the Cabinet Secretary for communications is not in the loop and has to hear that one of our priests has been relieved of his assignment from a Newsday reporter asking why." Suffolk County Supreme Court Special Grand Jury, May 6, 2002 Term ID, Grand Jury Report CPL §190.85(1)(C), January 17, 2003, p. 35.

[10] John Allen, Jr., "The Word from Rome" *The National Catholic Reporter*, (March 28, 2003) 2:31.

[11] *It Is So! (If You Think So!)* (*Così è, se vi pare!*) By Luigi Pirandello [1917] English version by Arthur Livingston (New York: E. P. Dutton, 1922) 70.

References

Catechism of the Catholic Church. 1994 (New York: Catholic Book Publishing Co.).

Pirandello, Luigi. 1917. *It is so! (If you think so!)* translated by Arthur Livingston. 1922. (New York: E.P. Dutton).

Part V
Concluding Note

Chapter 15

✦

WHITHER THE CATHOLIC ITALIAN VISION IN ITALIAN AMERICAN STUDIES?

In a previously published essay of mine, "The Saints in the Lives of Italian American Catholics: Toward a Realistic Multiculturalism" (1999), I concluded that "rebuilding the institutional integrity of Catholicism, recovering accurately the eternal truths found incarnate within the traditions of south Italy, and more fully appreciating the naturally positive dispositions and contributions of the Italian immigrants and their sons and daughters to American shores are the prerequisites for the development of a realistic Italian-American multiculturalism" (1999: 247). The focal point of that essay was on how the combination of two factors — the decomposition of Catholicism in the United States during the post-Vatican II years and the uncritical assimilation of most middle-class, post World War II Italian Americans into the contemporary secular worldviews of the American public sphere — severely attenuated classical Italian, Catholic, and Italian American world views in American society. No significant attention, however, was directed in that essay to a related matter, i.e., how the combination of these two factors affected the nature of what passes for, more or less, "official" scholarship in the area of Italian-American studies.[1]

The overall gestalt or "feel" of many of the essays in *Models and Images of Catholicism in Italian Americana: Academy and Society* is consistent with the claim that various "utopian" philosophies (e.g.subjectivist-/literary/social constructionist/deconstructionist; feminist; democratic capitalist; Marxist/socialist/anarchist, among others) have appealed, over the past four decades to both significant sectors of the Italian American academy or the formally educated, professional, and highly secularized Italian American middle and upper-middle classes. By a "utopian" imagination I mean a worldview that (a) emphasizes the ability of human beings to transcend through rational planning, limits set by biology, human nature, human finitude, and chance, (b) that defines happiness and success largely

283

through various forms of material and status acquisition and through the exercise of a radical individual autonomy, and (c) that denies that there is an objective reality in the nature of the social order that demands conformity in order for one, minimally, to survive to, maximally, flourish. Conversely put, contemporary Italian American studies are increasingly being under-represented by perspectives based on some philosophy of realism whether in "low culture" peasant and folk versions or in more fully articulated "high culture" natural law and Catholic traditions of thought that essentially reverse the baseline assumptions of the utopian imagination. A realist Catholic Italian worldview would include, contra the utopian imagination, the following postulates:

1) Spiritual/supernatural concerns are constitutive of what it means to be human;

2) Society and social relations are anchored by an unchanging human nature;

3) Family, neighborhood, and community are central to life;

4) There is a natural division of labor between men and women with gender relations seen as naturally complementary and organic;

5) There is a great respect given to tradition and tradition is viewed as both a dynamic reality and one relevant to the contemporary age;

6) There is a limited dependency on and expectations from the political process and even less from revolutionary movements;

7) Self-reliance, a sense of personal responsibility for one's actions, and duty to God are to be promoted; and

8) The maintenance of a healthy culture, rich interpersonal relations, well-earned respect, and a sense of honor and dignity are part and parcel of living out the "good life."

Put another way, the cognitive claim being made here is that the substantive content in the field of Italian American studies is more and more being explained by perspectives foreign to its natural baseline, i.e., the premodern southern Italian peasant and ethnic Italian American worldview — a baseline that shares a fundamental overlap with natural law and Catholic perspectives. The current direction of Italian American studies as I depict it, while not salutary overall, is not devoid of some substantial scholarly worth. Socialists and capitalists, Freudians and religionists, literary elitists and peasant populists, subjectivists and natural law advocates, and feminists and Catholics can and do regularly, and to some degree of useful intellectual profit, critique each other. There is, after all, a long tradition of scholarly inquiry that respects the usefulness of the "outsider's role" in the investigative process. Nonetheless it is curious, to say the least, to note the increasing dearth of Italian American scholars who have attempted to organically develop, update, and shape by way of transforming and coopting more modern analytical frameworks, an authentically southern Italian or Catholic realistic vision. Questions arise. Is the perception of most modern Italian American scholars toward their own heritage a hostile one? Are they victims of what Frederich Nietzche and Max Scheler have termed the reality of "resentment," of accepting the negative cultural evaluations of one's own oppressors? Are modern day Italian American scholars more and more ignorant of their own past or, conversely, trapped within the "tyranny of the present?" Or have they radically revised the past in a quite calculating fashion for immediate political and careerist purposes or to better "fit in" culturally and socially in the secular intellectual milieu? And, finally, do any contemporary Italian American Catholic scholars have the ability or the desire to articulate and develop cogently the incarnational truths found within their own tradition?

There are three reasons why some scholars within the Italian American academy might now want to attempt to craft an authentic Catholic Italian vision in Italian American studies at this particular juncture in time and space. First of all, the idea is given both plausibility and positive encouragement by the increasingly widespread contemporary acceptance of the philosophy of multiculturalism, within both the society-at-large and the academy. While some of the most ardent supporters of what I've termed "multicultural radicalism" (Varacalli, 1994), may, empirically speaking, see no personal, heartfelt need for the resurrection of philosophies and worldviews grounded in realism and in the cultural traditions of Western civilization, the logic of multiculturalism forces even some of the inwardly very unsympathetic to publicly support inclusiveness, both intellectually and morally. Secondly, there is the growing awareness, among both Italian American scholar and educated lay person alike, that, minimally, there are limits to the usefulness of hyper-modern worldviews and that, in many cases, they have actually brought about personal and social dysfunctions and pathologies. For those with the sentiments of the former, a Catholic

Italian worldview in Italian American studies would serve as an indispensable balance wheel to the hyper-modern. For those holding the latter judgment, the movement toward the reintroduction of a classical/realist/Catholic worldview should not stop short of it being granted a privileged paradigmatic status among Italian American scholars and citizens alike. Thirdly, and finally, at least since the publication of Thomas Kuhn's *The Structure of Scientific Revolutions* (1962), there is a slowly growing recognition of the inevitable and necessary role that images, models, and paradigms—whether consciously held or not—play in the intellectual research process. This greater recognition of the "groundedness" of scholarship makes the possibility of articulating and developing a Catholic Italian worldview in Italian American studies a more obvious and justifiable enterprise in the minds of its academic supporters and detractors alike.

JOSEPH A. VARACALLI

---

[1] A different version of this paper titled "The Contemporary Utopian Imagination and the Present State of Italian American Studies" was first presented on November 10, 2000 at the Thirty-Third Annual Conference of the American Italian Historical Association held in Lowell, Massachusettes.

References

Kuhn, Thomas 1962. *The Structure of Scientific Revolutions*. (Chicago: University of Chicago Press).

Varacalli, Joseph A. 1999. "The Saints in the Lives of Italian American Catholics: Toward a Realistic Multiculturalism," pp.231-249, *The Saints in the Lives of Italian Americans: An Interdisciplinary Investigation*, eds. Joseph A. Varacalli, Salvatore Primeggia, Salvatore J. LaGumina, and Donald J. D'Elia, (Stony Brook, NY: Forum Italicum).

Varacalli, Joseph A. 1994. "Multiculturalism, Catholicism, and American Civilization," *Homiletic and Pastoral Review*, 94:47-55.

CONTRIBUTORS

Linda Ardito is a member of the Music Department of Dowling College, Oakdale, New York.

Mary Jo Bona is a member of both the Department of European Languages, Literatures, and Cultures and Center for Italian Studies at Stony Brook University, Stony Brook, New York.

Michael P. Carroll is a member of the Sociology Department at the University of Western Ontario, London, Ontario.

Donald J. D'Elia is a member of the History Department at SUNY-New Paltz, New Paltz, New York.

Honorable Thomas P. DiNapoli, New York State Assemblyman, is a member of the Advisory Council of the Center for Italian Studies at Stony Brook University.

Fred Gardaphè is a member of both the Department of European Languages, Literatures, and Cultures and serves as Chair of American Studies at Stony Brook University, Stony Brook, New York.

Louis Gesualdi is a member of the Sociology Department and Director of the Center for Italian American Studies at St. John's University, Jamaica, New York.

Anthony L. Haynor is Chair of the Sociology Department at Seton Hall University, South Orange, New Jersey.

Salvatore J. LaGumina is Professor Emeritus of History and Director of the Center for Italian American Studies at Nassau Community College-SUNY, Garden City, New York.

Sister Margherita Marchione is Professor Emerita of Italian Language and Literature at Fairleigh Dickinson University and a member of the Religious Sisters Filippini.

287

Salvatore Primeggia is a member of the Sociology Department at Adelphi University, Garden City, New York.

John C. Rao is a member of the History Department at St. John's University, Staten Island, New York.

Frank A. Salamone is a member of the Sociology Department at Iona College, New Rochelle, New York.

Joseph A. Varacalli is a member of the Sociology Department and Director of the Center for Catholic Studies at Nassau Community College-SUNY, Garden City, New York.

Phyllis Zagano is a member of the Religious Studies Department at Hofstra University, Hempstead, New York.

INDEX

Abbreviations: IA for Italian American
    RC for Roman Catholic(s)
    RCC for Roman Catholic Church
    USA for United States of America
    SI for Southern Italian
    I for Italian